# Ecological Studies in the Coastal Waters of Mauritania

Developments in Hydrobiology 86

*Series editor*

H. J. Dumont

# Ecological Studies in the Coastal Waters of Mauritania

Proceedings of a Symposium held at Leiden, The Netherlands, 25–27 March 1991

*Edited by*

W.J. Wolff, J. van der Land, P.H. Nienhuis & P.A.W.J. de Wilde

*Reprinted from Hydrobiologia, vol. 258 (1993)*

Springer Science+Business Media, B.V.

**Library of Congress Cataloging-in-Publication Data**

Ecological studies in the coastal waters of Mauritania / edited by
W.J. Wolff ... [et al.]
p. cm. -- (Developments in hydrobiology ; 86)
"Proceedings of a symposium held at Leiden, The Netherlands, 25 -
27 March 1991."
Includes bibliographical references.
ISBN 978-94-010-4877-4 ISBN 978-94-011-1986-3 (eBook)
DOI 10.1007/978-94-011-1986-3
1. Marine ecology--Mauritania--Congresses. 2. Coastal ecology-
-Mauritania--Congresses. I. Series.
QH195.M479E25 1993
574.5'2636'09661--dc20 93-19531

ISBN 978-94-010-4877-4

*Printed on acid-free paper*

# Contents

*Hydrobiologia* **258**: vii–ix, 1993.
*W. J. Wolff, J. van der Land, P. H. Nienhuis & P. A. W. J. de Wilde (eds), Ecological Studies in the Coastal Waters of Mauritania.*



# Preface

Ultimately this volume on the ecology of Mauritanian coastal waters and in particular the Banc d'Arguin has its origin in a number of earlier Dutch projects. The pelagic system off the Mauritanian coast was studied during the NECTAR-77 and NECTAR-78 cruises. Benthic studies of the shelf, including the edge of the Banc d'Arguin and the Baie du Lévrier, were made during the CANCAP-III cruise in 1978. A visit of the Banc d'Arguin proper by four Dutch biology students followed in 1980. Their account (Altenburg *et al.*, 1982) of the millions of birds wintering in that tidal flat area and their description of the ecology of that area, provoked interest among other Dutch ornithologists and estuarine ecologists. This resulted in a larger Dutch–Mauritanian research project at the Banc d'Arguin carried out in 1985 and 1986. The ornithological studies of that project have been published in a special volume of Ardea (Ens *et al.*, 1990). However, such projects normally result in more questions than answers and this project proved no exception to that rule. At the same time the attention of some of the scientists involved shifted from purely ornithological to more general ecological questions, particularly the functioning of the Banc d'Arguin ecosystem. This led to a research project in which a large part of the Dutch oceanographic and ecological scientific community participated: the project Banc d'Arguin 1988.

The Banc d'Arguin is an area of tidal flats and shallows situated between roughly 19° 20′ and 20° 30′N along the Saharan coast of Mauritania (Fig. 1). It is characterized by water temperature of about 20–25 °C and salinities over 38. There are no rivers and rainfall is rare. Off the coast of Mauritania cold water masses from the northern Atlantic and warm water from tropical areas meet. Hence flora and fauna contain both temperate and tropical species. Examples are the temperate mudsnail *Hydrobia ulvae* and the seagrass *Zostera noltii*, and the tropical mangrove *Avicennia africana* and the bivalve *Anadara senilis*. Conspicuous features are extensive seagrass beds, relatively low biomasses of the zoobenthos and extremely high densities of birds (Wolff & Smit, 1990).

The project Banc d'Arguin 1988 endeavoured to understand the functioning of this remarkable coastal wetland. Major questions to be answered concerned the interaction between the Banc d'Arguin ecosystem on the one hand, and the adjoining systems of the open ocean with its intensive upwelling phenomena, and of the Sahara desert on the other. Is the Banc d'Arguin ecosystem relatively isolated from its surroundings and does it function as a tropical pocket in a more temperate surrounding as suggested by Sevrin-Reyssac (1983), or does it interact with both Sahara and Atlantic Ocean? It was decided beforehand that the interaction between the Banc d'Arguin and the open ocean should get the major part of the attention. However, at the Banc d'Arguin this is not an easy task, because the area between the estuarine parts of the area and the open ocean is extremely shallow and largely uncharted. Large research vessels cannot penetrate the shallows whereas small vessels are often unable to negotiate the outer part of the shallow area due to rough weather conditions. Hence, seperate teams were organized for work at the seaward side of this shallow sill and at the landward side. In addition other teams used the opportunity to address other questions such as the productivity of the seagrass beds and the benthic fauna, the number of birds spending the summer at the Banc d'Arguin, and the distribution of several taxonomic groups of flora and fauna.

The project was organized in cooperation with Mauritanian institutions, viz. the Banc d'Arguin National Park ('Parc National du Banc d'Arguin') and the National Centre of Oceanography and Fisheries ('Centre National de Recherches Océanographiques et de Pêches'), both at Nouadhibou.

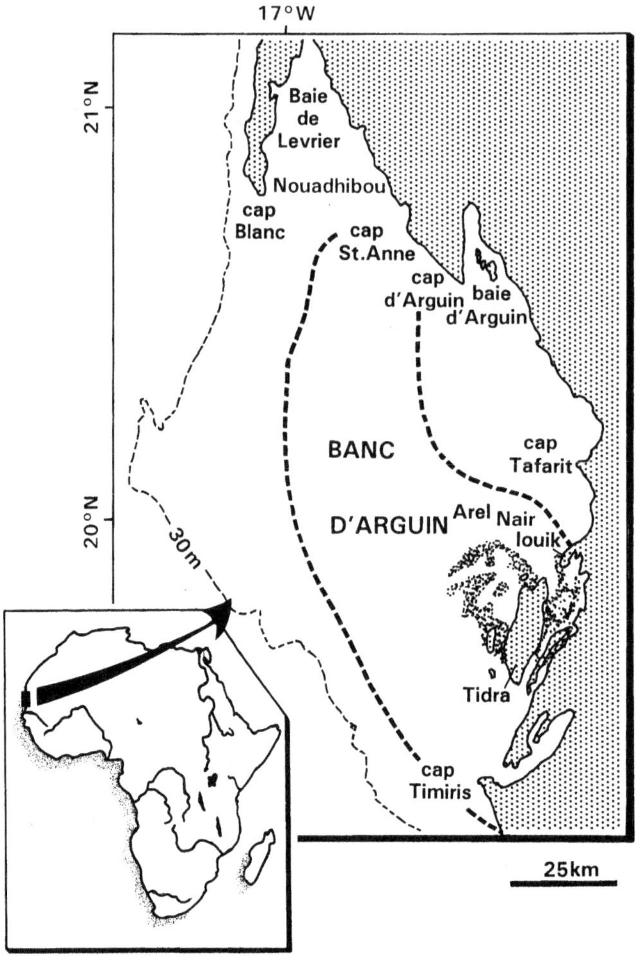


*Fig. 1.* The Mauritanian coast between Nouadhibou and Cap Timiris.

The participants from The Netherlands originated from Amsterdam University, the DLO – Institute for Forestry and Nature Research (formerly Research Institute for Nature Management), Groningen University, Leiden University, National Museum of Natural History, Netherlands Institute for Ecological Research (formerly Delta Institute for Hydrobiological Research), Netherlands Institute for Sea Research, Wageningen University and Working Group for International Waterfowl Research. In addition colleagues from Ghana and Germany took part.

To evaluate the results of the project Banc d'Arguin 1988 a symposium was held in the National Museum of Natural History at Leiden, The Netherlands, from 25–27 March 1991. At this symposium the results of the 1988 project have been placed in perspective by inviting a number of key scientists who reported on research at the Banc d'Arguin by Mauritanian and French institutions.

The symposium was opened by the Director of the Parc National du Banc d'Arguin, Mr Kane Hadya, and it was very appropiate that the first lecture could be given by the nestor of research at the Banc d'Arguin, Prof. Theodore Monod.

We are vere grateful to the Mauritanian authorities, in particular the Director of the Parc National du Banc d'Arguin, Mr Kane Hadya, for their permission to carry out our research in the coastal waters of Mauritania. We thank the Director of the Centre National de Recherches Océanographiques et de Pêches,

Dr Moctar Ba, as well as the Director of the Parc National, Mr Kane Hadya, for their cooperation and assistance, both by themselves and by their staff. We also want to acknowledge the help of Mr Tom van Spanje and his friends in Mauritania. The participants of the expedition based in Iouik during the project will not easily forget the kind help of the local population.

Financial support for either the work in Mauritania, or the symposium, or both was received from the Netherlands Marine Science Foundation (SOZ), which funded the lion's share, the Directorate-General for International Cooperation of the Netherlands Ministry of Foreign Affairs, UNESCO, the Netherlands Natural History Museum, and the Netherlands DLO – Institute for Forestry and Nature Research.

W. J. WOLFF
J. VAN DER LAND
P. H. NIENHUIS
P. A. W. J. DE WILDE

## References

Altenburg, W., M. Engelmoer, R. Mes & T. Piersma, 1982. Wintering waders on the Banc d'Arguin, Mauritania. Stichting Veth tot Steun aan Waddenonderzoek, Leiden, 283 pp.

Ens, B. J., T. Piersma, W. J. Wolff & L. Zwarts (eds), 1990. Homeward bound: problems waders face when migrating from the Banc d'Arguin, Mauritania, to their northern breeding grounds in spring. Ardea 78: 1–364.

Sevrin-Reyssac, J., 1983. Affinité biogéographique de la région des îles de Banc d'Arguin (Parc National du Banc d'Arguin, Mauritanie). Bull. Centre Natl Rech. Océanogr. Pêches (Nouadhibou) 11: 53–56.

Wolff, W. J. & C. J. Smit, 1990. The Banc d'Arguin, Mauritania, as an environment for coastal birds. Ardea 78: 17–38.

*Hydrobiologia* **258**: 1–8, 1993.
*W. J. Wolff, J. van der Land, P. H. Nienhuis & P. A. W. J. de Wilde (eds),*
*Ecological Studies in the Coastal Waters of Mauritania.*




# Hydrology and underwater climate of the Banc d'Arguin, Mauritania: a review

J. Sevrin-Reyssac
*Laboratoire d'Ichtyologie générale et appliquée, Muséum national d'Histoire naturelle, 43, Rue Cuvier, 75231 Paris, Cedex 05, France*

*Key words:* Banc d'Arguin, Mauritania, hydrology

## Abstract

The northern coast of Mauritania between Cap Blanc and Cap Timiris is characterized by a large variety of hydrological situations. At the western slope of the Banc d'Arguin, at a depth of about 20 m, the underwater climate is essentially of a temperate character. This is a highly unstable area due to the occurrence of upwellings and the movements of the intertropical front. Close to the coast the isolation of water masses by the very shallow Banc d'Arguin results in an increase of temperature and salinity. These factors show a gradient towards the South from Cap St. Anne to the islands. In this region the underwater climate is tropical, a feature which is most clearly expressed in land-locked bays such as the Baie d'Arguin, the area around the island of Tidra, and the Baie de St. Jean. In the latter bay the conditions are most extreme (salinity $> 80‰$). Due to the coastal topography the area of the islands around Tidra is highly isolated. This area receives no fertilization from the upwellings offshore.

## Introduction

The Mauritanian coast between Cap Blanc and Cap Timiris is a very interesting region because of its hydrological complexity and its climatological and biogeographical pecularities. The richness of the upwelling system in the coastal zone beyond 25 m is known worldwide and it has been the goal of numerous international oceanographic expeditions.

The littoral zone, however, and in particular the Banc d'Arguin, has not been explored by the large expeditions because of its very shallow nature making it inaccessible for the large vessels. The small trawler 'Almoravide' of the Centre National de Recherches Océanographiques et des Pêches at Nouadhibou has enabled the first studies of the shallow water area. In this paper the major conclusions from this research will be reviewed.

### *The situation of the Mauritanian coast in a West African framework*

The northern part of the Atlantic Ocean in front of the African coast belongs to the Atlantic–Mediterranean province (Ekman, 1953) including the Mediterranean and the adjoining parts of the Atlantic. Within this area the African coast between Gibraltar and Cap Blanc constitutes the Mauritanian subprovince (Fig. 1). In this area the waters are relatively cold. This is due to the presence of the cold Canary Current penetrating South until the environs of Cap Blanc, and the forming of upwellings between this current and the coast. The frequency and intensity of these phenomena increase from Cape Bojador towards the South. Maximum frequency occurs near Cap Blanc (Margalef, 1975; Schemainda *et al.*, 1975; Bas & Cruzado, 1976).

Between Cap Blanc and Cape Verga in Guinea

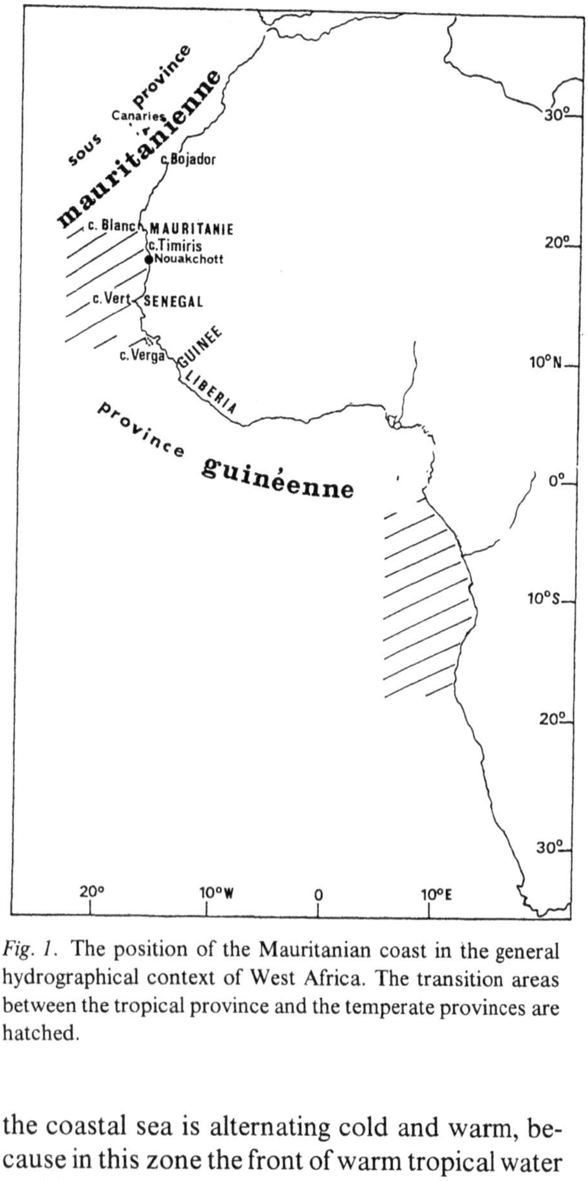


*Fig. 1.* The position of the Mauritanian coast in the general hydrographical context of West Africa. The transition areas between the tropical province and the temperate provinces are hatched.

the coastal sea is alternating cold and warm, because in this zone the front of warm tropical water oscillates. During the boreal summer and autumn the coast is bathed by 'tropical' warm and salt ($T > 24$ °C; $S > 35‰$) water, characteristic of the tropical region, as well as by 'Guinean' warm, but less salt ($T > 24$ °C; $S < 35‰$) water originating from the Liberian region where it is formed in an area with heavy precipitation. The remainder of the year the coast is under the influence of 'Canarian' cold, but salt ($T < 24$ °C; $S > 35‰$) water originating from the region of the Canary Islands (Berrit, 1952; Rossignol, 1973).

The transitional character of the Mauritanian coast is underlined by many authors. It is most clearly expressed between Cap Blanc and Cap Timiris (Figs 1 and 2). More to the South, between Cap Timiris and Cape Verga, the marine climate becomes warmer. Although upwellings still occur locally (for example near Nouakchott, and near Cape Verde), their frequency diminishes from North to South. Near Cape Verga winds causing upwellings occur only early in the year (Schemainda *et al.*, 1975), whereas near Cap Blanc the wind directions and strength are nearly always able to cause upwelling.

## Hydrology of the northern coast of Mauritania

The coast between Cap Blanc and Cap Timiris not only is situated in the area where the tropical thermal front oscillates, but it is also an area with strong hydrological heterogeneity. From Cap Blanc in the North the cold Canary Current leaves the coast and takes a south-westerly direction. Its speed is 1–1.5 knot in summer and 0.5–1 knot in winter (Miro Orell, 1973). In between this current and the coast upwellings occur during the larger part of the year, but with maximal intensity in the periods February–June and October–December, bringing water from a depth between 60 and 200 m at the surface. This is due to the tradewinds which are particularly stable and strong along the northern coast of Mauritania. The winds can change in direction and force during the day, but winds from the North, NNW and NNE dominate. A study of 10 years of wind data showed a high constancy of direction, but important annual and seasonal variations in start, duration and intensity of wind periods (Maigret, 1972; see also Ould Dedah, this volume). It has to be stressed that between 20 and 24°N, in an average year, 70–80% of the time northerly winds favour upwelling phenomena (Schemainda & Nehring, 1975).

The presence of the Canary Current and the coastal upwellings cause the underwater climate to be cold for this latitude. Quite a number of authors have shown that the fauna in these waters has numerous affinities to that of Iberian

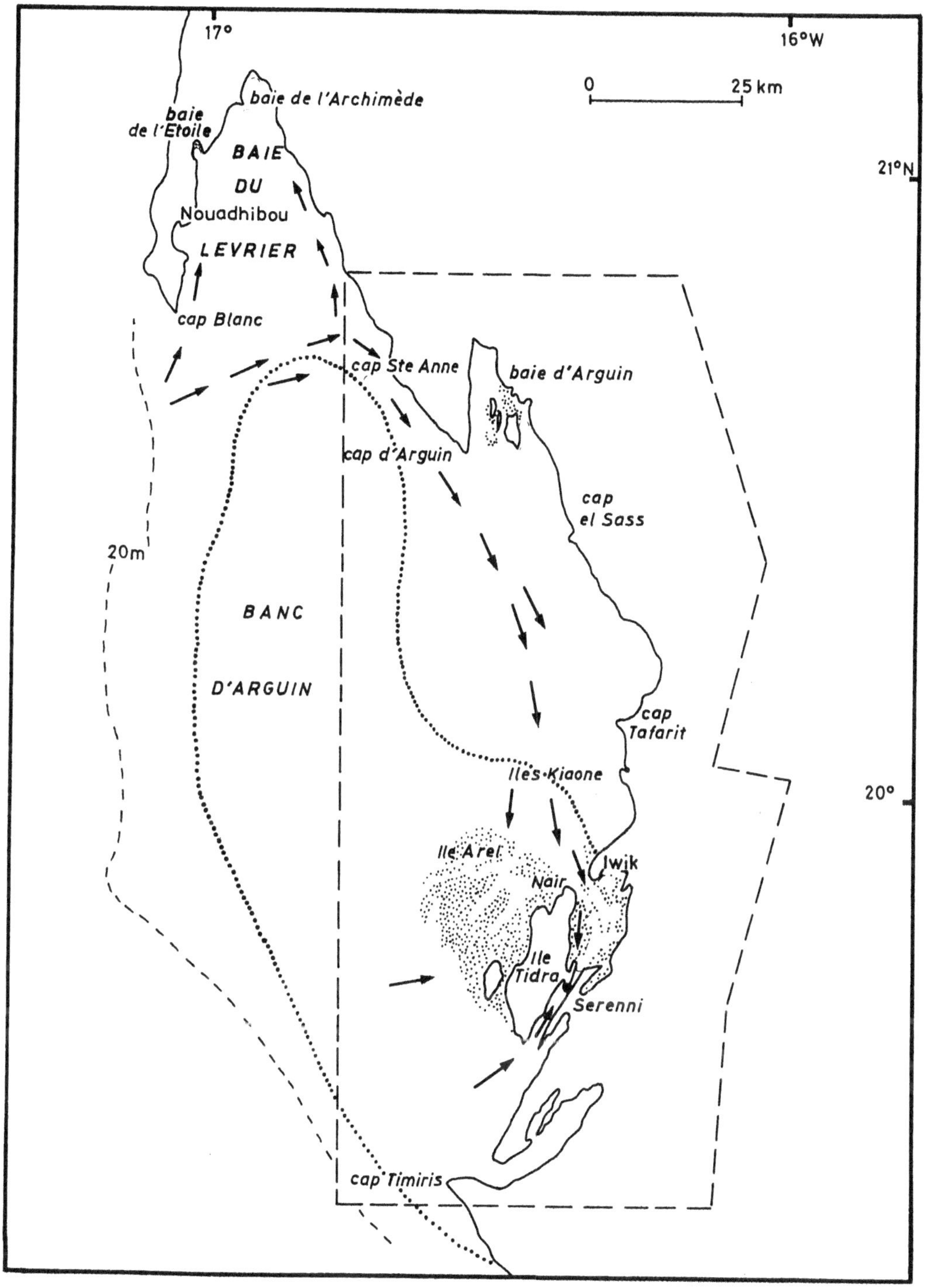


*Fig. 2.* The coast of the northern part of Mauritania between Cap Blanc and Cap Timiris. The arrows denote the main currents during flood tide. The straight broken line denotes the boundaries of the Parc National du Banc d'Arguin; the stippled line indicates the position of the Banc d'Arguin proper. The stipled areas denote tidal flats.

coastal waters, *i.e.* its biogeographical character is dominantly temperate (Reyssac, 1978).

A warmer and less unstable area is situated South of Cap Timiris although here deep water may well up locally due to eddies between currents or between a current and the shore (Margalef, 1975).

Between these two different areas a thermal front moves North during summer and South during winter. It reaches its northermost position near Cap Blanc in July–August (Ingham, 1970; see also Cuq, this volume).

Maigret (1972) defines three major water masses along the northern Mauritanian coast:

- cold Canarian water (Canary Current) is year round present North of Cap Blanc and moves periodically South; it reaches the Banc d'Arguin in January–February;
- warm water from the Liberian area following the northward movement of the intertropical front in summer; this water occurs along the western edge of the Banc d'Arguin especially from July to September–October;
- cold upwelling water, but less salt than the Canarian water, especially near Cap Blanc but also further South, although the upwellings diminish in intensity in the direction of Cap Timiris.

## Hydrology of the Banc d'Arguin

### *Bottom topography and currents*

To understand the hydrological pecularities of the Banc d'Arguin, one has to study the bottom topography.

The Banc d'Arguin is a vast body of accumulated sediments. East of the longitude of Cap Blanc the depth is nowhere larger than 20 m and at the latitude of the islands the 20 m isobath occurs 30 km offshore. At the latitude of Iwik (Iouik) (Fig. 2) sand banks (depth about 5 m) occur up to 50 km from the coast (Sevrin-Reyssac, 1988). Closer to the mainland along the eastern edge of the Banc depths are larger than at the Banc itself, but do not reach 20 m.

In the southern part of the Banc d'Arguin the desert and the marine environment meet, with islands of sand and sandstone, mudflats emerging at low tide and intersected by complex network of channels with depths between 2 and 6 m. An extraordinary area is the Baie de Saint Jean (Fig. 3). This landlocked bay with a length of about 30 km and a maximum depth of 5 m is a dune valley invaded by the sea. However, it seems to have been the estuarine mouth of a freshwater stream in the early Quaternary (Faure *et al.*, 1982). These past climatic conditions are still reflected by the presence of an abundant population of the cichlid fish *Sarotherodon melanotheron* which is one of the principal elements of the estuarine relict fauna of the Banc d'Arguin (Sevrin-Reyssac & Richer de Forges, 1985).

The shallows of the Banc d'Arguin cause the landward tidal wave to pass North of the Banc during flood. Following the deeper areas the tide then moves towards Cap Sainte Anne. Here the water mass splits. One part goes North towards the Baie du Lévrier, the other one moves southwards following the deeper areas along the coast (Fig. 2). During ebb tide the currents move the other way (Reyssac, 1977).

Next to the tidal currents the observations of Schaffer (1974), Peters (1976) and Reyssac (1977) seem to show a residual current from North to South driven by the wind. The offshore water penetrates into the Banc d'Arguin area at its northwestern border and leaves the Banc at its southwestern border after a time of residence of 18 to 34 days depending on the force of the wind. The water leaving the Banc has become quite saline, and hence quite dense; therefore it leaves the Banc d'Arguin along submarine canyons off Cap Timiris (Peters, 1976).

The average tidal range is about 1.5 m.

### *The hydrology and underwater climate in the different parts of the Banc d'Arguin*

According to Margalef (1975) the Banc d'Arguin influences the hydrology of the entire coastal area between Cap Blanc and Cap Timiris. It behaves

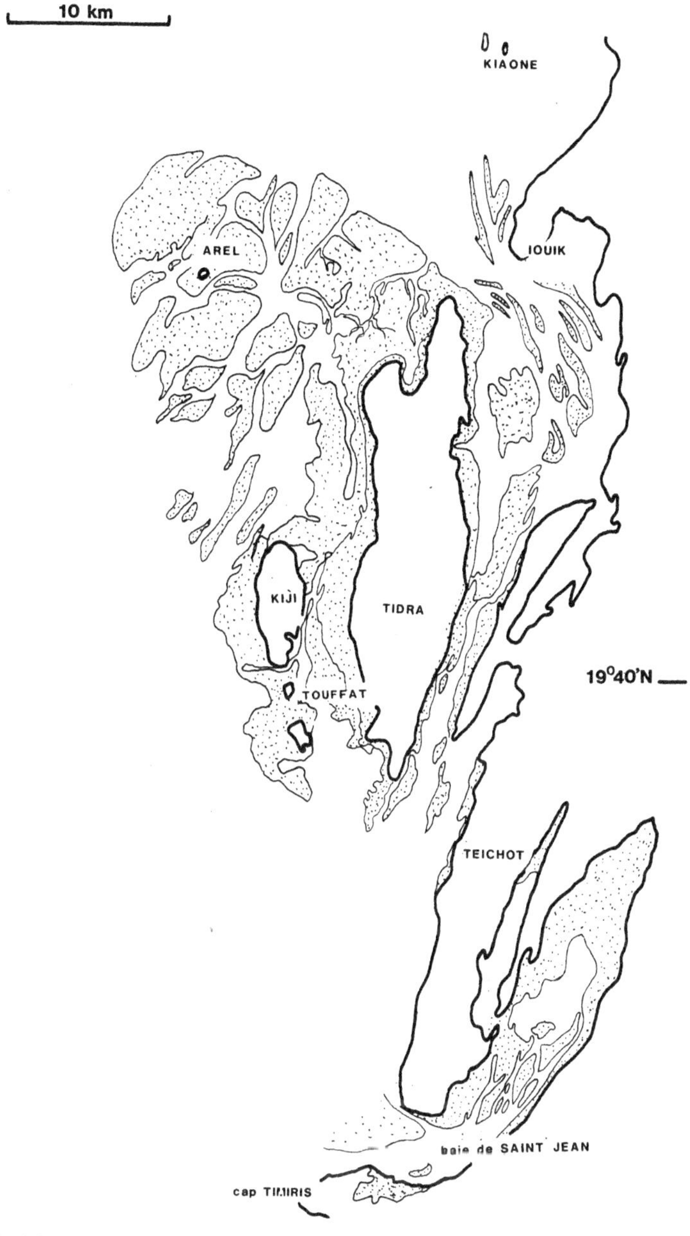


*Fig. 3.* The area of the isles of the Banc d'Arguin and the Baie de Saint Jean. The stippled areas denote tidal flats.

in fact as a negative estuary. The absence of any freshwater discharge from the mainland (precipitation is less than 30 mm per year), the strong insolation, the evaporation and the confinement of the water due to the coastal topography cause a strong increase of temperature and salinity in the area between 20–25 m depth and the coast. There exist, however, large differences between the various parts of the Banc d'Arguin.

*Western slope of the Banc d'Arguin*

South of Cap Blanc the western slope of the Banc d'Arguin is subject to the influence of upwelling water giving it a cold marine climate. From one series of observations to another, the upwellings vary in strength and cover larger or smaller areas. In April–May 1972 and in September–October 1973 Reyssac (1977) observed cold water (16–17 °C) following the 20 and 25 m isobaths from Cap Blanc till off the Kiaone Isles. Salinity was 35.5 to 36.5‰.

More to the South West of the island of Tidra temperature and salinity increase rapidly along an West-East transect when the tidal flats are approached. Sevrin-Reyssac (1984) also observed cold and less saline water (15.4 °C; 35.3‰) at depths of 20 m, but such water masses are rapidly replaced by warmer and more saline water. At the edge of the tidal flats temperature amounts already to 20 °C and the salinity surpasses 38‰. Near the isle of Touffat (Fig. 3) temperature is 21 °C and salinity nearly 40‰. Also the transparency of the water decreases on approaching the islands. The euphotic layer is 7–8 m thick at depths of 20 m, but near the mudflats where the tidal currents resuspend the sediment, the layer is under 2.5 m.

It should be emphasized that, although the highest salinities occurring at the western edge of the Banc d'Arguin correspond to low tide periods, the corresponding values during high tide are also quite high: 38.7‰ near Arel (Fig. 3).

These results lead to the hypothesis that the area West of the isles, although facing offshore, is continuously washed by very saline water (see also Peters, 1976). Due to the large difference in density this saline water does not mix very well with the cold and less saline coastal water. The small tidal range and the coastal topography contribute to the poor mixing. Hence the area West of the isles, although in close vicinity of an area of upwelling, appears to be rather isolated from the fertilizing influence of this phenomenon. This also applies to the area around the isles further to the East (Sevrin-Reyssac, 1984).

When the phytoplankton productivity of this area is compared to that of the Baie du Lévrier which is not only situated close to the upwellings off Cap Blanc, but also shows a bottom topography favourable to the penetration of oceanic water, phytoplankton productivity in the Baie du Lévrier appears at least ten times higher (Sevrin-Reyssac, 1984).

*Eastern slope of the Banc d'Arguin*

This area is well protected against oceanic influences by the shallows of the Banc d'Arguin. Consequently it has a warm underwater climate. Temperatures here can be ten centigrades higher than those on the western slope of the Banc d'Arguin. Therefore, this environment shows many analogies to coastal areas much farther to the South. This phenomenon repeats itself along other parts of the West-African coast. Such estuaries, lagoons, and networks of tidal channels locally develop conditions which exaggerate their coastal character (Postel, 1963). This is, for example, also the case for the estuary of the Casamance, the complex Saloum-Gambia, the mouth of the Senegal river and, also in Mauritania, in the Baie de l'Archimède, the northernmost part of the Baie du Lévrier (Fig. 2).

In the eastern part of the Banc d'Arguin the coastal underwater climate remains warm all around the year as shown by remote sensing (Domain, 1980; see also Ould Dedah, this volume). It appears very clearly that after the warm water front (isotherm of 20 °C) has retreated to the Gulf of Guinea and the waters off the Banc d'Arguin have changed to temperate conditions, tropical conditions continue to exist near the coast North of Cap Timiris.

Like temperature also salinity increases strongly from West to East (just as is the case for the area West of the isles), but starting from Cap Sainte Anne it increases also from North to South. Near this cape temperature and salinity show minimum values due to the fact that the tidal wave with oceanic water enters the shallows right here. Since the water masses are increasingly isolated going to the South salinity increases from 36.5–36.7‰ near Cap Sainte Anne to 39‰ off Cap Tafarit and to nearly 40‰ near the Kiaone islands.

Also water temperatures increase from North to South, the difference approaching 7 °C in some periods (Sevrin-Reyssac, 1982). Between the same two points Secchi disc observations show the thickness of the euphotic layer to be 5–8 m.

Hence a major difference exists between the hydrology of the western and that of the eastern part of the Banc d'Arguin. In addition to that, both areas are very heterogeneous. In the western area the heterogeneity in time and space is due to the irregularity of the upwelling, both in intensity and in place as well as to the seasonal displacement of the thermal front. In the eastern part it is due to the coastal topography leading to an increasing isolation of water masses from North to South.

*Baie d'Arguin and area East of the island of Tidra*
The Baie d'Arguin (Fig. 2) and the area East of Tidra are far removed from oceanic influences. This results in an very warm and highly salines environment.

The Baie d'Arguin is doubly isolated from offshore waters by the shallows of the Banc d'Arguin proper and by the sill at Cap d'Arguin.The observations of Sevrin-Reyssac (1982) in November 1980 showed that water temperatures in the Baie were 6 °C higher than those at Cap Blanc. Also salinities increases strongly, from 36.5‰ near Cap d'Arguin till over 41‰ in the innermost part of the Baie. Being quite isolated from the upwelling influences, the water of the Baie d'Arguin is very clear due to the poverty of the plankton (Sevrin-Reyssac, 1982). The depths being 10–12 m at maximum the entire water layer receives sunlight.

North of the Isle of Tidra as well as between this isle and the mainland we find water masses from the eastern part of the Banc d'Arguin. Being already warm and saline, these features become even more prominent. The salinity is over 41‰ in the more landward parts of the tidal channels. It is already near 40‰ off Iouik (Iwik) (Fig. 3). Even higher values (43‰) have been observed on the flood tide South of Tidra near the village of Teichot. Wolff & Smit (1990) registered 54‰ near Serenni in the most landward part of a tidal channel.

Extremely high temperatures and salinities have been found in the Baie de Saint Jean: salinities range from 52‰ at the entrance of the Baie to over 80‰ at the most landward station. Measurements made half way the Baie at high and at low tide demonstrate that the tides do not result in important changes of salinity: 55.4‰ at high tide and 57.4‰ at low tide. At the same time (March) the waters of the Baie were very warm: 23.5 °C at the entrance of the Baie and 26.5 °C at the most landward part. At that time the water temperature South of Cap Blanc is 16–17 °C.

As in the entire tidal flat area the turbidity in the Baie de Saint Jean is high because the tides keep a large amount of silt and organic matter suspended in the water. This situation contributes to the low phytoplankton stocks in the Baie (Sevrin-Reyssac, 1988).

## Discussion

The interesting character of the northern part of the coast of Mauritania in general and of the Banc d'Arguin in particular is the very large diversity of aquatic environments in a relatively small area. This abiotic situation results in a large diversity of flora and fauna. In the areas with a tropical underwater climate the composition of the biocenosis is similar to those of areas much farther to the South. Other parts of the Banc d'Arguin, however, with a underwater climate which is relatively cold for this latitude, show a more temperate biocenosis.

The area around the islands of the Banc d'Arguin probably is the most characteristic part of the Banc because of its 'coastal' character. The contrast between this area of the coastal zone and the adjacent ocean is very large and these two environments seem to be independent from each other.

The Baie de Saint Jean may considered as a tropical estuary in a desert region. The environmental change due to the isolation of the water of this Baie creates an exceptional habitat.

The Banc d'Arguin deserves scientific research as well as protection of its environments. The complexity of this system warrants the continuing interest of hydrologists, geologists, biologists and other scientists. In a time when numerous wetlands disappear or at least improve, the Banc d'Arguin still seems to be doing well. Thus it is one of the last wetlands where natural conditons still prevail.

## References


Bas, C. & A. Cruzado, 1976. Campana oceanografica 'Altor V' (C. Bojador-C. Blanco, abril–mayo 1974). Caracteristicas y algunos resultados preliminares. Res. Exp. Cient. B/O Cornide 5. Suppl. Invest. Pesq.: 113–122.

Berrit, G. R., 1952. Esquisse des conditions hydrologiques du plateau continental du Cap Vert à la Gambie (températures et salinités). Bull. Inst. fr. Afrique noire 14: 735–761.

Cuq, F., this volume. Remote sensing of sea surface and coastal features in the area of the Golfe d'Arguin, Mauritania.

Domain, F., 1980. Contribution à la connaissance de l'écologie des poissons démersaux du plateau continental sénégalo-mauritanien. Les ressources démersales dans le contexte général du golfe de Guinée. Thèse Doc. Fac. Sci. Paris VI, 10 décembre 1980. I: 342 pp; II 68 pp.

Ekman, S., 1953. Zoogeography of the sea. Sidgwick & Jackson, London, 417 pp.

Faure, H., P. Pirazzoli & J. Monteillet, 1982. Où était situé le niveau de la mer le long du littoral Ouest africain à 7000 ans BP? Ass. sénégal. Etude quatern. Afr. Bull liais. Sénégal 66/67: 51–57.

Ingham, M. C., 1970. Wind and sea surfaces off Mauritania and Sierra Leone. Mar. Techn. Soc. J. 4: 55–57.

Maigret, J., 1972. Campagne expérimentale de pêche des sardinelles et autres espèces pélagiques. Juillet 1970–Octobre 1971. I. Observations concernant l'océanographie et la biologie des espèces. Lab. Pêches Nouadhibou, Soc. centr. Equip. Territ. intern. 23–51.

Margalef, R., 1975. Composition y distribucion del fitoplancton marino en la region de afloramiento del NW de Africa en marzo de 1973 (campaña 'Altor II' del 'Cornide de Saavedra'. Res. Exp. Cient. B/O Cornide, 4 Suppl. Invest. Pesq.: 145–170.

Miro Orell, M. de, 1973. Sedimientos recientes del margen continental de Mauritania (expedicion Sahara II). Res. Exp. Cient. B/O Cornide, 2 Suppl. Invest. Pesq.: 1–12.

Ould Dedah, this volume. Wind, surface water temperature, surface salinity and pollution in the area of the Banc d'Arguin, Mauritania.

Peters, H., 1976. The spreading of water masses of the Banc d'Arguin in the upwelling area off the northern Mauritanian coast. Meteor Forsch. Ergebnisse A 18: 78–100.

Postel, E., 1963. Hydrologie et biogéographie marine dans l'Ouest africain. In: Atlas international de l'Ouest africain. Organisation de l'unité africaine. Comm. sci. tech. rech., Dakar, Inst. fond. Afrique noire: 13–16.

Reyssac, J., 1977. Hydrologie, phytoplancton et production primaire de la Baie du Lévrier et du Banc d'Arguin. Bull. Inst. fond. Afrique noire 39: 487–554.

Reyssac, J., 1978. Principales caractéristiques du phytoplancton marin dans les différentes provinces biogéographiques de l'Atlantique africain. Vie Milieu 28/29: 353–370.

Rossignol, M., 1973. Contribution à l'étude du 'complexe guinéen'. Rép. Sénégal, Serv. Océanogr. Pêches ORSTOM, Océanographie, 017. Diffusion: Centre ORSTOM, Cayenne: 1–143.

Schaffer, G., 1974. On the Northwest African coastal upwelling system. Thesis Univ. Kiel.

Schemainda, R. & D. Nehring, 1975. The annual cycle of the space temporal dislocation of the North West African upwelling region. Third internat. symp. Upwelling Ecosystems, 25–28 August 1975, Kiel, Fed. Rep. Germany, 7 pp.

Schemainda, R., D. Nehring & S. Schulz, 1975. Ozeanologische Untersuchungen zum Produktionspotential der nordwestafrikanischen Wasserauftriebsregion, 1970–1973. Nat. Komm. Geod. Geoph. Akad. Wissenschaften D.D.R. IV/16: 4–84.

Sevrin-Reyssac, J., 1982. Etude comparée de quelques secteurs du littoral mauritanien entre le Cap Blanc et le Cap Timiris. Bull. Inst. fond. Afrique noire, 44: 39–56.

Sevrin-Reyssac, J., 1984. Conditions hydrologiques et phytoplancton dans la partie méridionale du Parc National du Banc d'Arguin (Mauritanie) en mars et avril 1983. Bull. Centre nat. Rech. Océanogr. Pêches, Nouadhibou (Mauritanie) 12: 107–113.

Sevrin-Reyssac, J., 1988. Hydroclimat et biogéographie de la région des îles du Banc d'Arguin (Mauritanie). Particularités du réseau trophique. Ann. Biol. 26: 185–198.

Sevrin-Reyssac, J. & B. Richer de Forges, 1985. Particularités de la faune icthyologique dans un milieu sursalé du Parc National de Banc d'Arguin (Mauritanie). Océanogr. tropic. 20: 85–90.

Wolff, W. J. & C. J. Smit, 1990. The Banc d'Arguin, Mauritania, as an environment for coastal birds. Ardea 78: 17–38.

*Hydrobiologia* **258**: 9–19, 1993.
*W. J. Wolff, J. van der Land, P. H. Nienhuis & P. A. W. J. de Wilde (eds), Ecological Studies in the Coastal Waters of Mauritania.*




# Wind, surface water temperature, surface salinity and pollution in the area of the Banc d'Arguin, Mauritania

Sidina ould Dedah
*Centre National de Recherches Océanographiques et de Pêches, B.P. 22, Nouadhibou, Mauritania*

*Key words:* Banc d'Arguin, Mauritania, wind, water temperature, salinity, pollution

## Abstract

Based on meteorological observations at Nouadhibou Airport, Mauritania, over the period 1953–1990 frequency distributions and averages are computed for wind speed and direction. Average wind speed reaches a maximum in May–June (9 m $s^{-1}$) and a minimum in November–December (6 m $s^{-1}$). About 85% of the time winds blow from northerly directions. Based on a data set collected since 1952 maps of surface water temperatures are constructed. Based on these maps and on observations on salinity an hypothetical current pattern for the shallow area between Nouadhibou and Cap Timiris, Mauritania, is proposed.

## Introduction

Coastal areas are considered to be highly productive. The Baie du Lévrier and especially the area of the Banc d'Arguin (Fig. 1) in Mauritania are no exception to that rule. In the latter two areas the combination of water masses of different origins creates a highly diverse mixture of tropical and temperate plant and animal species. Moreover the shallow seagrass beds attract numerous species for their reproduction. This results in these areas being a nursery and it has even been suggested that these waters function as reservoirs for some West-African fish populations (Maigret & Ly, 1986). Also bird numbers are extremely high; with over 2 million wintering waders the Banc d'Arguin ranks among the principal wintering sites of the world (Trotignon *et al.*, 1980; Altenburg *et al.*, 1982).

The richness of the ecosystems of the Banc d'Arguin is due to the interaction of a great number of factors. To protect and manage these sys-

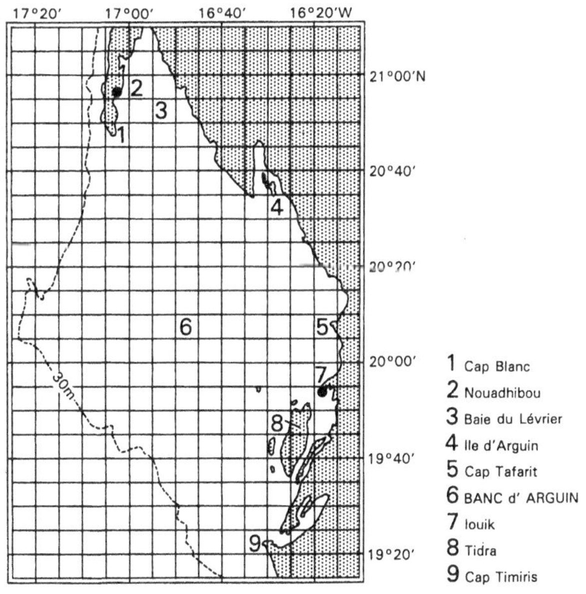


*Fig. 1.* The area of the Baie du Lévrier and the Banc d'Arguin in Mauritania. The rectangles have been defined for the analysis of temperature and salinity data.

*Table 1.* Frequency distribution of hydrographical rectangles (Fig. 1) according to the number of observations of surface water temperature per rectangle.

| Number of observations | 0 | 1–20 | 21–50 | 51–100 | 101–200 | > 200 | Total |
|---|---|---|---|---|---|---|---|
| Number of rectangles | 34 | 98 | 26 | 22 | 19 | 13 | 212 |
| Percentage | 16.0 | 46.2 | 12.3 | 10.4 | 9.0 | 6.1 | 100 |

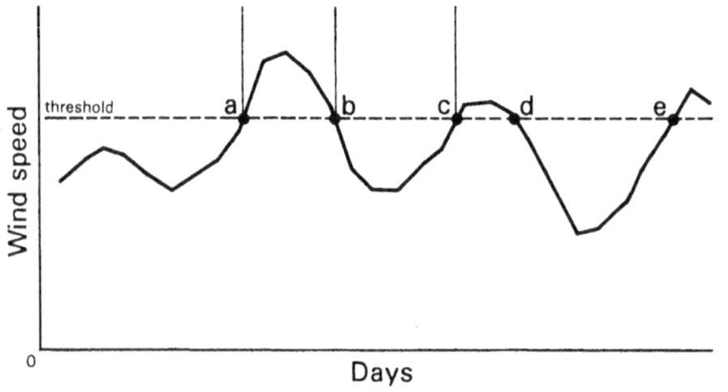


*Fig. 2.* Example of changes of wind speed and definition of periods of high wind (a-b, c-d), windows of quiet weather (b-c, d-e) and wind speed cycles (a-c, c-e).

tems in a sustainable way intimate knowledge of these factors is required. This paper is a contribution to the knowledge of the area of the Baie du Lévrier and the Banc d'Arguin and concerns a description of three major factors: wind, water temperature and salinity. In addition some data on local pollution are presented.

## Material and methods

Wind data have been obtained from the meteorological station of Nouadhibou Airport for the period 1953–1990 and were made available by the Agence pour la Sécurité de la Navigation Aerienne en Afrique (ASECNA). In addition wind data for Iouik have been derived from Smit *et al.* (1989) for two spring periods in 1985 and 1986.

For the analysis of wind data we have calculated monthly frequency distributions of wind speed and wind direction. For the analysis of wind stability on a time scale of days we have chosen the approach of Matuschevsky & Nadeev (1987). Figure 2 shows an example how wind speed may change over time. Having fixed an arbitrary threshold value of 8 m $s^{-1}$, we can distinguish periods of high winds: a-b, c-d, and windows of quiet weather: b-c, d-e. Moreover we can define the length of a wind cycle: a-c, c-e. These data have been presented as frequency distributions. The threshold value has been based on Roy (1990) who observed a declining phytoplankton productivity at wind speeds over 8 m $s^{-1}$.

The data on water temperatures and salinity have been extracted from the data base of the Centre National de Recherches Océanographiques et de Pêches (CNROP). These data have different origins. Part of these derive from measurements made by the Laboratoire des Pêches (now CNROP) since 1952. Another part is derived from measurements during scientific expeditions and by vessels engaged in industrial

*Table 2.* Comparison of wind speeds (m $sec^{-1}$) and standard deviations at Nouadhibou Airport and Iouik. Data from ASECNA and Smit *et al.* (1989).

| Period of observations | Number of days | Nouadhibou | | Iouik | | Correlation coefficient |
|---|---|---|---|---|---|---|
| | | Mean | s.d. | Mean | s.d. | |
| March-May 1985 | 34 | 9.50 | 1.83 | 9.07 | 2.03 | −0.15 |
| Febr.-April 86 | 69 | 9.58 | 2.13 | 7.46 | 1.65 | 0.45 |
| Total | 103 | 9.55 | 2.03 | 7.99 | 2.93 | 0.22 |

*Table 3.* Wind speed (m sec$^{-1}$) frequency distribution (1984–1990), average monthly wind speed with standard deviation (1960–1988) and maximum wind speed ever observed (1953–1982) for Nouadhibou Airport. Data from ASECNA.

| Wind speed | Jan | Feb | Mar | April | May | June | July | Aug | Sept | Oct | Nov | Dec |
|---|---|---|---|---|---|---|---|---|---|---|---|---|
| 0 | 2.08 | 1.50 | 2.65 | 0.36 | 0.17 | 0.62 | 1.77 | 2.26 | 0.97 | 1.68 | 1.80 | 3.71 |
| 1 | 0.56 | 0.12 | 0.17 | 0.06 | 0.00 | 0.07 | 0.00 | 0.47 | 0.21 | 0.20 | 0.41 | 0.97 |
| 2 | 2.53 | 2.17 | 2.71 | 0.53 | 0.17 | 0.55 | 1.13 | 2.20 | 2.22 | 1.88 | 3.11 | 4.60 |
| 3 | 5.63 | 4.16 | 4.61 | 1.18 | 0.63 | 0.62 | 2.34 | 3.86 | 4.86 | 4.17 | 7.94 | 10.00 |
| 4 | 7.60 | 4.97 | 8.82 | 2.54 | 1.15 | 2.00 | 5.16 | 5.99 | 6.11 | 7.19 | 10.32 | 11.45 |
| 5 | 8.72 | 11.18 | 9.57 | 5.39 | 3.05 | 2.83 | 8.06 | 9.98 | 11.39 | 11.29 | 15.56 | 14.19 |
| 6 | 8.27 | 11.12 | 11.31 | 6.57 | 4.15 | 2.62 | 8.55 | 8.92 | 8.33 | 11.29 | 16.30 | 14.44 |
| 7 | 13.90 | 11.24 | 10.44 | 8.93 | 6.34 | 5.86 | 8.06 | 9.31 | 12.15 | 17.00 | 14.91 | 12.74 |
| 8 | 11.93 | 12.67 | 9.86 | 12.78 | 11.35 | 11.79 | 8.15 | 10.85 | 13.89 | 14.65 | 11.55 | 9.84 |
| 9 | 10.19 | 13.35 | 8.48 | 11.95 | 12.73 | 13.93 | 9.19 | 9.45 | 11.88 | 10.01 | 6.55 | 7.10 |
| 10 | 14.07 | 13.17 | 10.27 | 16.99 | 22.07 | 22.28 | 14.68 | 11.71 | 13.40 | 11.56 | 6.55 | 4.35 |
| 11 | 7.20 | 5.90 | 8.07 | 11.12 | 16.24 | 16.69 | 11.61 | 10.58 | 8.68 | 6.25 | 2.62 | 4.19 |
| 12 | 4.90 | 5.03 | 6.52 | 8.64 | 13.25 | 11.72 | 9.11 | 8.45 | 3.96 | 2.02 | 1.39 | 1.53 |
| 13 | 1.52 | 2.05 | 3.69 | 6.39 | 6.05 | 6.28 | 6.45 | 3.26 | 1.46 | 0.67 | 0.82 | 0.73 |
| 14 | 0.56 | 0.87 | 1.79 | 3.49 | 2.25 | 1.79 | 3.06 | 2.13 | 0.35 | 0.07 | 0.16 | 0.08 |
| 15 | 0.34 | 0.50 | 1.04 | 2.07 | 0.40 | 0.28 | 2.33 | 0.60 | 0.07 | 0.07 | | 0.08 |
| 16 | | | | 0.65 | | 0.07 | 0.16 | | 0.07 | | | |
| 17 | | | | 0.18 | | | 0.08 | | | | | |
| 18 | | | | 0.12 | | | 0.08 | | | | | |
| 19 | | | | 0.06 | | | | | | | | |
| Mean | 6.68 | 7.14 | 7.96 | 8.57 | 9.00 | 9.06 | 8.30 | 7.67 | 7.39 | 6.93 | 6.10 | 6.02 |
| s.d. | 1.75 | 1.29 | 1.70 | 1.73 | 1.40 | 1.76 | 1.57 | 1.34 | 1.27 | 1.47 | 1.09 | 1.24 |
| Max | 31 | 45 | 50 | 36 | 41 | 32 | 55 | 65 | 33 | 27 | 34 | 31 |

*Table 4.* Frequency distribution of wind directions for Nouadhibou Airport for the period 1984–1990. Data from ASECNA.

| Direction | Jan | Feb | Mar | Apr | May | June | July | Aug | Sept | Oct | Nov | Dec | Year |
|---|---|---|---|---|---|---|---|---|---|---|---|---|---|
| 231–250 | | 0.9 | 0.5 | 0.1 | | | 0.7 | 1.2 | 1.0 | | 0.3 | 0.7 | 0.4 |
| 251–270 | 0.1 | 0.5 | 1.0 | 0.3 | | 0.7 | 1.6 | 1.2 | 1.7 | | 0.7 | 0.4 | 0.7 |
| 271–290 | 0.8 | 0.3 | 3.0 | 1.0 | 0.2 | 0.1 | 2.8 | 2.2 | 2.2 | 0.6 | 2.7 | 1.3 | 1.4 |
| 291–310 | 1.9 | 4.0 | 4.3 | 1.2 | 0.1 | 1.5 | 5.5 | 5.3 | 4.2 | 2.4 | 4.6 | 3.0 | 3.0 |
| 311–330 | 5.4 | 7.2 | 11.2 | 7.3 | 8.9 | 10.0 | 19.2 | 18.0 | 15.8 | 14.8 | 13.4 | 7.8 | 11.5 |
| 331–350 | 8.8 | 15.8 | 17.7 | 27.8 | 35.5 | 33.0 | 36.1 | 31.1 | 27.0 | 26.3 | 15.1 | 9.0 | 23.5 |
| 351–10 | 17.2 | 24.3 | 27.0 | 37.0 | 43.5 | 38.0 | 24.5 | 28.1 | 26.4 | 30.8 | 24.5 | 16.1 | 28.1 |
| 11–30 | 20.8 | 22.7 | 16.8 | 18.7 | 8.8 | 14.8 | 5.1 | 7.6 | 12.3 | 16.3 | 18.7 | 11.6 | 14.6 |
| 31–50 | 21.4 | 11.8 | 10.3 | 4.6 | 2.6 | 1.3 | 0.3 | 1.0 | 2.5 | 4.9 | 10.2 | 15.7 | 7.3 |
| 51–70 | 11.4 | 4.8 | 2.8 | 0.6 | 0.2 | | 0.3 | 0.2 | 0.7 | 1.1 | 2.4 | 14.4 | 3.3 |
| 71–90 | 5.9 | 2.6 | 1.9 | 0.5 | | | 0.1 | 0.2 | 0.4 | 0.9 | 1.5 | 9.1 | 2.0 |
| 91–110 | 3.0 | 3.5 | 1.2 | 0.1 | 0.1 | 0.2 | | 0.1 | 0.6 | 0.4 | 0.8 | 5.0 | 1.3 |
| 111–130 | 1.0 | 0.4 | 0.3 | 0.1 | 0.1 | 0.2 | 0.2 | 0.2 | 0.6 | 0.2 | 0.4 | 2.6 | 0.5 |
| 131–150 | 1.0 | 0.4 | 0.3 | 0.2 | | | 0.3 | 0.6 | 1.3 | 0.3 | 0.9 | 1.0 | 0.5 |
| 151–170 | 0.6 | 1.0 | 0.7 | 0.1 | | 0.2 | 0.7 | 0.6 | 0.6 | 0.4 | 1.5 | 1.5 | 0.6 |
| 171–190 | 0.6 | 1.1 | 0.3 | 0.1 | | | 1.7 | 1.1 | 0.9 | 0.3 | 1.0 | 0.4 | 0.6 |
| 191–210 | 0.1 | 0.7 | 0.6 | 0.2 | | | 0.7 | 0.8 | 0.3 | 0.3 | 0.4 | 0.3 | 0.4 |
| 211–230 | 0.1 | 0.5 | 0.3 | 0.1 | | 0.1 | 0.3 | 0.7 | 1.5 | 0.2 | 1.2 | 0.2 | 0.4 |

*Table 5.* Length and standard deviation of periods of high wind, windows of quiet weather and wind cycles for Nouadhibou Airport. A wind speed of 8 m.sec$^{-1}$ is considered to discriminate between high winds and quiet weather.

| | Jan–March | | April–June | | July–Sept | | Oct–Dec | |
|---|---|---|---|---|---|---|---|---|
| | Mean | s.d. | Mean | s.d. | Mean | s.d. | Mean | s.d. |
| Period of high winds (>8m sec$^{-1}$) | 4.5 | 3.5 | 14.3 | 10.1 | 3.6 | 3.1 | 3.9 | 3.2 |
| Quiet weather window (<8m sec$^{-1}$) | 4.7 | 4.5 | 2.7 | 2.0 | 2.8 | 2.3 | 8.7 | 9.2 |
| Cycle length | 8.7 | 4.7 | 16.3 | 8.9 | 6.4 | 3.6 | 12.0 | 9.4 |

fishing. The CNROP data base now contains about 65% of all available data. In total we have used 8590 observations on surface water temperature and 4220 observations on surface salinity from the area within the 30 m isobath (Fig. 1).

For the analysis we have divided the sea area between 19°20′ and 21°20′ N and East of 17°25′ W into rectangles of 5 degrees latitude and longitude each. Only rectangles containing seabed shallower than 30 m have been considered (Fig. 1). The data show an inequal distribution in time and in space (Table 1). Many data are available for the regions West and North-East of the Banc d'Arguin proper as well as for the Baie du Lévrier. On the other hand, due to difficult access to some areas, we have no data for 16% of the rectangles, especially West of the Cap Blanc peninsula, in the Baie d'Arguin and around the island of Tidra.

For each rectangle we have computed the mean temperature and mean salinity per month. Through these monthly means for each rectangle we have fitted manually a curve and the monthly values thus obtained have served as the basis for all further calculations. The temperature maps have been based also on other available information such as some satellite pictures, the coastal and submarine topography of the area, and the temperature regime at the continental shelf, which is fairly well known. Based on the maps we have developed a subdivision of water masses and computed temperature characteristics per region.

*Table 6.* Annual development of average surface water temperatures for six subdivisions of the area of the Baie du Lévrier and the Banc d'Arguin (compare Fig. 13), for the entire area of the Baie du Lévrier and the Banc d'Arguin (BdA), and for the Mauritanian continental shelf between 19° 20′ and 21° 00′N (shelf).

| Area | 1 | 2 | 3 | 4 | 5 | 6 | BdA | shelf |
|---|---|---|---|---|---|---|---|---|
| January | 17.5 | 17.6 | 17.7 | 18.3 | 18.4 | 18.7 | 18.0 | 18.4 |
| February | 18.0 | 18.5 | 18.6 | 18.5 | 19.0 | 18.2 | 18.4 | 18.1 |
| March | 18.2 | 18.7 | 19.6 | 19.4 | 20.3 | 19.1 | 19.1 | 18.2 |
| April | 17.8 | 18.5 | 20.1 | 19.8 | 21.5 | 18.4 | 19.1 | 17.9 |
| May | 17.6 | 18.5 | 21.2 | 20.8 | 22.7 | 18.4 | 19.5 | 17.9 |
| June | 18.7 | 19.3 | 22.1 | 22.0 | 23.7 | 20.3 | 20.8 | 18.8 |
| July | 22.0 | 21.5 | 23.8 | 24.6 | 25.4 | 23.8 | 23.5 | 21.4 |
| August | 23.3 | 23.4 | 24.9 | 25.6 | 26.9 | 26.1 | 24.9 | 23.7 |
| September | 23.4 | 23.8 | 25.1 | 26.1 | 27.3 | 25.7 | 25.6 | 23.0 |
| October | 21.5 | 22.0 | 23.3 | 24.4 | 25.4 | 23.1 | 23.2 | 21.0 |
| November | 19.3 | 19.2 | 20.3 | 21.8 | 22.3 | 21.0 | 20.7 | 19.6 |
| December | 18.4 | 18.1 | 18.3 | 19.4 | 19.9 | 19.6 | 19.1 | 18.8 |

The subdivision has taken into account the pattern of seasonal development of temperature, the shape of the temperature curves, the maximum and the minimum values.

Because of time constraints the salinity data have not been put on maps. We did, however, establish temperature – salinity diagrams for each month.

## Results

### *Wind data*

Table 2 shows a comparison of wind data for Nouadhibou Airport and Iouik for two periods in which meteorological observations were made at both stations. Average wind speeds for the two

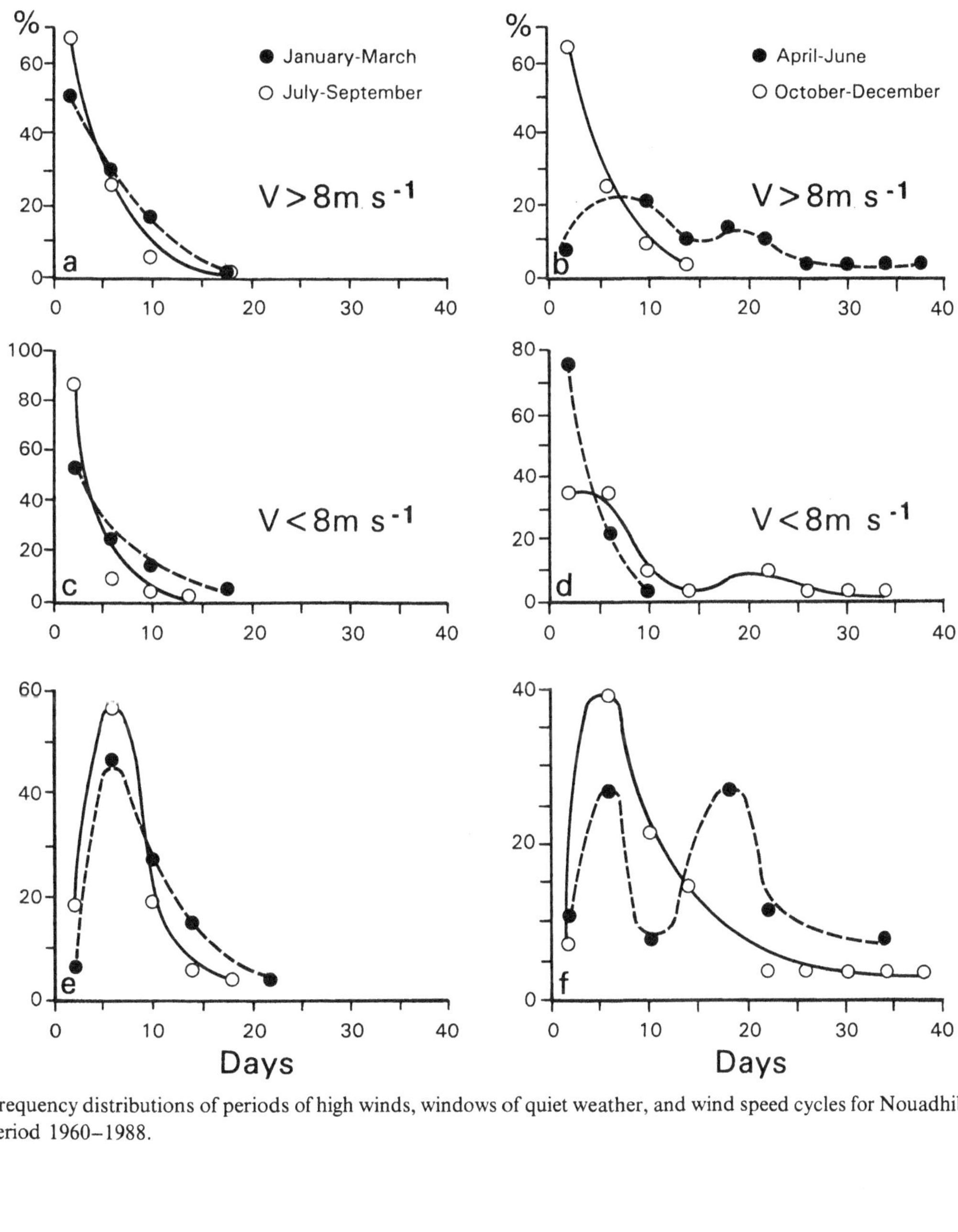


*Fig. 3.* Frequency distributions of periods of high winds, windows of quiet weather, and wind speed cycles for Nouadhibou Airport for the period 1960–1988.

stations do not differ significantly for the periods considered, so it may tentatively be concluded that the wind climates of Nouadhibou and Iouik are similar. On the other hand wind speeds on a day to day basis were not correlated significantly for the same periods.

Table 3 gives the wind speed frequency distribution per month for Nouadhibou Airport as well as the average wind speed per month and the maximum speed ever observed. It appears that there is a clear annual pattern with highest wind speeds in May–June and lowest values in November–December. The variability of the wind speed shows a similar pattern, but less clearly.

Table 4 shows the frequency distribution of wind directions per month for Noauadhibou Airport. Wind comes from the North (NW-N-NE) during nearly 85% of the year. In the period December–February northeasterly winds predominate. In spring the dominating winds shift to the North and in April, May and June northerly directions dominate. In July–September a northwesterly tendency becomes visible, but from October onwards a shift to the North-East develops again.

Table 5 shows the results of the analysis of wind stability. During the larger part of the year periods of high wind ($>8\ \mathrm{m\ s^{-1}}$) last only for 3.5–4.5 days on average. Only in the period April–June, when the trade winds reach maximum strength, the periods of high wind last for about 14 days on average. The windows of quiet weather ($<8\ \mathrm{m\ s^{-1}}$) are short from April to September and have their maximum duration in October–December. Figure 3 shows the frequency distribution of the various aspects of wind stability.

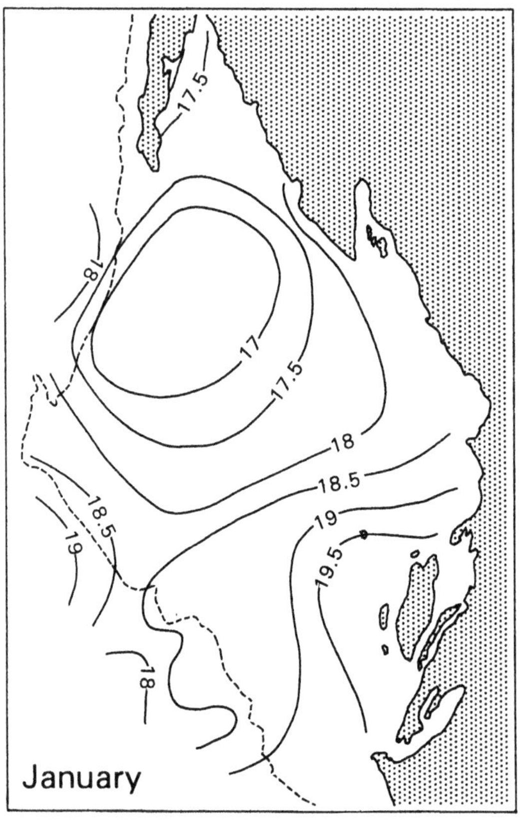


*Fig. 4.* Distribution of surface water temperature in January.

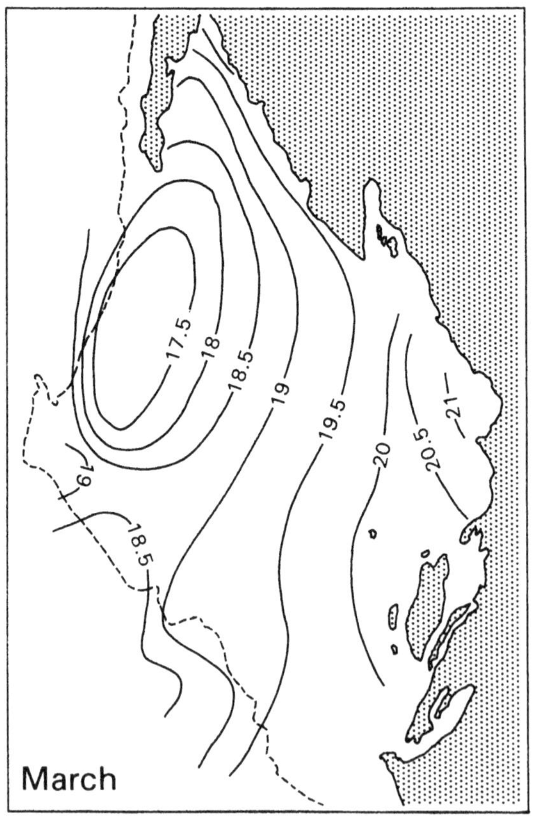


*Fig. 5.* Distribution of surface water temperature in March.

*Surface water temperature*

Figures 4–9 show the surface water temperatures in the course of the year. In January temperature conditions are rather homogeneous (Fig. 4), but in March on the one hand upwelling becomes more dominant whereas on the other hand insolation becomes more important (Fig. 5). Thus the temperature gradient across the area becomes stronger. The strongest gradient, however, is observed in May (Fig. 6). Lowest temperatures occur always South of Cap Blanc whereas the highest values occur in the coastal areas such as around the island of Tidra and inside the Baie du Lévrier.

The same pattern occurs in the annual average values of temperature (Fig. 10). The annual fluctuations of temperature are rather small with an increasing trend from the vicinity of Cap Blanc towards the area around the island of Tidra (Fig. 11).

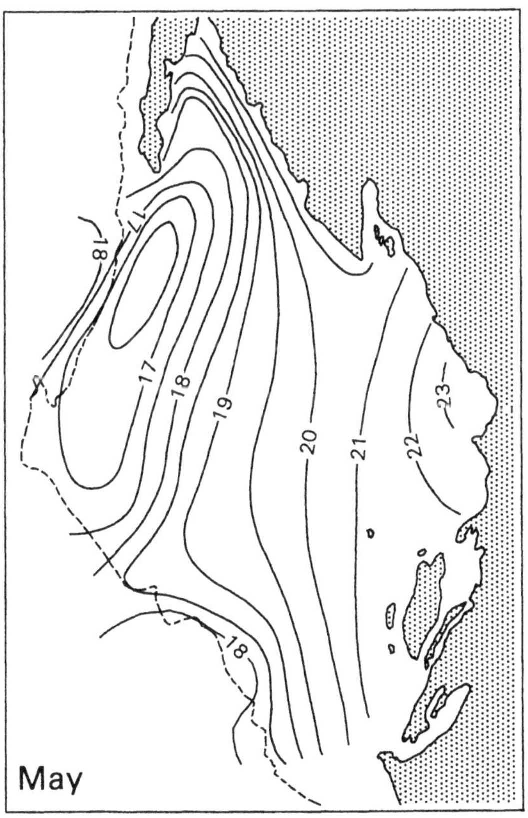


*Fig. 6.* Distribution of surface water temperature in May.

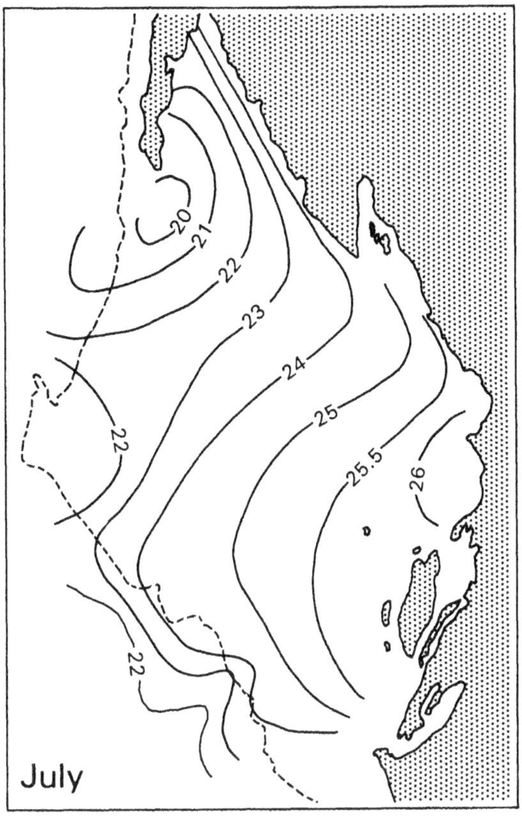


*Fig. 7.* Distribution of surface water temperature in July.

*Surface salinity*

The T-S diagrams (Fig. 12) demonstrate the existence of two different water types. One type with salinities between 35.5 and 37 is found at the seaward part of the area and in the entrance of the Baie du Lévrier. The other type with salinities over 38, occurs inshore. Both types change in the course of the year, apparently following the yearly cycle of insolation.

*Subdivision of the study area*

Our temperature data allow a subdivision of the area studied (Fig. 13). Table 6 gives the various

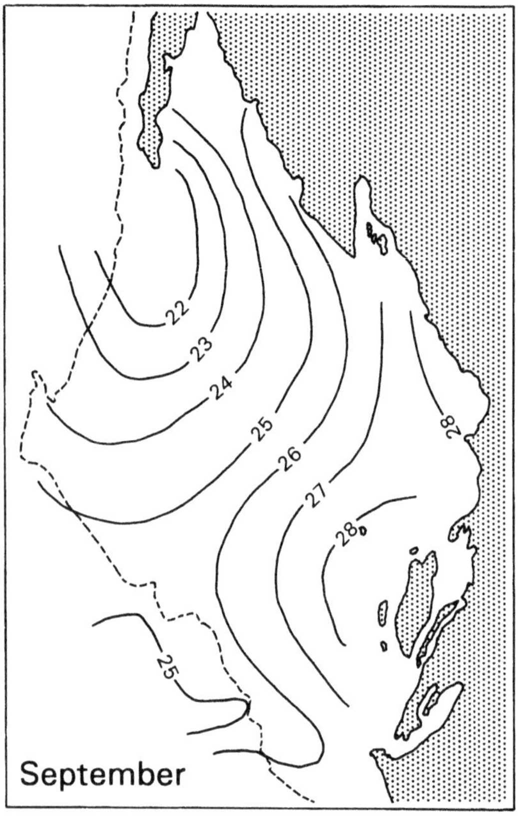


*Fig. 8.* Distribution of surface water temperature in September.

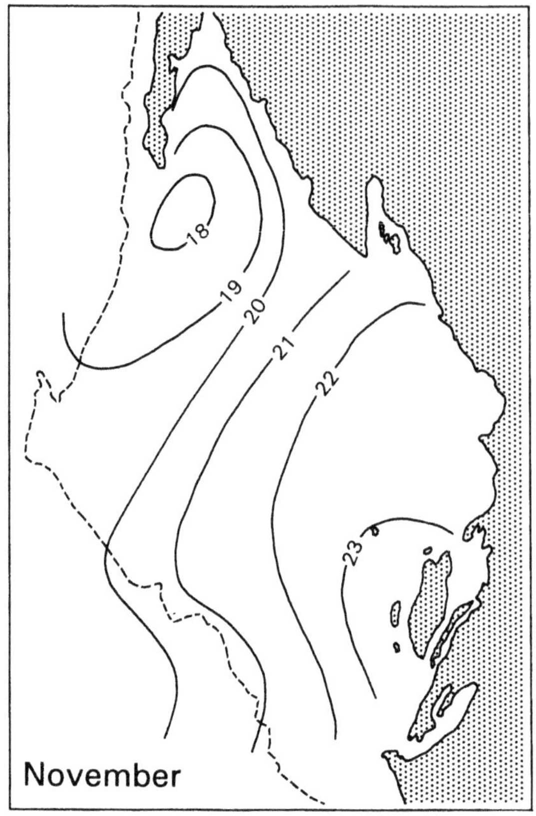


*Fig. 9.* Distribution of surface water temperature in November.

characteristics of these subareas and compares these with the Mauritanian continental shelf. Based on these data a pattern of water circulation is proposed for the Baie du Lévrier and the region of the Banc d'Arguin (Fig. 13).

*Pollution*

All domestic and industrial sewage of the city of Nouadhibou (about 50000 inhabitants) are discharged untreated into the Baie du Lévrier. Particularly the fish processing industry discharges large amounts of oxygen demanding sewage.

Another source of pollution is the iron ore loading facility at Cansado. Dust from this facility is blown into the Baie du Lévrier by the strong northerly winds of the area. Quantitative data are lacking, however.

A major source of pollution are the ships on the roads of Nouadhibou. It is estimated that annually over 8000 tons of oil are being lost or deliberately discharged into the Baie du Lévrier. The same ships also are the source of large quantities of solid waste. Because of the prevailing winds and currents floating wastes may be expected to drift towards the Parc National du Banc d'Arguin in the South. Indeed, according to observations of the Coast Guard during three periods in 1990 huge quantities of this material wash ashore in several bays near Cap Tafarit. Also South of Iouik solid waste obviously originating from ships is found along the shores (W. J. Wolff, pers. comm.).

Oil pollution may also result from accidents. Especially the vessels provisioning the country itself or the fishing fleet in Nouadhibou are likely candidates for collisions or strandings with serious consequences. Recently two vessels collided near La Guerra and another one ('Rosso II') with 500 tons of gasoil ran ashore on the Banc d'Arguin. Also the large number of abandoned vessels and wrecks (120 in 1989) in Nouadhibou roads constitutes a potential danger of pollution.

Finally oil pollution may result from the offshore transport of very large quantities of crude oil from the Arabian Gulf to Europe, but sofar no signs of this source of pollution have been noticed in the area between Cap Blanc and Cap Timiris. Operational discharges from these tankers are the most likely source of pollution, but the recent large oilspill off the coast of Morocco shows that the Mauritanian coast is also vulnerable to large-scale oil pollution resulting from accidents of oil tankers passing offshore.

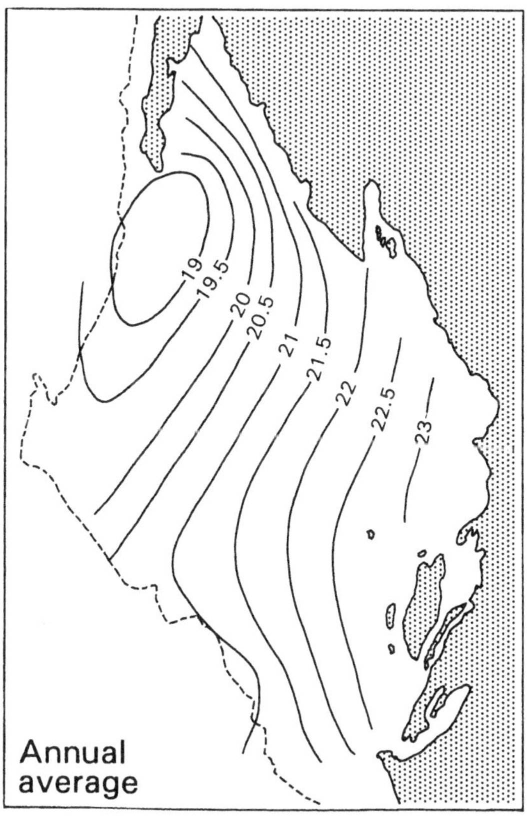


*Fig. 10.* Distribution of annual mean surface water temperature.

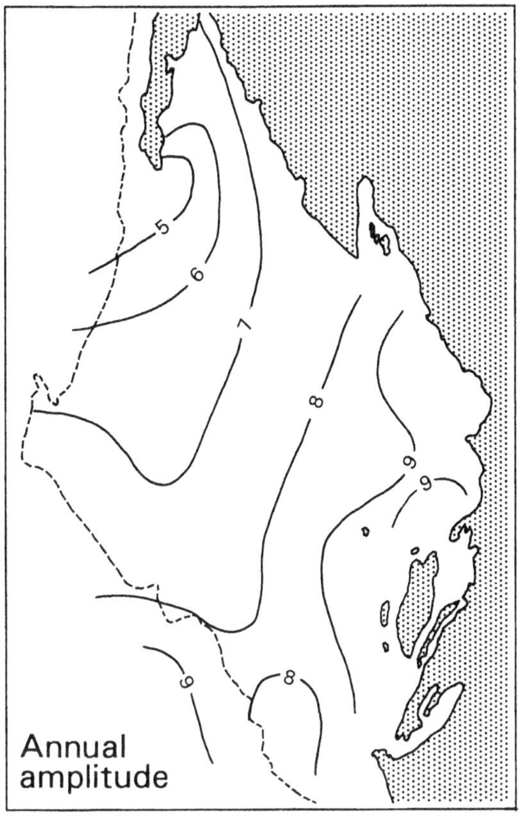


*Fig. 11.* Distribution of the annual surface water temperature amplitude.

## Discussion

The major interest of the CNROP is vested in the large-scale fisheries off the Mauritanian coast. However, data have also been collected in the Baie du Lévrier and in the area of the Banc d'Arguin. Notwithstanding the shallow nature of this area and the fact that large parts are uncharted, it has proven possible for the first time to produce maps for an environmental factor for the entire area between Nouadhibou and Cap Timiris.

The analysis of wind data may be compared to the analysis of a similar data series from the same source by Smit (Smit *et al.*, 1989; Wolff & Smit, 1990). Although this author analysed data from periods different from ours his results are very similar to those of our analysis.

Our analysis of the hydrology of the Baie du

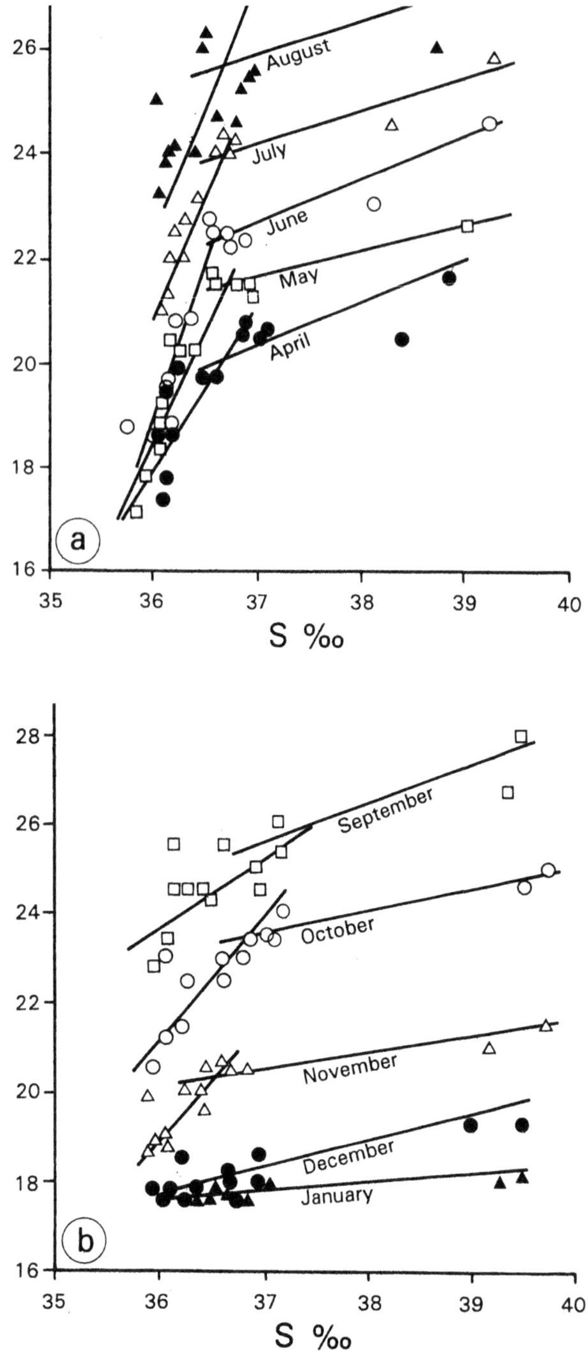


*Fig. 12.* T-S diagrams for the area of the Baie du Lévrier and the Banc d'Arguin per month.

Lévrier and the Banc d'Arguin has to depart from the observation that this area is influenced by at least four different water types (Tixerant, 1966; Maigret & Ly, 1986; Loktionov, 1993).

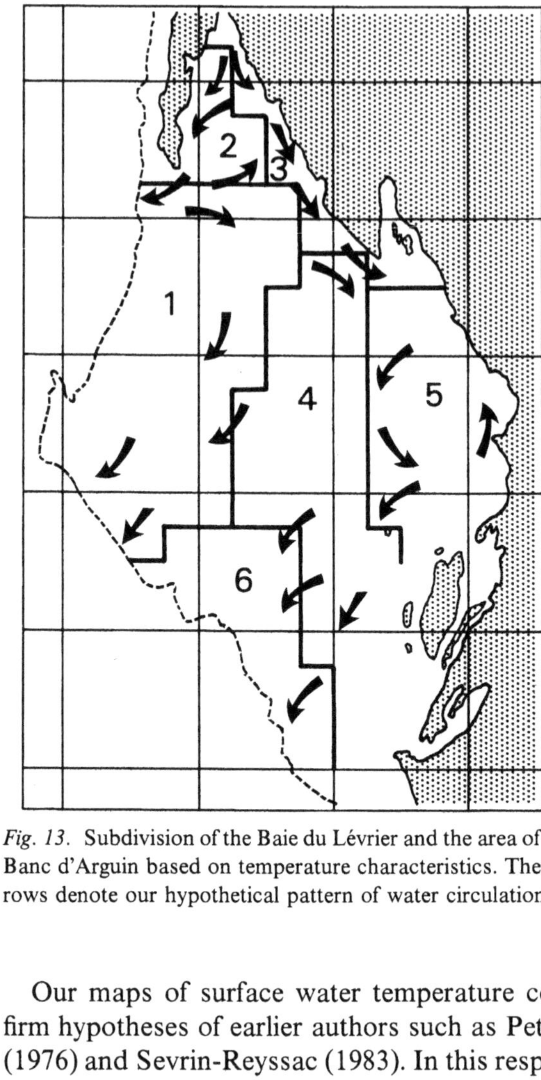


*Fig. 13.* Subdivision of the Baie du Lévrier and the area of the Banc d'Arguin based on temperature characteristics. The arrows denote our hypothetical pattern of water circulation.

Our maps of surface water temperature confirm hypotheses of earlier authors such as Peters (1976) and Sevrin-Reyssac (1983). In this respect the paper of Cuq (1993) more or less confirms our results.

Our work has also shown how many gaps in our knowledge still exist, particularly for the region of the shallow Banc d'Arguin. The collection of such data will form a major task for the years ahead.

## References

Altenberg, W., M. Engelmoer, R. Mes & T. Piersma, 1982. Wintering waders on the Banc d'Arguin, Mauritania. Stichting Veth tot Steun aan Waddenonderzoek, Leiden, 283 pp.
Cuq, F., 1993. Remote sensing of sea surface and coastal

features in the area of the Golfe d'Arguin, Mauritania. Hydrobiologia 258: 33–40.
Loktionov, Y., 1993. Hydrographical observations west of the Banc d'Arguin, Mauritania, in May 1988. Hydrobiologia 258: 21–32.
Maigret, J. & B. Ly, 1986. Les poissons de mer de Mauritanie. Centre National de Recherches Océanographiques et de Pêches, Nouadhibou, 213 pp.
Matushevsky, G. V. & V. V. Nadeev, 1987. Statistical laws for the distribution of wind-wave persistence characteristics. Meteorol. Hydrol. 9: 70–75. (In Russian).
Peters, H., 1976. The spreading of water masses of the Banc d'Arguin in the upwelling area off the northern Mauritanian coast. Meteor Forsch. Ergebnisse A 18: 78–100.
Roy, C., 1990. Reponses des stocks de poissons pélagiques à la dynamique des upwellings en Afrique de l'Ouest. Analyse et modélisation. Thèse Doct. Etat, Univ. Bretagne Occ., 149 pp.
Sevrin-Reyssac, J., 1983. Quelques particularités de la chaîne alimentaire marine dans la région des îles du Banc d'Arguin (Parc National du Banc d'Arguin, Mauritanie). Bull. Centr. Natl Rech. Océanogr. Pêches, Nouadhibou 11: 41–52.
Smit, C. J., A.-M. Blomert & L. Zwarts, 1989. Meteorological measurements. In: B. J. Ens, T. Piersma, W. J. Wolff & L. Zwarts (eds), Report of the Dutch-Mauritanian project Banc d'Arguin 1985–86. Report WIW0 25/RIN-rapport 89/6: 16–42.
Tixerant, G., 1966. Régime hydrologique de la Baie du Lévrier. Doc. Lab. Pêches, Nouadhibou, 62 pp.
Trotignon, E., J. Trotignon, M. Baillou, J.-F. Dejonghe, L. Duhautois & M. Lecomte, 1980. Recensement hivernal des limicoles et des autres oiseaux aquatiques sur le Banc d'Arguin (Mauritanie) (Hiver 1978/79). L'Oiseau et R.F.O. 50: 323–343.
Wolff, W. J. & C. J. Smit, 1990. The Banc d'Arguin, Mauritania, as an environment for coastal birds. Ardea 78: 17–38.

*Hydrobiologia* **258**: 21–32, 1993.
*W. J. Wolff, J. van der Land, P. H. Nienhuis & P. A. W. J. de Wilde (eds), Ecological Studies in the Coastal Waters of Mauritania.*




# Hydrographical observations west of the Banc d'Arguin, Mauritania, in May 1988

Y. Loktionov
*Centre National de Recherches Océanographiques et de Pêches, B.P. 122, Nouadhibou, Mauritania; Present address: Institute for Fisheries and Oceanography Atlantniro, vl. Frunze 91 KW 19, 236016 Kaliningrad, Russia*

*Key words:* Mauritania (West Africa), Banc d'Arguin, shelf hydrography

## Abstract

The present paper discusses the distribution of hydrographical parameters in Mauritanian waters between 19° 20′ N and 20° 40′ N. Data were collected during the first leg of the Dutch Mauritania Expedition in May 1988 with the R/V 'TYRO' and comprise measurements of water temperature (T), salinity (S) and dissolved oxygen contents ($O_2$) in the area off the Banc d'Arguin. The time of investigation fell in the transitional period between the cold and warm season, which in this area is characterised by a peak in the trade wind speed and progressive heating of waters. Wind speed and sea surface temperature distribution in May 1988 were close to the seasonal averages which suggests that the observed hydrographical conditions were normal for this period as well. The highest spatial variability of T, S and $O_2$ was found near the southern end of the Banc d'Arguin, where the combined effect of coastal upwelling and seasonal heating made the thermocline upper limit rise towards the surface and upwelling waters reach the shallowest parts of the shelf. In the northern part of the study area upwelling phenomena were observed in the subsurface layers below 50 m depth; upwelling water did not reach beyond the outer edge of the shelf. The analysis of the TS diagram showed that water masses in the upper 300 m layer were mainly composed of South Atlantic Central Water with an admixture of warm, low-salinity surface water of Guinean origin. Traces of North Atlantic Central Water were found in the subsurface layer and at the deepest levels of measurements. Almost no traces of warm, saline water originating from the shallow Banc d'Arguin were registered during the survey.

## Introduction.

The spatio-temporal variability in the waters on the shelf and continental slope on the western side of the Banc d'Arguin is the result of a complex superposition of several phenomena, the most important of which are:

1. The area can be defined as a zone of transition from tropical conditions (present south of Cap Timiris) to waters with a greater degree of mixing and well pronounced seasonal fluctuations of hydrographical characteristics. Highest spatial variation of salinity and temperature is observed in the sector between 20° 00′ N and 20° 30′ N (Loktionov, 1989).
2. The upper layer down to a depth of 300 m in the investigated region consists of a mixture of South Atlantic Central Water (SACW) and warm, low salinity water of Guinean origin. It is periodically intruded by tongues of North Atlantic Central Water (NACW) coming from

the west (Mittelstaedt, 1972). An additional complication is the frequently observed 'dumps' of warm, saline and dense water originating from the Banc d'Arguin shallows, where it is formed by evaporation and solar irradiation (Peters, 1976).

3. The cold and the warm season, each characterised by a relatively stable spatial distribution of water properties and circulation pattern, alternate here in an annual cycle. The cold season (January–April) is separated from the warm season (July–October) by a transitional period (May–June) during which progressive heating of the water leads to a high temporal variability of its physical parameters (Loktionov, 1989).
4. The peak in trade wind speed, normally registered in May–June (Loktionov, 1988), creates most favourable conditions for the development of coastal upwelling, leading to an even more complicated distribution of hydrographical and chemical parameters in the transitional period of May–June.

The foregoing shows that in the waters along the western edge of the Banc d'Arguin 'spring' is the time of year when the highest level of variability can be observed. The survey with R/V 'TYRO' in May 1988 was aimed at describing the hydrographic features occurring during this specific period. Special attention was given to the interaction of waters with dissimilar properties over the outer shelf, which is of great biological significance (Maske, 1976).

## Materials and methods

The data were collected with R/V 'TYRO' in Mauritanian waters between 10 and 29 May 1988. The CTD data consist of two sets collected in consecutive periods. The first set was obtained in the period from 10 to 13 May during a hydrographical survey of the study area. The stations were situated on 8 transects (Fig. 1a), which extended from the inner shelf (10–15 m depth) to the waters over the continental slope where the deepest measurements were made at 300 m.

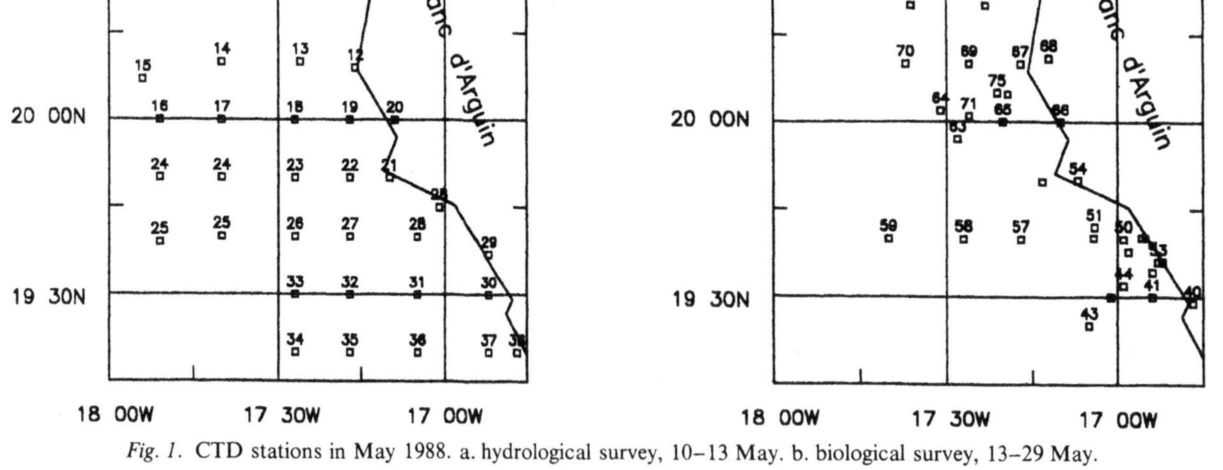


*Fig. 1.* CTD stations in May 1988. a. hydrological survey, 10–13 May. b. biological survey, 13–29 May.

The second dataset consists of incidental CTD-casts made during the biological survey, which took place during the remaining part of the cruise, *i.e.* from 13–29 May. The positions of the stations are shown in Fig. 1b. Some of these stations, especially in the northern part of the study area, had also been visited during the hydrographical survey. The time interval between duplicate readings ranged from 10–18 days.

For the analysis of the vertical distribution of water characteristics the CTD readings from the international standard depths were used. The upper limit of the thermocline was defined as the level where the value of the vertical temperature gradient was equal or higher than 0.2 °C $m^{-1}$. To be more precise in the estimation of this parameter the CTD readings from 1 or 1.5 m depth intervals were used. The maps and sections shown in this paper were computer-made whereby some of the smallest details, especially in the horizontal distribution of the parameters, may have been lost.

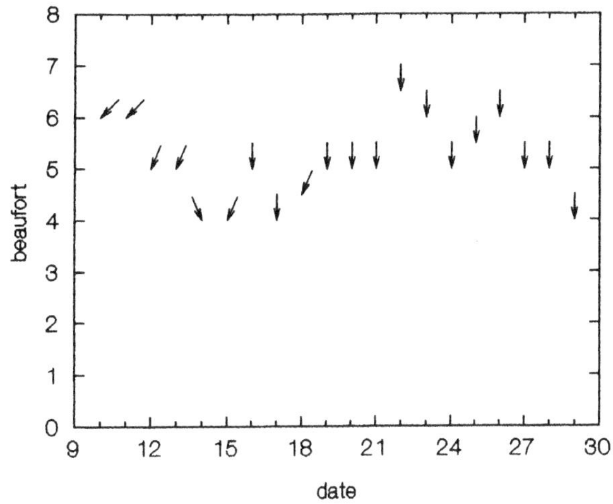


*Fig. 2.* Shipboard measurements of wind speed (Beaufort scale) and direction (denoted by arrows).

## Results and interpretation.

### *Wind conditions*

In the annual variation of the trade wind speed, the maximum wind speed in the region situated between Cap Blanc and Cap Timiris normally occurs from May to June. The long-term monthly mean value for May is 9.9 m $s^{-1}$ (Loktionov, 1988). The same series of data showed that in this period the winds are northerly. As the onboard wind force measurements during the cruise were made in the Beaufort scale, it was not possible to follow the wind speed in details. These measurements (Fig. 2), nevertheless, show that the wind was moderate to strong from northerly directions throughout the whole cruise. For a more precise appraisal of wind conditions it was possible to use the monthly averaged wind speed, based on

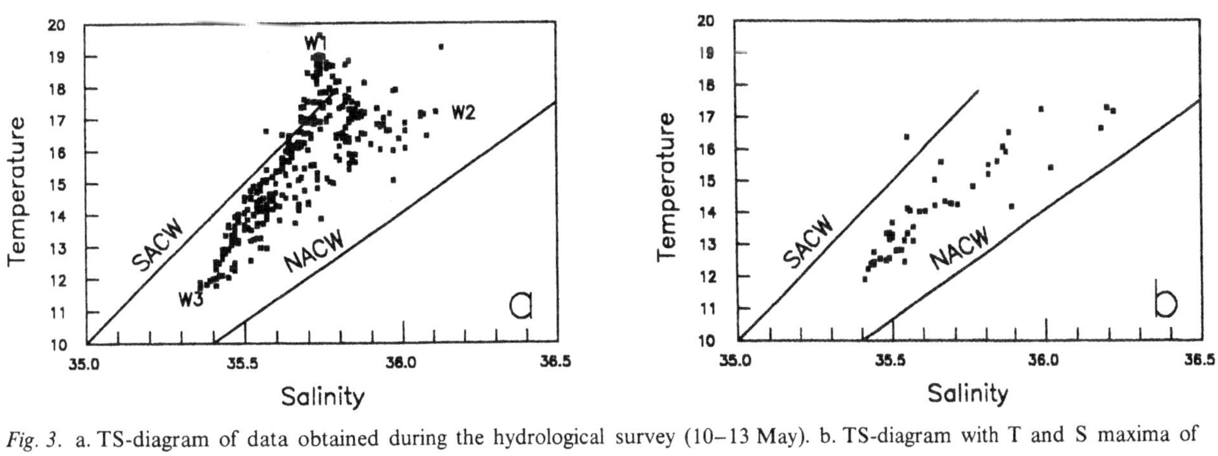


*Fig. 3.* a. TS-diagram of data obtained during the hydrological survey (10–13 May). b. TS-diagram with T and S maxima of important intrusions.

continuous 3 hour observations by the Meteorological Service at Nouadhibou airport. The value of this parameter in May 1988 was 10.6 m $s^{-1}$, which means that the trade winds during the survey were slightly stronger than normal. Moreover, the constant northerly direction of the wind caused a positive monthly anomaly (1.2 m $s^{-1}$) of the meridional component of the wind speed. The importance of this for the development of nearshore upwelling was shown by Barton *et al.* (1977), Brink *et al.* (1981) and Brink (1983).

Because the wind registered during the expedition was close to normal, it can be expected that those hydrographical features that mainly depend on wind conditions are normal for this period of the year.

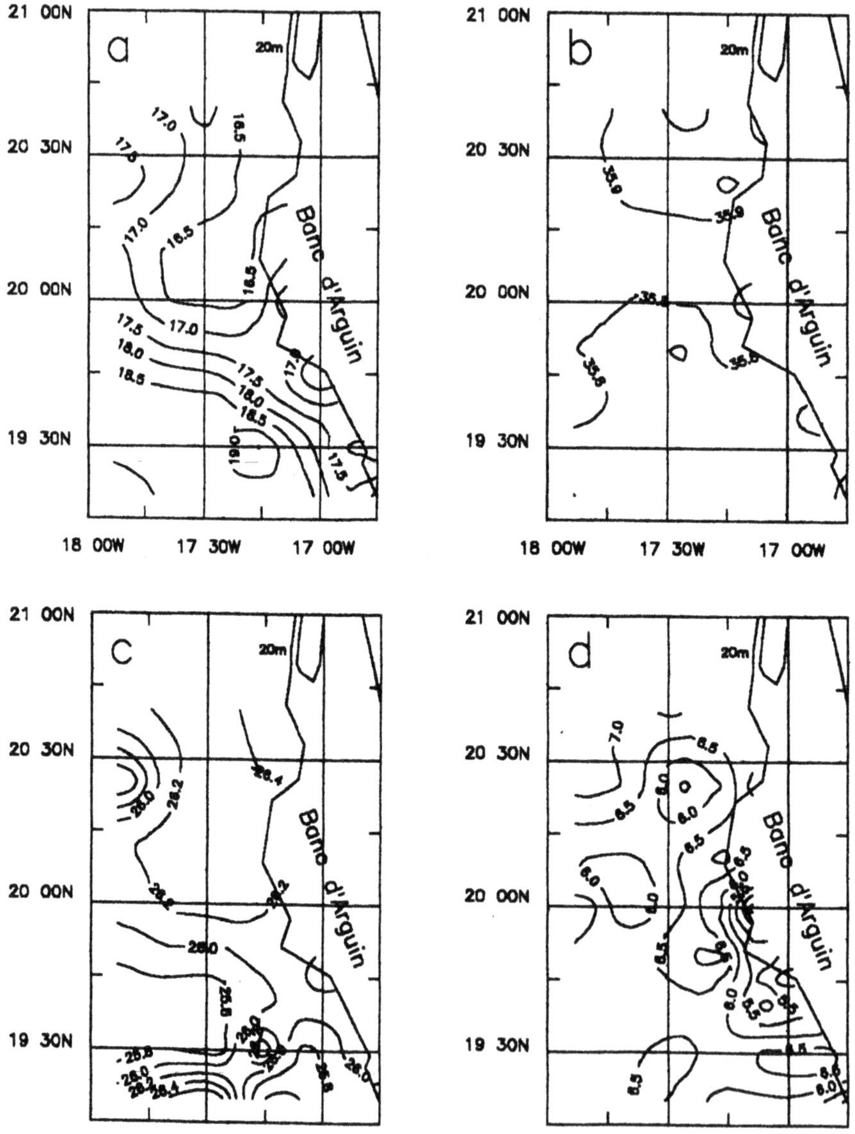


*Fig. 4.* Distribution of the principal characteristics of the surface water in the period of 10–13 May (20 m depth is indicated) a. temperature °C. b. salinity. c. σt. d. dissolved oxygen in ml $l^{-1}$.

*Hydrography.*

Before analyzing the distribution of water characteristics during the survey, the types of water found in the area of investigation will be defined. With the well known TS-diagram approach, taking into account the thermohaline indices of principal water masses of the region, the TS-values of the watermasses measured during the 'Tyro' survey have been plotted in Fig. 3a, together with the mixing lines of SACW and NACW (Sverdrup *et al.*, 1942). The TS-points in the diagram cover a triangular area with the majority of the points situated near the SACW line and to the left of it. Taking into account the depth of the observations, the diagram shows that in the upper layer a mixture of SACW with the warm and low-salinity waters dominated. The second corner of

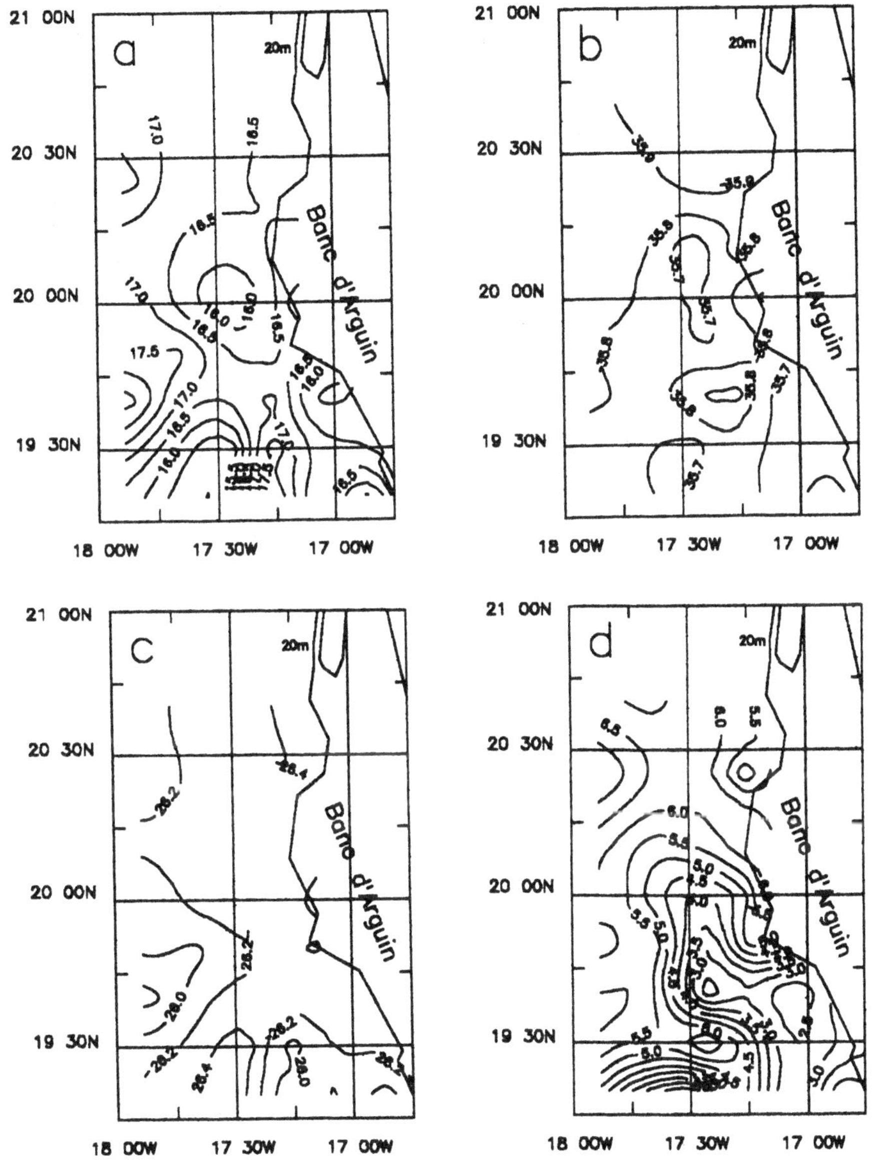


*Fig. 5.* Distribution of the principal characteristics of the water at 30 m depth in the period 10–13 May. a. temperature °C. b. salinity. c. $\sigma t$. d. dissolved oxygen in ml $l^{-1}$.

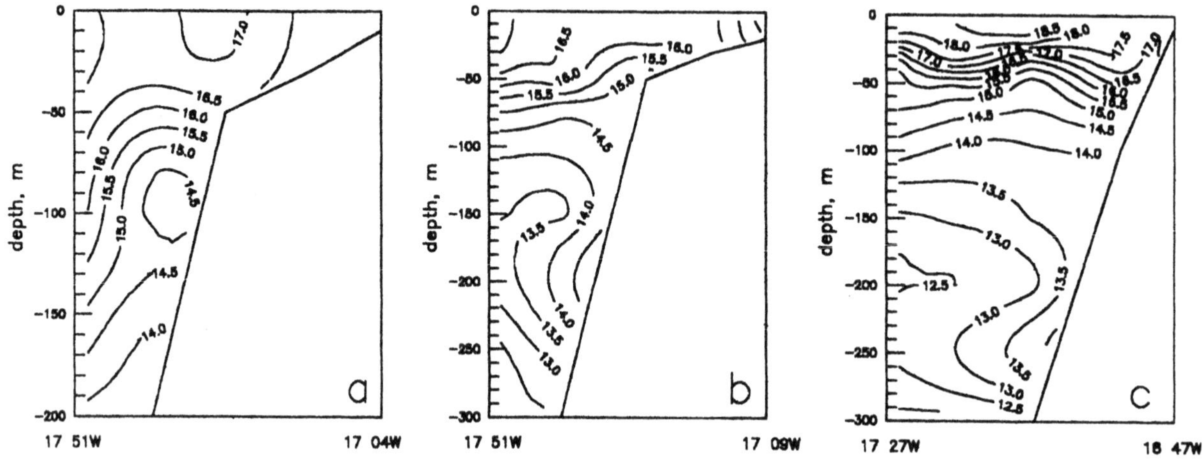


*Fig. 6.* T° C-distribution along three sections at different latitudes. a. at 20° 40′ N on 10 May. b. at 20° 00′ N on 11 May. c. at 19° 20′ N on 13 May.

the TS-triangle is consistent with the influence of NACW on subsurface depths. On the deepest levels of measurements water was found with TS indices intermediate between those of NACW and SACW. In general, all three types of water encountered during the survey represented the product of mixing of SACW, NACW and surface water coming from the south. Here these waters are named as W1 (19.2 °C, 35.73), W2 (17.0 °C, 36.20) and W3 (11.2 °C. 35 33).

A special point of interest in our watermass analysis was to determine whether the water from the Banc d'Arguin ('Banc water') was present in the area during the expedition. Due to the particular conditions during the formation of this superficial water over the shallows of the Banc d'Arguin its TS indices should normally be found in the part of TS-diagram where the salinity and temperature values are much higher than 36.5–37.0 and 19–20 °C (Peters, 1976). Outside the shelfbreak 'Banc water' has frequently been encountered at depths ranging from 75 to 350 m in

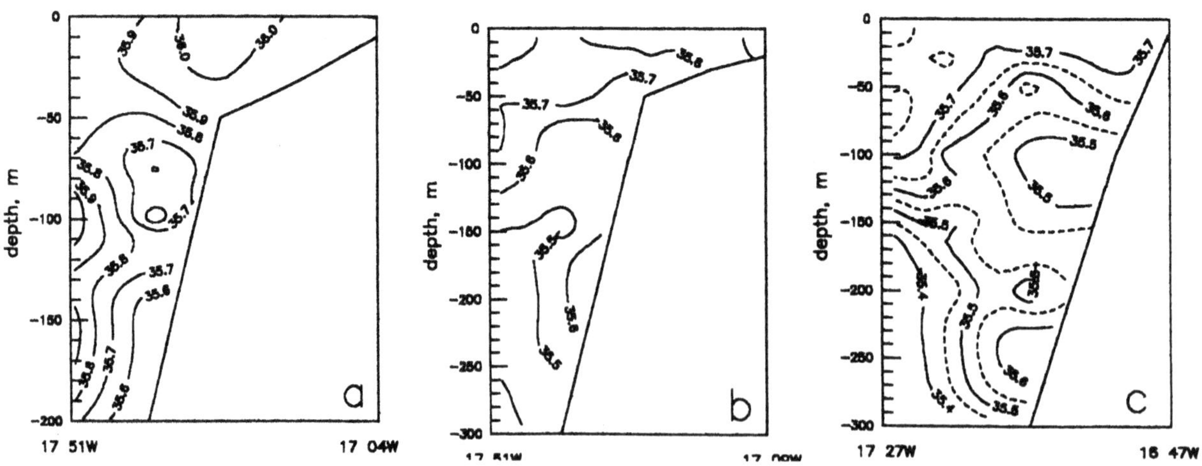


*Fig. 7.* Salinity-distribution along three sections at different latitudes. a. at 20° 40′ N on 10 May. b. at 20° 00′ N on 11 May. c. at 19° 20′ N on 13 May.

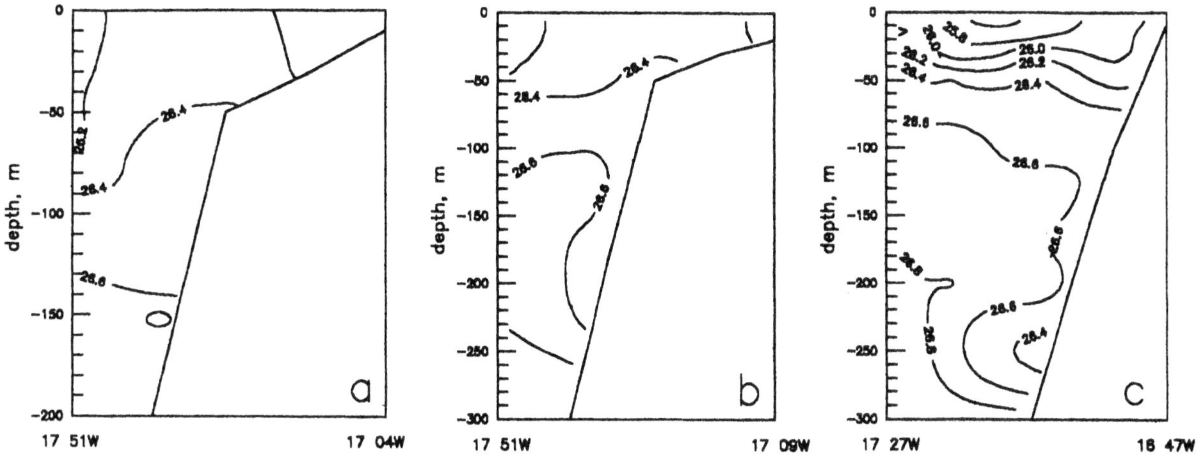


*Fig. 8.* σt-distribution along three sections at different latidues. a. at 20° 40′ N on 10 May. b. at 20° 00′ N on 11 May. c. at 19° 20′ N on 13 May.

the form of cells where temperature and salinity values deviate from surrounding conditions by up to 1 °C and 1 respectively (Peters, 1976).

Judging from the TS-diagram in Fig. 3a, no TS-pair of the surface water in the investigated area was high enough to be associated with the presence of 'Banc water'. As to deeper layers, almost everywhere outside the shelfbreak evidence was found of water cells with higher temperature and salinity. The values of these parameters measured in the cores of important intrusions are represented in Fig. 3b. Such bodies were encountered by earlier observations at depths ranging from 50 to 280 m. Cells, which by their σt values (> 26.8) might be related to water from the Banc d'Arguin (Peters, 1976), were discovered only at depths between 230 and 280 m in the southern part of the survey. They differed only slightly by 0.2 °C and 0.5 from surrounding waters. The intrusion registered at lesser depths in the northern sections revealed thermohaline characteristics close to those of NACW.

The above mentioned observations lead to the conclusion that the discharge of 'Banc water' was at least slowed down in the period of the survey. The possible cause might be an effect of strong winds, which induced the nearshore upwelling preventing the downflow of the 'Banc water'. It is also possible that the wind activity and the predominantly cool and clouded weather prevented in some way the formation of 'Banc water' over the Banc d'Arguin.

The extensive hydrographical survey, which was realised in four days time from 10 to 13 May, provided a practically synoptic picture of the spatial variability of the water characteristics in the whole investigated area (Figs 4–9).

At the surface an increase of water temperature in the seaward direction was observed (Fig. 4a). Another important feature of the distribution of surface temperatures is the tongue of water with temperatures equal to or lower than 16.5 °C extending from the vicinity of Cap Blanc to 20° 00′ N, 17° 30′ W. The tongue was surrounded by a thermic front, which had the greatest temperature contrasts in the south of the area. These phenomena are normal for this season (Loktionov, 1989). Apart from this, intermittent patches of cold and warm water were observed during the survey at the southern end of the Banc d'Arguin. They had a diameter of a few miles or less.

The salinity at the surface (Fig. 4b) showed minor differences, from 36.07 near Cap Blanc to 35.70 in the southern part of the study area, without any sharp irregularities. Due to this the σt distribution reflected the spatial variability of temperature (Fig. 4c).

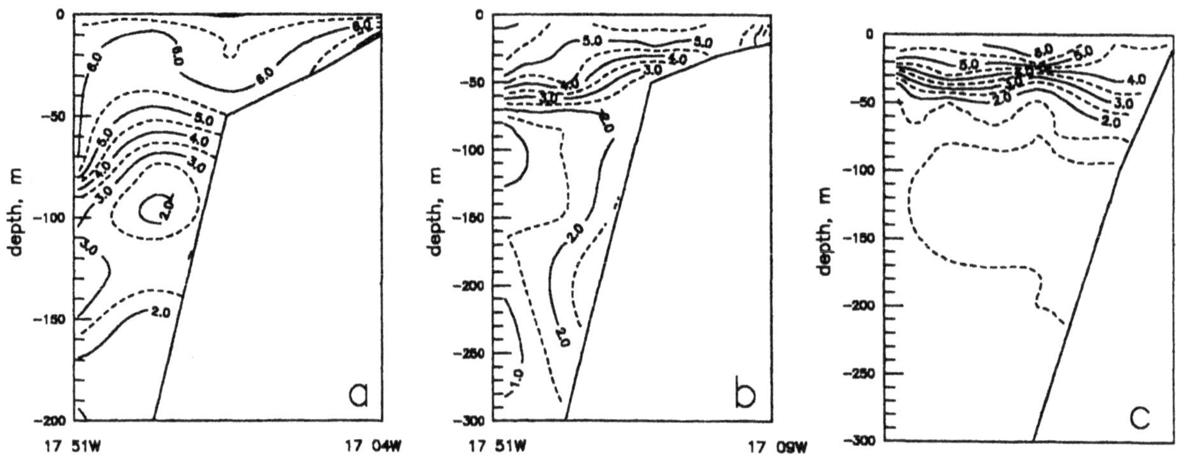


*Fig. 9.* Distribution of dissolved oxygen (ml $l^{-1}$) along three sections at different latitudes. a. at 20° 40′ N on 10 May. b. at 20° 00′ N on 11 May. c. at 19° 20′ N on 13 May 1988.

The dissolved oxygen contents were rather high at the surface as a result of aeration due to the strong wind (Fig. 4d). The minimum value of 4.0 ml $l^{-1}$ was registered over the western edge of the Banc d'Arguin (20° 00′ N, 17° 09′ W). In the immediate surroundings values showed a sharp increase to 6.0–6.5 ml $l^{-1}$. The tongue of water with relatively low temperatures (< 16.5 °C) had also a relatively low oxygen content (<6.0 ml $l^{-1}$).

As the distribution of hydrographical parameters at the surface is subject to a short-term variability, probably induced by wind fluctuations, description of the distribution pattern at 30 m depth (Fig. 5) offers several advantages:

- to a certain degree the effect of the short-term variations due to wind is excluded;
- parameters measured at this level and close to the bottom on the shallower stations could be

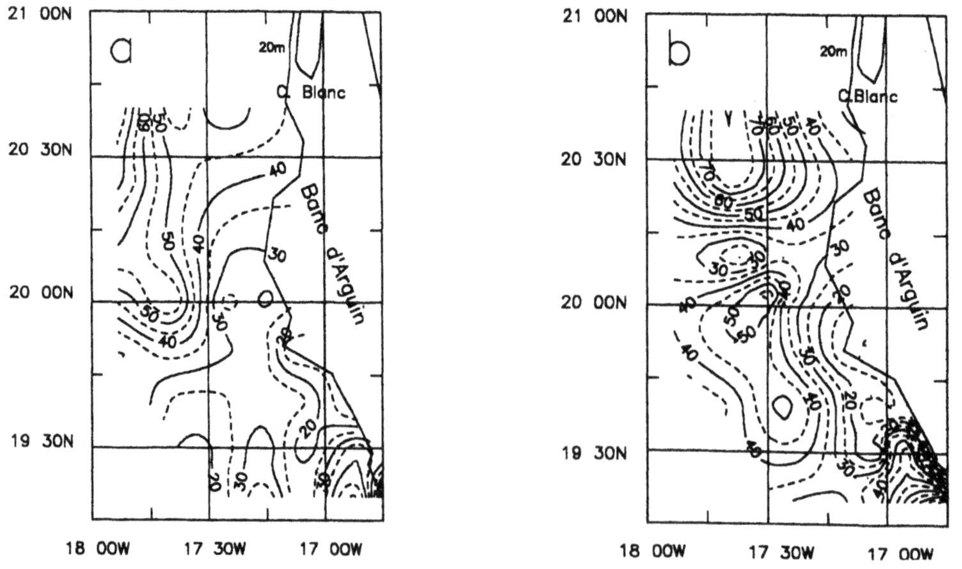


*Fig. 10.* Depth of thermocline upper limit (m). a. 10–13 May 1988. b. 13–25 May 1988.

analysed together as if measured at the same horizontal;

- discontinuity zones existing in the surveyed area are better manifested in the subsurface layer (Arkhipkin *et al.*, 1987);
- the level is close to the thermocline upper limit.

The distribution of hydrographical parameters at 30 m depth (Fig. 5) displays a high degree of spatial variability south of 20° 15′ N as a result of the existence of a patches of water with a low temperature (< 16.0 °C), salinity (< 35.7) and oxygen saturation (< 3.5 ml $l^{-1}$). This water is present at the outer shelf, closely following the isobaths. In an offshore direction temperature, salinity and dissolved oxygen content increase again at the 30 m level. The sharpest contrasts were observed in the distribution of the oxygen

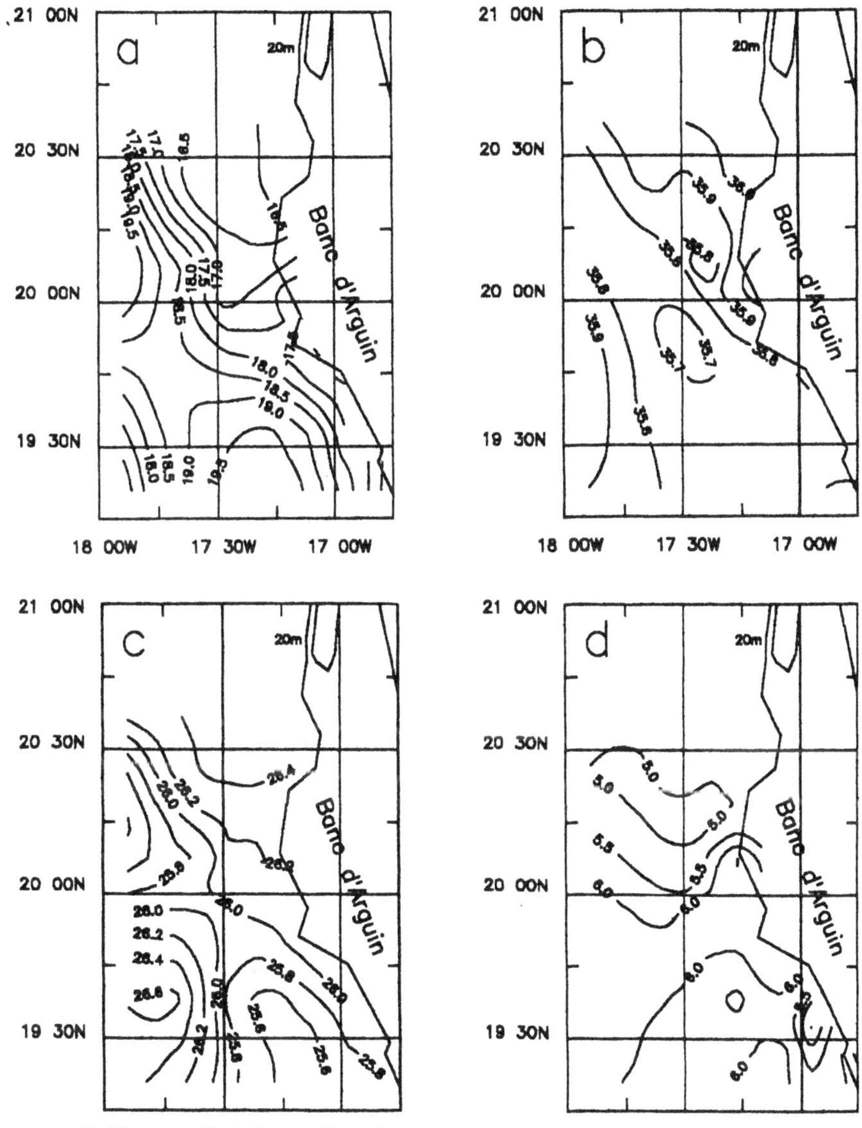


*Fig. 11.* Distribution of the principal characteristics of the surface water in the period 13–29 May. a. temperature °C. b. salinity. c. σt. d. dissolved oxygen in ml $l^{-1}$.

content (Fig. 5d). The results of earlier investigations show the presence of nutrient rich upwelling water in nearly the same position (19° 38′ N, 17° 00′ W), at the end of a shelfbreak canyon (Maske, 1976). The horizontal distribution of water properties in Fig. 5 suggests the presence of a chain of upwelling phenomena around the south-western part of the Banc d'Arguin shelf slope.

This is confirmed by the distribution of water parameters along vertical sections at 20° 40′ N, 20° 00′ N and 19° 20′ N (Figs 6–9). These sections were chosen to represent the different conditions found in the northern, central and southern parts of the survey. In all three sections the presence of upwelling can be very clearly seen from the position of isopleths. However, the degree to which the upwelling influenced the shelf

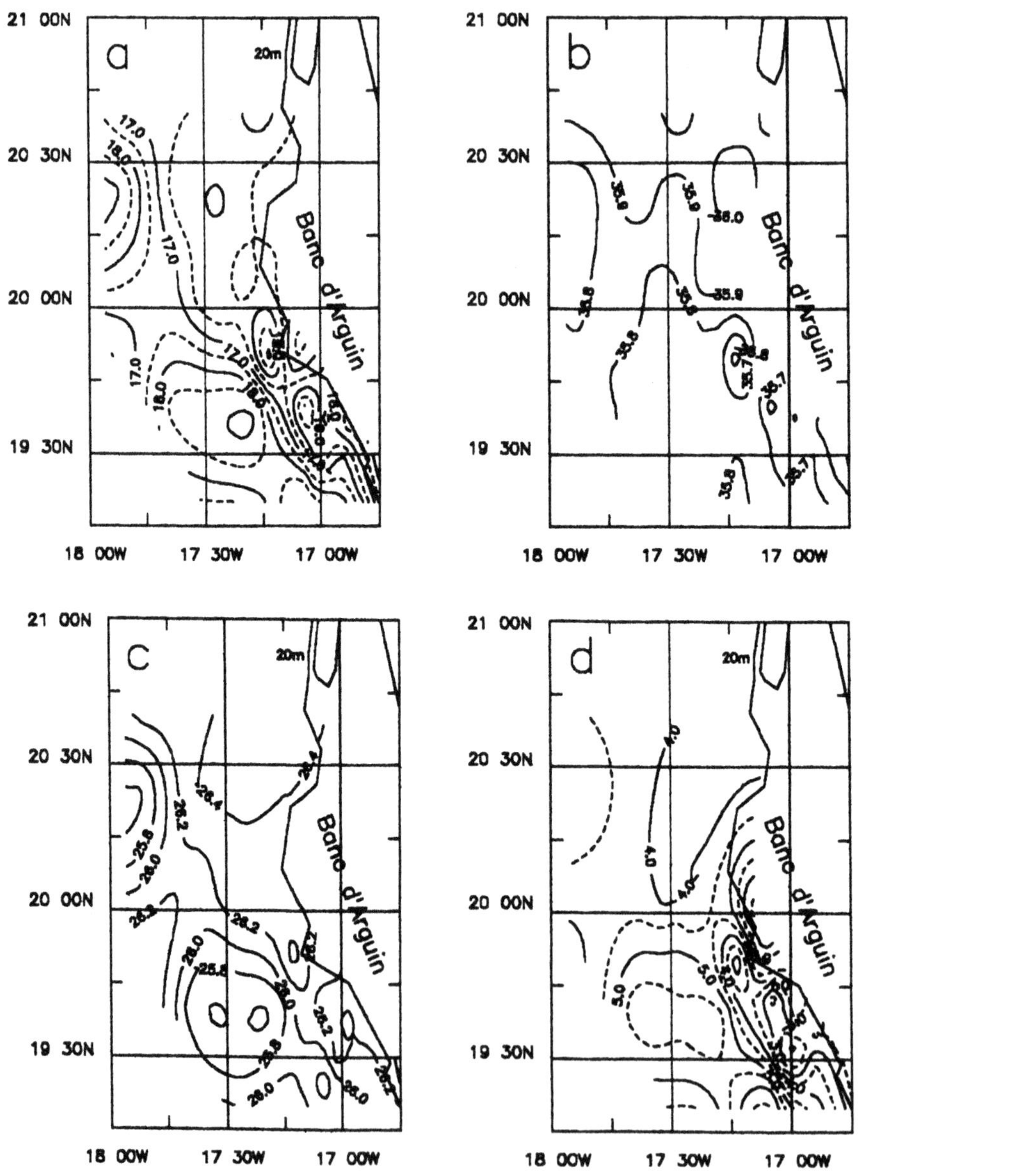


*Fig. 12.* Distribution of the principal characteristics of water at 30 m depth in the period 13–29 May. a. temperature °C. b. salinity. c. σt. d. dissolved oxygen in ml l⁻¹.

zone changed from north to south. Along the section at 20° 40′ N (Fig. 6a–9a) the steepest gradients of temperature, salinity and oxygen, created by the water rising from the deeper layers, were registered at the depth of 50–60 m. This upwelling water did not reach beyond the 50 m isobath on the shelf. Along the section at 20° 00′ N (Fig. 6b–9b) upwelling water was found nearer to the surface and reaching depths at the shelf of less than 30 m. The sharpest decrease in oxygen content, *i.e.* from 6 to 2.5 ml $l^{-1}$, was observed at this latitude over the depths of 30–50 m (Fig. 9b). Along the southernmost section at 19° 20′ N (Fig. 6c–9c) a well developed thermocline, associated with high vertical gradients of other properties, could be observed as a result of the combined effect of upwelling and surface heating. Evidence for upwelling in the form of rising isopleths was found below the 200 m isobath at approximately 17° 10′ W. In between the upwelling water and the shelf slope, water of a relatively high temperature, salinity and oxygen content seems to plunge down.

The spatial distribution of the thermocline upper limit can be used as a supplementary indicator of hydrographical processes in the area. This parameter is plotted separately for the hydrological survey of 10–13 May and for the subsequent observations in the period 13–29 May (Fig. 10). During the hydrographical survey a sharp decline of the thermocline upper limit is seen at 19° 45′ N–20° 00′ N (Fig. 10a), dividing the area in two parts differing in upwelling intensity. The thinnest homogeneous layer (< 20 m) existed over the southern end of the Banc d'Arguin, between 19° 25′ N and 19° 50′ N, which is consistent with a greater upwelling intensity. The distribution of the thermocline upper limit during the subsequent period is shown in Fig. 10b. In general, the thermocline seems to have descended in comparison to the first period. Nevertheless, during the second period there is still a clear difference between the northern and southern part of the study area with regard to the upper limit of the thermocline.

In searching for evidence of temporal changes in the hydrography of the area, a comparison is made between the horizontal distribution of water properties at the surface and 30 m depth in the periods 10–13 May (Figs 4–5) and 13–29 May (Figs 11–12). During the second period a well-pronounced, North-South oriented thermic front was formed along the western limit of the surveyed area with temperatures reaching 19.5 °C on its outer side (Fig. 11a–12a). This front ap-

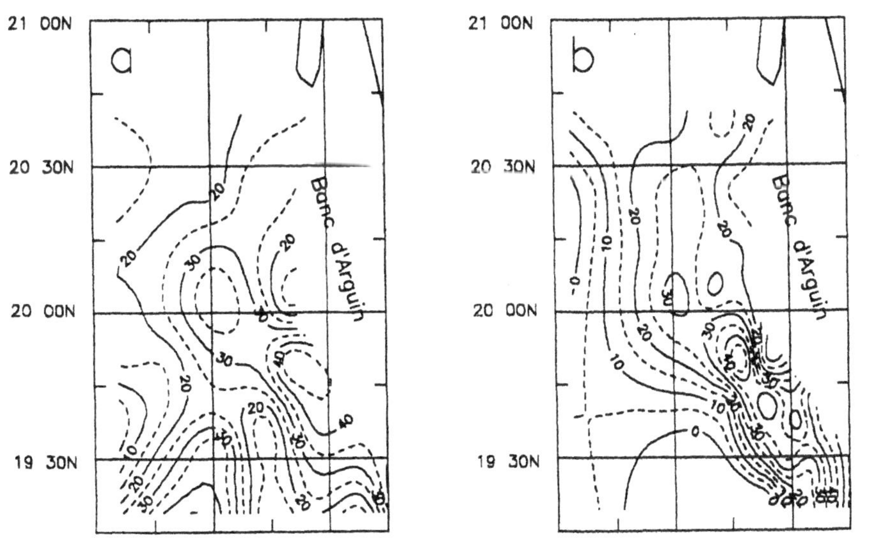


*Fig. 13.* The proportion of W3 water at a depth of 30 meters. a. 10–13 May 1988. b. 13–29 May 1988.

parently pressed the tongue of relatively cold and saline surface water, observed in the first period (see Fig. 4a–b), back to the north. Concurrently, the oxygen content of the water at the 30 m depth level on the shoreward side of the front, notably north of 20° 00′ N, seemed to have dropped from more than 6 to 4 ml $l^{-1}$ (Figs 4d–5d and 11d–12d).

The changes that took place during May 1988, can also be illustrated by the distribution of the proportion of W3 water at the 30 m depth level in the two periods. Since the W3 water characteristics are near to those of SACW at a depth of 300 m, and SACW was defined as an original source of upwelling in this region (Maske, 1976), the percentage of W3 water would indicate the degree of upwelling. The W3 content at 30 m was calculated using the TS-parameters of W1, W2 and W3 waters as points of the triangle of mixing. The distributions of the W3 proportion (Fig. 13) show that throughout May 1988 the highest contributions of upwelling water at 30 m depth were consistently found along the slope and edge of the shelf with maximum values in the southern part of the study area. The belt of the water with a high percentage of W3 (>30%) changed its width during the investigation. Notably in the central part of the area (20° 00′ N), the percentage W3 water decreased during the second period of the cruise, which corresponds with the descent of the thermocline upper limit (Fig. 10).

## Conclusions.

The observations lead us to conclude that, although the biological investigations during the cruise took place in the transitional period, the main hydrographic features distinguishing the southern from the northern part of the study area remained relatively stable during the period of investigation. This conclusion is corroborated by the results of the zooplankton study by Kuipers *et al.* (this volume) who found no significant difference between the distribution of copepod species collected during the hydrographical survey and the subsequent period.


## Acknowledgements

I would like to thank the technicians and the crew of R/V 'TYRO' for their help in collecting the data.


## References


Arkhipkin, V. S., V. N. Veremeyev & V. A. Ivanov, 1987. Upwelling in the Boundary Areas of the Ocean. Academy of Sciences of the USSR, Sea Hydrographical Institute, Sevastopol, 45 pp.(In russian).

Barton, E. D., A. Huyer & R. L. Smith, 1977. Temporal Variation Observed in the Hydrographic Regime Near Cabo Corveiro in the North West African Upwelling Region, February to April 1974. Deep Sea Res. 24: 7–23.

Brink, K. H., 1983. The Near Surface Dynamics of Coastal Upwelling. Progr. Oceanogr. 12: 223–257.

Brink, K. H., B. H. Jones, J. C. Van Leer, C. N. K. Mooers, D. W. Stuart, M. R. Stevenson, R. C. Dugdale & G. W. Heburn, 1981. Physical and Biological Structure and Variability in an Upwelling Center off Peru Near 15°S. In F. A. Richards (ed.), Coastal Upwelling. Am. Geophys. Union, Washington D.C.: 473–496.

Kuipers, B. R., H. J. Witte & S. R. Gonzalez, 1993. Zooplankton distribution in the coastal upwelling system along the Banc d'Arguin, Mauritania. Hydrobiologia 258: 133–149.

Loktionov, Y., 1988. Evolution de l'Hydroclimat et les Conditions Hydrologiques dans la ZEE Mauritanienne en 1979–1987. In Rapport du Groupe de Travail CNROP-CRODT/ISRA sur les Ressources Pelagiques Cotieres (Mauritanie-Senegal), 8–16 Juin 1988, Nouadhibou – Mauritanie, Annexe V: 55–73.

Loktionov, Y., 1989. Analise des Variations de la Temperature de l'Eau de Surface le Long des Cotes Mauritaniennes. In Les Ressources Halieutiques de la ZEE Mauritanienne: Description, Evaluation et Amenagement. Rapport du 2 Groupe de Travail CNROP/FAO/ORSTOM, Nouadhibou, Mauritanie, 12–22 novembre 1988 114–126.

Maske, H., 1976. Nutrients and Oxygen in the Upwelling Area off NW-Africa. 'Meteor' Forsch.-Ergebnisse A 17: 73–87.

Mittelstaedt, E. D., 1972. Der hydrographische Aufbau und die zeitliche Variabilität des Schichtung und Strömung im nordwest-afrikanischen Auftriebsgebiet im Frühjahr 1968. 'Meteor' Forsch.-Ergebnisse. A 11: 1–57.

Peters, H., 1976. The Spreading of the Water Masses of the Banc d'Arguin in the Upwelling Area off the Northern Mauritanian Coast. 'Meteor' Forsch.-Ergebnisse. A 18: 78–100.

Sverdrup, H. U., M. W. Johnson & R. H. Fleming, 1942. The Oceans, their Physics, Chemistry and General Biology. Prentice-Hall, Englewood Cliffs. New York. 1087 pp.

*Hydrobiologia* **258**: 33–40, 1993.
*W. J. Wolff, J. van der Land, P. H. Nienhuis & P. A. W. J. de Wilde (eds),*
*Ecological Studies in the Coastal Waters of Mauritania.*




# Remote sensing of sea surface and coastal features in the area of the Golfe d'Arguin, Mauritania

F. Cuq
*Laboratoire Géosystèmes, URA 1518 CNRS, Université de Bretagne Occidentale, B.P. 452, 29275 Brest Cedex, France*

*Key words:* Banc d'Arguin, Mauritania, remote sensing, sea surface temperature, organic matter

## Abstract

The analysis of a series of NOAA AVHRR satellite data recorded between July and September 1986 shows the interaction between northerly and southerly water masses during summer as well as the northern limit of the northward advance of water masses originating from the South Equatorial Current. Two upwelling situations analysed by means of NOAA data show the intrusion of suspended matter from the ocean into the Golfe d'Arguin and illustrate the water circulation within the coastal zone. This circulation pattern is detailed by LANDSAT and SPOT data. A SPOT picture from October 1987 shows the structure of the intertidal and coastal geomorphology.

## Introduction

The region of the Banc d'Arguin lies in the northern part of Mauritania between Cap Blanc and Cap Timiris (Fig. 1). Within this area the Banc d'Arguin proper forms a remarkable structure. It is a vast semi-circular accumulation of sediments with an average depth of about 5 m. It was here that the frigate 'La Méduse' wrecked in 1816. The Banc shelters a coastal zone with an irregular underwater morphology, an average depth of about 10 m, and partly covered with intertidal and subtidal seagrasses. Following Savigny & Corréard (1818) we name this area between Cap Blanc and the Baie du Lévrier in the North and Cap Timiris in the South and between the Banc d'Arguin in the West and the mainland coast in the East, the Golfe d'Arguin.

The region of the Banc d'Arguin is a tropical coastal desert at the interface between northern and southern climatic and oceanic influences. The first studies of this area were focussed principally on vertebrates, in particular on marine mammals and birds. These studies led to the creation of the Parc National du Banc d'Arguin in June 1976, but did not result in an overall picture of the functioning of the ecosystem. Present coordination of research by the Banc d'Arguin International Foundation recognizes the need to address many more topics to arrive at a more accurate description of the ecological system of the area and of the exchanges of matter with surrounding systems.

Our research aims to survey the coastal landscape, to describe the physical and biological exchanges between the Golfe d'Arguin and the marine and terrestrial areas around it, and to throw light on the evolution of the environment by analyzing the impact of the geodynamic factors that form it.

In this paper we report the first results based on analysis of satellite images and describing the

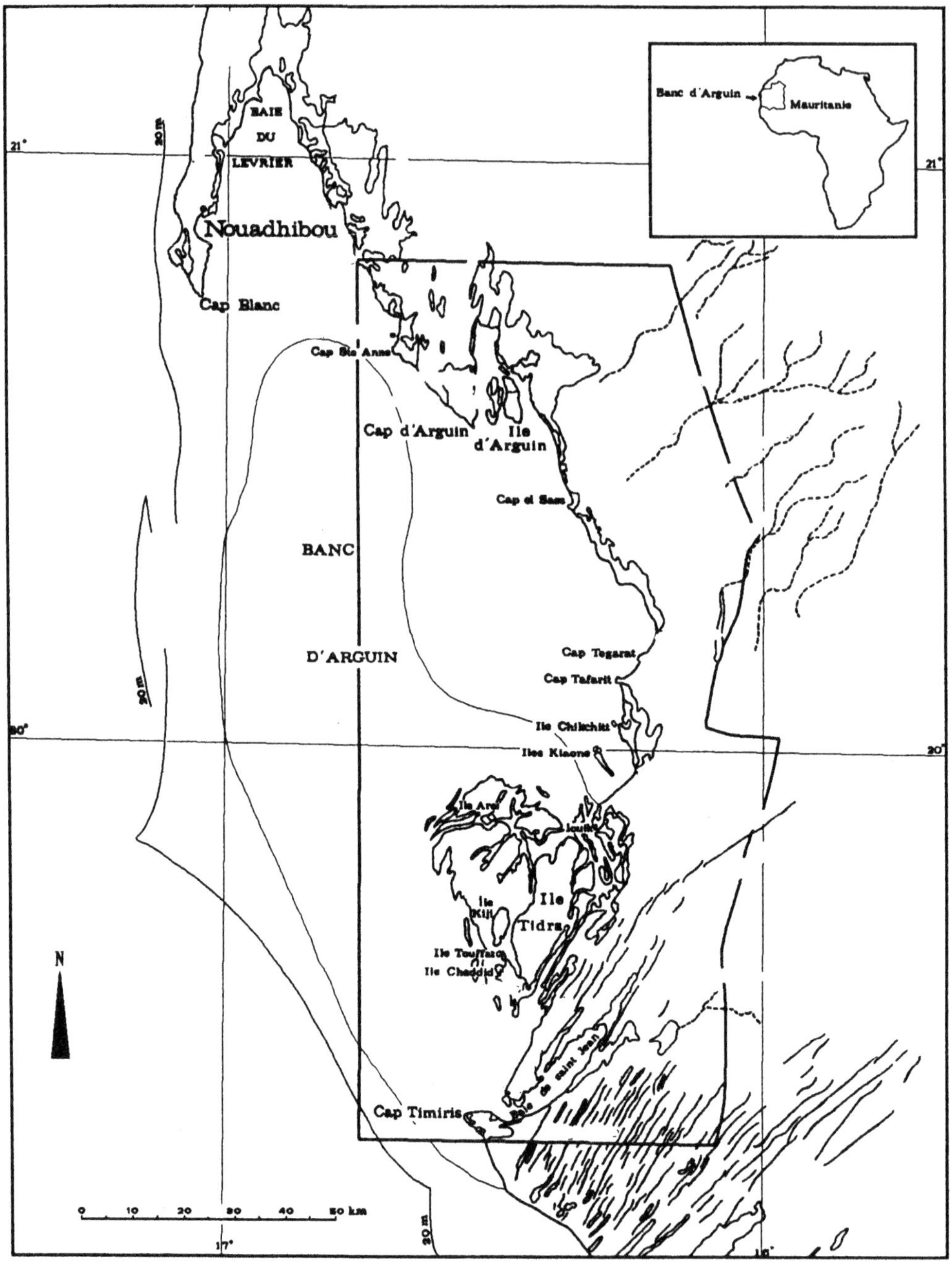


*Fig. 1.* The area of the Golfe d'Arguin. The thick line denotes the boundaries of the Parc National du Banc d'Arguin.

movements of water masses and suspended matter in the open sea and along the coast, as well as the typology of the coastal landscapes.

## Material and methods

We used data from the NOAA Advanced Very High Resolution Radiometer (AVHRR), the

LANDSAT Multi Spectral Scanner (MSS) and the SPOT Haute Résolution Visible (HVR) satellites. We analysed a set of 26 NOAA-09 images recorded in LAC mode (Local Area Coverage ($1.1 \times 1.1$ km at Nadir)). The AVHRR sensor carried on board the NOAA satellites collects data in five wavebands, *viz.* band 1: 0.58–0.68 μm; band 2: 0.72–1.10 μm; band 3: 3.53–3.93 μm; band 4: 10.30–11.30 μm, and band 5: 11.50–12.50 μm. Bands 1 and 2 belong to the visible and near-infrared part of the spectrum, band 3 to the middle infrared, and bands 4 and 5 to the thermal infrared.

Image processing included data calibration and geometric rectification in the form of geographic coordinates, temperature calibration and correction of the atmospheric effects using bands 4 and 5 by application of a split-window algorithm, and 10-days synthese of images, selecting maximum temperature in order to eliminate most of the clouds. The split-window algorithm takes the form

$$T_{corr} = T_{11\mu} + a(T_{11\mu} - T_{12\mu}) + b\,,$$

in which coefficients $a$ and $b$ are those recommended by Deschamps & Phulpin (1980).

The LANDSAT-MSS sensor operates in four bands of the visible and near-infrared part of the spectrum, *viz.* band 1: 0.5–0.6 μm; band 2: 0.6–0.7 μm; band 3: 0.7–0.8 μm, and band 4: 0.8–1.1 μm. Resolution of each MSS pixel is $80 \times 60$ m. Data processing included calibration and illumination correction, geometric rectification in accordance with the UTM reference frame and sampling at the resolution of $80 \times 80$ m, and image classification.

We have analyzed the intertidal zone with a SPOT image in multi-band mode, recorded on 26 October 1987, when we were working on the ground (Cuq, 1989). The SPOT multi-band HRV sensor operates in the following three bands, *viz.* band 1: 0.50–0.59 μm; band 2: 0.61–0.68 μm, and band 3: 0.79–0.89 μm. The processing applied was similar to that applied to the MSS images, retaining the HRV pixel resolution of $20 \times 20$ m. In some cases the Normalized Difference Vegetation Index (NDVI) was calculated: NDVI = (XS3 − XS2)/(XS3 + XS2). The geographic area covered by the image extends from Cap Tafarit in the North to the island of Gibene in the South, and from Arel in the West to the desert in the East.

## Results

### *Movements of water masses and suspended load*

#### *Sea-surface temperatures*

Based on 24 NOAA-09 images covering eight 10-days periods Fig. 2 describes the reversal of the northern and southern influences in summer. We see the gradual intrusion of the Guinea Current in July with the boundary between water masses clearly marked by cyclonic eddies (Fig. 2b). From the end of July to mid-September the water circulation along the coast is dominated by the South-Equatorial Current. Between the end of August and the end of September the mode of the currents becomes reversed as shown by the gradual advance of cold water of the Canary Current. The boundary between water masses in that period is marked by anticyclonic eddies (Fig. 2f). The dislocation of the thermal structure of the South-Equatorial Current is carried out by the temporary insulation of water cells that are hotter and less dense than the surrounding water (Figs 2g, 2h).

This analysis reveals that:

- Cap Blanc seems to be the boundary of the northern and southern influences during the maximum advance of the Inter Tropical Convergence Zone;
- there is always cold water to the North of Cap Blanc;
- the South-Equatorial influence lasts for a relatively short time: about 1.5 months;
- the position and the direction of the swirling shapes describe the confrontation of the water masses and their respective forces.

Two NOAA-AVHRR images of 19 February and 9 March 1987, respectively, illustrate two upwelling situations that are confirmed by vertical

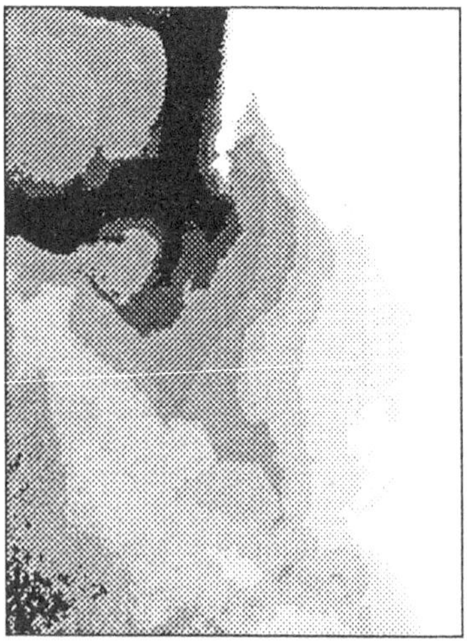

A - First decade of July

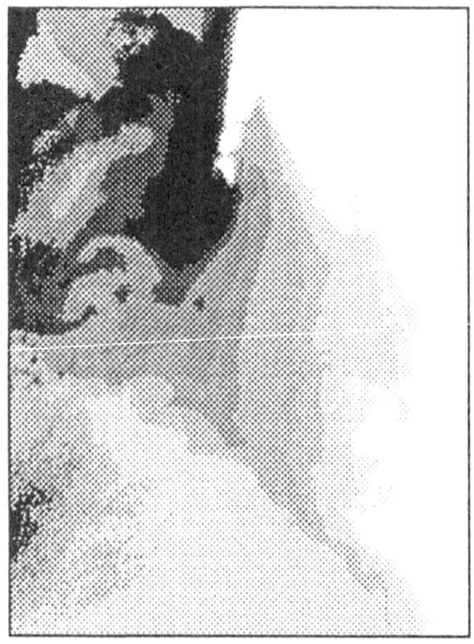

B - Second decade of July

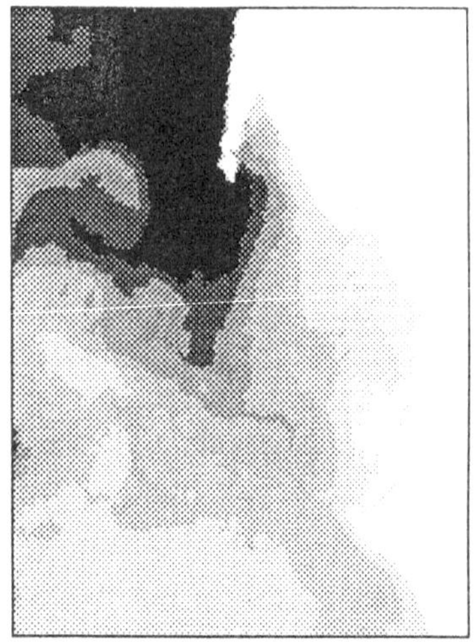

C - Third decade of July

D - First decade of August

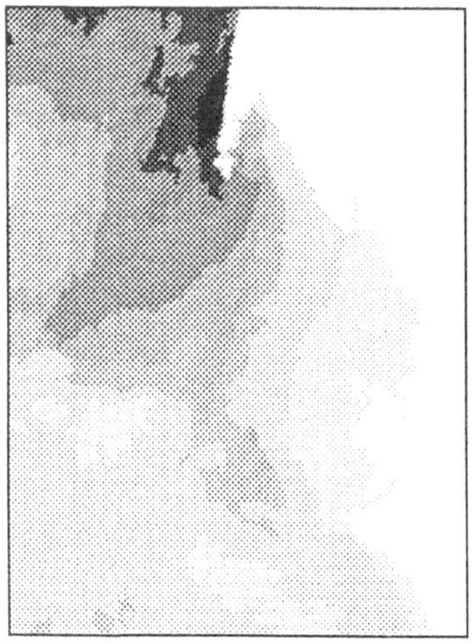

E - Second decade of August

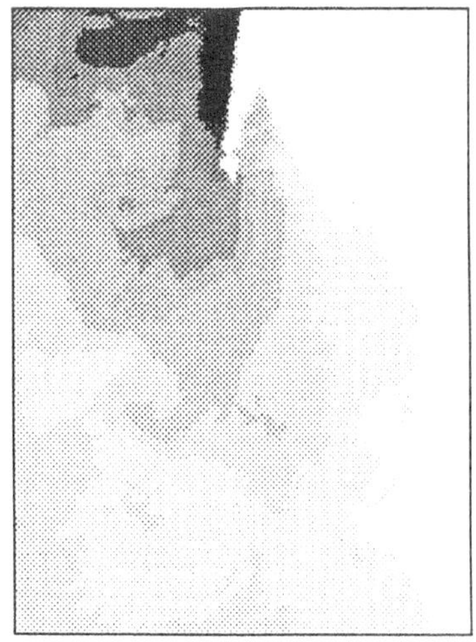

F - Third decade of August

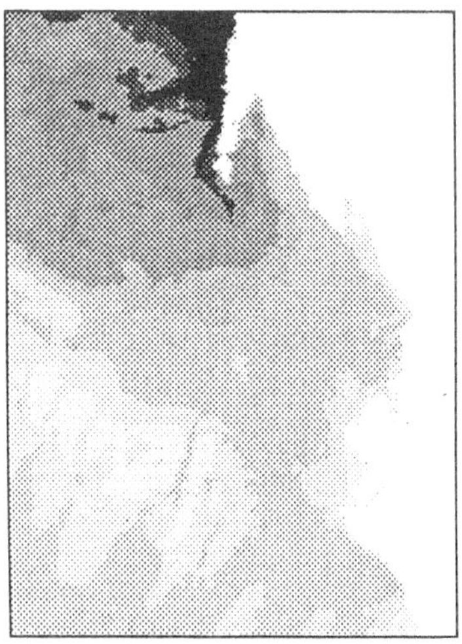

G - First decade of September

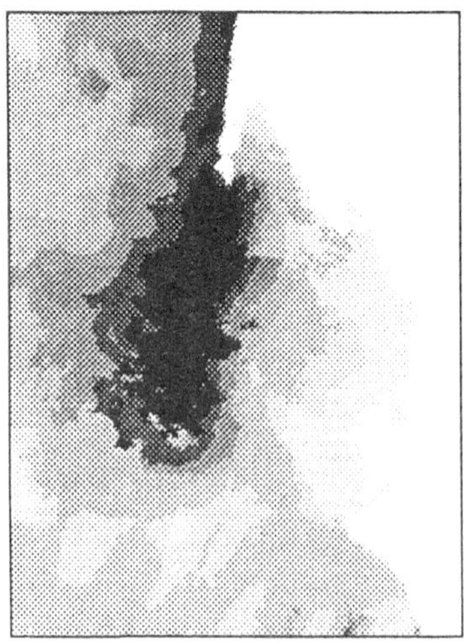

H - Second decade of September

*Fig. 2.* Decadal syntheses of sea-surface temperatures obtained from NOAA-AVHRR data for the area off the northern Mauritanian coast during July–September 1986. The sequence describes the reversal of the northern and southern influences during the summer.

thermal profiles taken from ships at that time (Fig. 3a, b).

*Suspended load*

Analysis of the two visible bands of the NOAA-AVHRR image of 9 March 1987 reveals a significant concentration of suspended matter that is partly carried towards the coast (Fig. 3c). The radiometric characteristics of the NOAA-AVHRR bands 1 and 2 do not allow analysis of the composition of the suspended load. However, the position of the turbidity is directly associated with the upwelling cells.

To produce more detailed information on the suspended material than obtained from the NOAA data, we processed a LANDSAT MSS image taken on 20 February 1987, illustrating the upwelling situation already observed on the NOAA image of 19 February.

Analysis of the results allows three water types to be distinguished qualitatively (Figs 4 and 5):

- type 1: water with a low content of suspended matter;
- type 2: water with a high content of suspended matter, characterized by a low radiometric deviation between MSS bands 1 and 2;
- type 3: water with a high content of suspended matter, characterized by a high radiometric deviation between bands 1 and 2.

Type 1 water occurs in the open sea; the boundary between this water type and the other ones shows the boundary between the open sea and the coastal water. Water of type 2 coincides with seagrass beds at shallow depths and coastal turbid water. Type 3 water has spectral characteristics indicating high reflectance in band 1 (green) and a more marked absorption in band 2 (yellow-orange). These optical properties seem to indicate a concentration of chlorophyll pigments illustrating the penetration into the coastal zone of phytoplankton associated with upwelling. Note that the waters of high suspended load content on the seaward edge of the Banc correspond to the position of an upwelling cell on the NOAA image during the same period (Fig. 3a).

The waters of type 3 follow the contour of the seaward slope of the Banc d'Arguin. Particularly in the southern section of the image the effect of the swell and of the rising tide allow the suspended load held up by the Banc to partially cross it. Thus a system of sandy ridges with hydraulic dunes in NW–SW direction of 6–10 m height and 4 km distance from crest to crest is clearly revealed. The fact that suspended matter enters the Golfe d'Arguin in the SW–NW direction is unexpected since the Canary Current runs N–S. It has to be explained by the effect of the rising tide together with refracted long-wavelength swell.

*The littoral zone*

*Coastal waters*

The SPOT image classification furnishes a typology of the coastal waters divided into three distinct groups (Fig. 6).

Waters of group 1 are found in the tidal flat area West of the island of Tidra, and are characterized by a low turbidity, most probably because the suspende matter is trapped by the seagrass beds.

Waters of group 2 are found in the tidal inlet of Tidra. Their suspended load is relatively low with its concentration increasing towards the North into the direction to which the inlet drains. We note the presence of lens-shaped volumes of water carrying high turbid loads as indicated by the high reflectancy recorded in band 2. These turbid waters were discharged four hours previously during maximum ebb. The variations of the turbid load together with the shallow depths, indicate immersed banks.

Waters of group 3 correspond to waters of the rising tide and their optical properties are very different from those of group 1 and 2. A fairly strong absorption in band 2 (yellow-orange) shows that the seston content is low whereas these waters are highly refective in band 1 (green). Such optical properties seem to characterize the green waters from the open sea we observed on the ground (Cuq, 1989). These waters have a chlorophyll content greater than that of the warmer and less oxygenated waters of groups 1 and 2. As

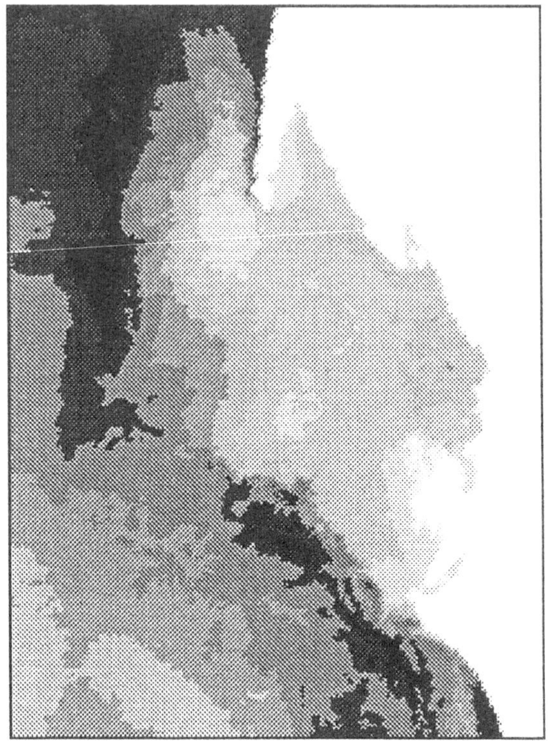

A - 19 February 1987 : thermal band

**C - 9 March 1987 : visible band**

*Fig. 3.* Upwelling situations. Images A and B (thermal band NOAA-AVHRR) from 19 February and 9 March 1987, respectively, show the sea-surface temperature when upwelling occurs off Cap Blanc. Image C (visible bands NOAA-AVHRR) shows turbidity carried from the open sea towards the Golfe d'Arguin.

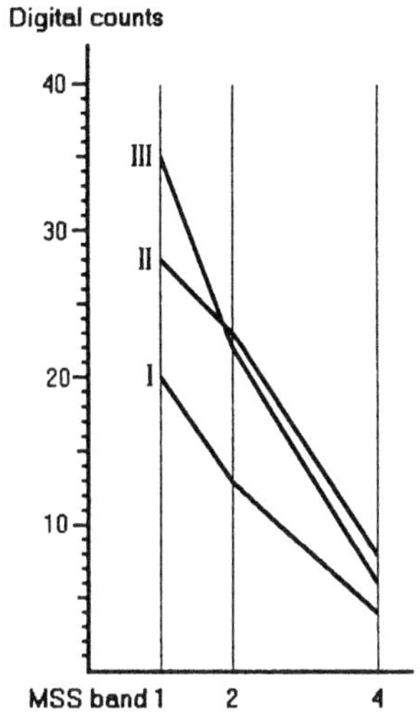


*Fig. 4.* Spectral characteristics of water masses distinguished on LANDSAT MSS data.

was the case for the LANDSAT image we can only analyze the water colour on the SPOT image qualitatively. Figure 6, however, shows that this leads to consistent results.

*Seagrass coverage*

The classification of the SPOT image throws light on two types of plant cover, *viz.* 1) dense seagrass beds and 2) sparse or dispersed seagrasses, *Spartina* beds, *Avicennia* mangroves and diatoms.

Dense seagrass beds (Fig. 7) are represented by four classes, the first including *Zostera* and *Cymodocea* immersed at this stage of the tide. The second class comprises *Zostera* beds that are partly emerged. The two remaining classes include *Zostera* beds at different states of drainage.

This typology also presents a description of the morphology and the hypsometry of the tidal flats. It is thus possible to distinguish between dome-shaped and basin-shaped tidal flats, the latter type being the most widely represented. In particular we note a bar facing the flood tide on the basin-shaped flats, illustrating how sediments in suspension are trapped. The total area on this image occupied by emerging seagrasses and those identified at shallow depths is 132 km$^2$ representing

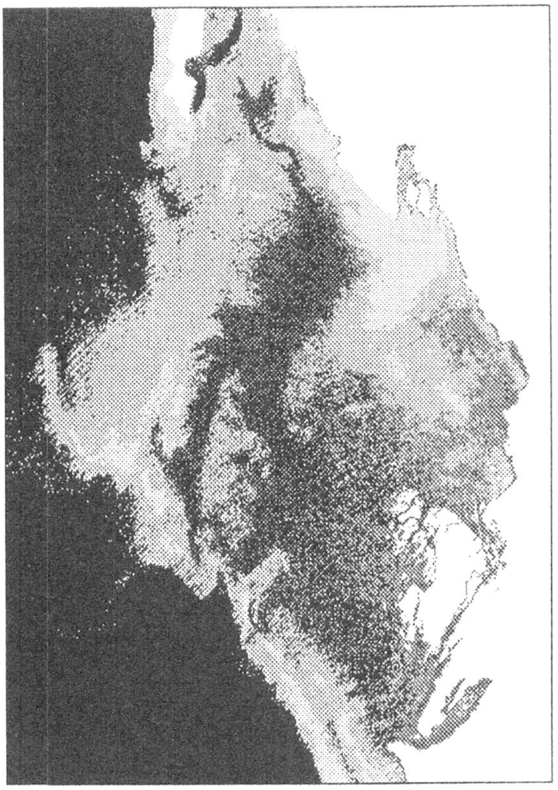

*Fig. 5.* Spatial distribution of water masses on the LANDSAT MSS image of 20 February 1987.

50.5% of the intertidal zone analyzed on the image.

It is very difficult to discriminate between the sparse vegetation types because the bare ground interferes with the spectral response. We have distinguished two classes in relation to their NDVI value. These two classes differ in chlorophyll density and activity. They include sparse *Zostera* beds, *Spartina* stands, relict *Avicennia* mangroves near the islands of Nair and Niroumi, and also areas covered with diatoms. These diatoms are due to the presence of former ponds high on the tidal flats (e.g. in the centre of the island of Niroumi) filled during the spring tide on 8 October.

The area covered by sparse vegetation types is 58.14 km or 22.5% of the part of the intertidal zone analyzed.

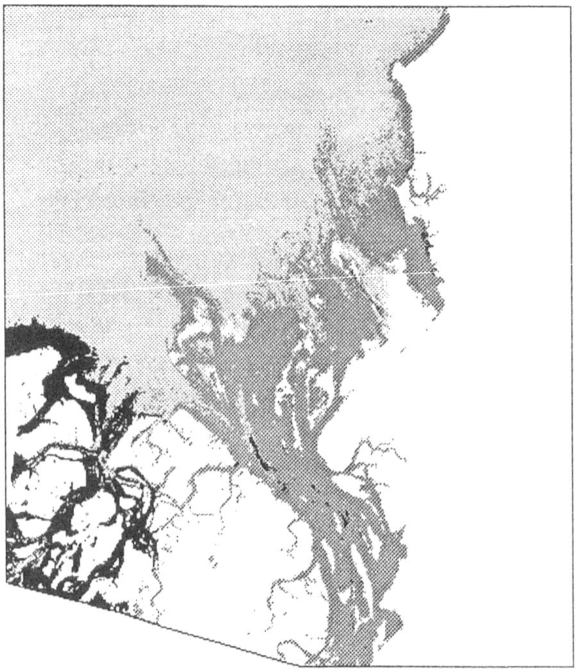

*Fig. 6.* Spatial distribution of coastal water masses on the SPOT image of 26 October 1987 at low tide.

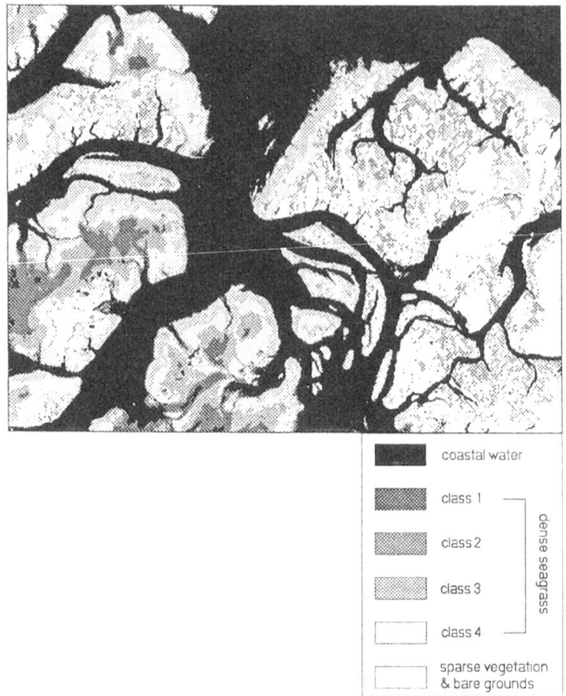


*Fig. 7.* Classification of seagrass beds on the tidal flats West of the island of Tidra. Based on a SPOT image of 26 October 1987.

## Discussion

The satellite data not only allow to draw conclusions regarding the ecology of the Golfe d'Arguin, but also provide insight into its geomorphology and the processes behind it. These data will be published elsewhere, however. Thus, our results reveal all the advantages of processing satellite images for understanding a complex and relatively unknown area such as the Golfe d'Arguin.

The study of sea-surface temperature and suspended matter distribution shows that blooms of phytoplankton due to upwelling supply organic material to the Golfe d'Arguin during winter and part of the spring. This contradicts the earlier held belief that the Banc d'Arguin is isolated from marine inputs by the Canary Current.

The spectral resolution and the choice of the bands of the sensors used cannot furnish a quantitative analysis of the optical properties of the water. Hence we are considering the use of a sequence of images of the Coastal Zone Color Scanner (CZCS) carried on the Nimbus-7 satellite for further studies in order to estimate the transfer of organic matter towards the Golfe d'Arguin.

The description of the intertidal environment illustrates the importance of the seagrass beds, thus confirming the conclusions of Wolff & Smit (1990) who estimated that about 84% of the tidal flats of the entire Banc d'Arguin area is covered by dense or sparse seagrass beds.

## References

Cuq, F., 1989. Processus côtiers de la région du Banc d'Arguin (Mauritanie). Actes du colloque du groupe français de géomorphologie. Bull. Centre Géomorphol. Caen 36: 43–46.
Deschamps, P. Y. & T. Phulpin, 1980. Atmospheric correction of infrared measurements of sea surface temperature using channels at 3–7, 11 and 12 $\mu$m. Bound. Layer Met. 18: 131.
Savigny & Corréard, 1818. Récit d'un voyage au Sénégal. Paris.
Wolff, W. J. & C. J. Smit, 1990. The Banc d'Arguin (Mauritania) as an environment for coastal birds. Ardea 78: 17–38.

*Hydrobiologia* **258**: 41–58, 1993.
*W. J. Wolff, J. van der Land, P. H. Nienhuis & P. A. W. J. de Wilde (eds),*
*Ecological Studies in the Coastal Waters of Mauritania.*



# Heavy metals (Cu, Zn, Pb, Cd) in sediment, zooplankton and epibenthic invertebrates from the area of the continental slope of the Banc d'Arguin (Mauritania)

J. M. Everaarts, R. Heesters & C. V. Fischer
*Netherlands Institute for Sea Research, P.O. Box 59, 1790 AB Den Burg, Texel, The Netherlands*

*Key words:* heavy metals, sediment, zooplankton, epibenthic invertebrates, Banc d'Arguin

## Abstract

The concentrations of copper, zinc, lead and cadmium in the surface sediment (upper 5 mm) were generally higher in the silt fraction than in the bulk sediment. No significant geographical trend in the metal concentrations of the surface sediments was found, nor a correlation between concentrations in bulk sediment as well as in the silt fraction and the % silt could be established. In general, the metal concentrations in both bulk sediment and silt are lower, when compared to marine environments in other climatological regions.

In zooplankton, the metal concentrations were relatively high: expressed in $\mu g\ g^{-1}$ on a dry weight (D.W.) basis, they ranged from 15–90 for copper, 70–580 for zinc, 12–55 for lead and 4–10 for cadmium.

In epibenthic invertebrate species, both in crustaceans and bivalve molluscs, the concentrations of copper, zinc, and lead were in the same order of magnitude as compared to corresponding species from other geographical latitudes. Cadmium concentrations were relatively low, ranging from 0.13–0.42 $\mu g\ g^{-1}$ D.W. in the bivalve molluscs *Pitaria tumens* and from 0.04–0.27 $\mu g\ g^{-1}$ D.W. in the shrimp *Processa elegantula*. Also in the crab species *Ilia spinosa*, *Inachus* sp. and *Pagurus* sp., the cadmium concentrations were low, varying between 0.1 and 0.2 $\mu g\ g^{-1}$ D.W.

No significant relation between the metal concentration in whole-body samples and sediment (either bulk or silt) was present. Also no gradient was apparent in concentrations in organisms sampled at different depths (5 to 200 m) along two off-shore transects perpendicular to the Banc d'Arguin. Data indicated lower metal concentration in epibenthic organisms from sampling stations along a northern transect (southwest of Cap Blanc) than in organisms from the southern transect, off Cap Timiris.

Evidence was obtained for a considerable atmospheric input of heavy metals, in particular zinc and lead, in a certain area along the continental slope of the Banc d'Arguin.

## Introduction

Heavy metals are natural constituents of the marine environment, sometimes present at very low concentration levels. A number of these elements are biologically essential. In sea-water, trace metals can occur dissolved in ionic form, bound to organic substances, associated with humic acids, chelated with colloids or adsorbed to organic and inorganic particulate matter (Stumm & Bilinski, 1972). The chemical form of a metal may not only determine its availability and the rate of uptake by organisms, but also its toxicity. Aquatic organisms may accumulate metals from the surrounding water by absorption and diffusion across the body surfaces, and by ingestion of food and particulates (Phillips, 1977; cf. Moore & Ramamoorthy, 1984). Uptake from sediment fractions does not occur or the availability of sediment-bound metals to benthic biota is very restricted (Samant *et al.*, 1990). In many marine organisms uptake of the predominant soluble metal form of copper, zinc and cadmium, mainly occurs from the waterphase, and the uptake is proportional to the metal ion activity in the water (Engel *et al.*, 1981, 1985; Sanders *et al.*, 1983).

Trace metals introduced into the marine environment by domestic, industrial and mining activities are ultimately absorbed onto deposits and incorporated into marine sediments. Many investigations were carried out to evaluate the role of geochemical and biological processes in marine sediments in order to understand the pathways of trace metals of anthropogenic origin, transported by rivers into the marine environment. Most of the world's estuarine systems are characterized by a substantial input of river water. Since rivers discharge all kind of chemical substances into the marine system, both in solution and in colloidal or particulate form, the impact of this outflow will be most direct and intensive in the estuarine and adjacent coastal areas.

The objective of the present study was to determine the quality of a coastal marine environment not influenced by rivers, in terms of the concentration of the heavy metals copper, zinc, lead and cadmium in sediment, zooplankton and epibenthic invertebrates. Surveys were carried out along two transects over the continental slope of and perpendicular to the Banc d'Arguin (Mauritania), in order to establish the metal concentration levels in sediments and the bioaccumulation capacities of organisms of two levels of the foodchain, zooplankton and invertebrates. Furthermore, a comparison was made with concentration levels in the particular environmental compartments of marine and estuarine areas from other geographical latitudes.

## Material and methods

Samples were obtained during a cruise with the research vessel 'Tyro' (Netherlands Marine Research Foundation, SOZ) from 7 May through 29 May, 1988.

Surface sediment samples were taken by means of sub-sampling the upper 5 mm layer of sediment cores that were obtained with a Reineck box corer. The sub-samples were stored in polypropylene bags at $-20\ ^{\circ}C$ up to processing. The grain-size distribution in the sediments was determined according to the following procedure. About 20 gram dry sediment was wet-sieved with demineralized water (to disintegrate aggregates of fine material) over a sieve with a pore size of 63 $\mu m$, and subsequently diluted to 1 $dm^3$ and filtered over a preweighed 0.45 $\mu m$ filter. The remaining sediment suspension was dried at 70 °C and sieved (pore sizes 100, 200, 315 and 500 $\mu m$) for 10 minutes at 50 Hz. After weighing the different fractions, grain-size distribution was calculated.

Zooplankton samples were collected using a 300 $\mu m$ vertical plankton net for hauls from 200 m depth or from 1 m above the sea-bed (at shallow stations) to the surface. The zooplankton species were mainly copepods and their numbers were dominated by the species *Calanoides carinatus*, *Paracalanus parvus*, *Temora turbinata*, *Metridia lucens* and *Centropagus chierchiae*.

The macrobenthic organisms, collected with a 5 m beam-trawl and a trawling time of 5–10 minutes, represented the phyla Arthropoda (Crustacea Natantia, shrimp – *Processa elegantula*, and Crustacea Reptantia, spidercrabs – *Ilia spinosa*, *Inachus* sp. and the hermitcrab *Pagurus* sp.) and Mollusca (Bivalvia, bivalve mollusc – *Pitaria tumens*).

To avoid contamination of the samples, all chemicals used were suprapure (Merck) and all materials used were rinsed for 12 hours with 6N HCl and subsequently rinsed five times with ultra clean water (obtained by nanopure Barnstead filtering device).

The metal content of the sediment was determined after decomposition of about 500 mg, measured to the nearest 0.1 mg, with a mixture of 1 $cm^3$ aqua regia (1 part $HNO_3$ + 3 parts HCl) and 5 $cm^3$ HF in a teflon destruction bomb (Loring & Rantala, 1977) for 2 hours at 120 °C. Subsequently the decomposed samples were diluted with 30 $cm^3$ saturated $H_3BO_3$-solution and ultra clean water to 50 $cm^3$. From the samples of organisms about 400 mg, measured to the nearest 0.1 mg, was decomposed with 3 $cm^3$ of 65% $HNO_3$ in a teflon bomb for 2 hours at 120 °C (Paus, 1972). The decomposed samples were transferred quantitatively to polypropylene tubes and diluted with ultra-clean water to 10 $cm^3$.

For the analysis of the metals both flame- and flameless atomic absorption spectrophotometry (AAS) was applied, using a Perkin Elmer 403 and a heated graphite-furnace atomizer HGA 500 coupled to Perkin Elmer 5000, respectively. To calculate the metal concentrations, the standard addition method was applied and calibration curves were made on standard solutions of copper, zinc, lead and cadmium, resp. The AAS-measurements were carried out in duplicate on each sample.

*Table 1.* The geographical position of the sampling stations along a Northern- and Southern transect of a survey on the continental slope of the Banc d'Arguin (Mauritania); the depth (m), the salinity of the surface water (‰), the temperature of the bottom and surface (in italic) water (°C) and the texture of the surface layer (upper 5 mm) of the sediment are listed.

| Sampl. station | Geographic position | Depth (m) | Sal. (‰) | Temp. (°C) | % grain-size fraction (μm) | | | | | |
|---|---|---|---|---|---|---|---|---|---|---|
| | | | | | > 500 | 315–500 | 200–315 | 100–200 | 63–100 | < 63 |
| Northern transect | | | | | | | | | | |
| B5A | 19° 54′ 3 N | 1080 | 35.7 | 7.1 | 2.7 | 5.9 | 10.0 | 22.6 | 19.7 | 39.0 |
| | 17° 40′ 9 W | | | *17.5* | | | | | | |
| B5 | 19° 57′ 3 N | 112 | 35.8 | 14.0 | 5.1 | 7.6 | 9.3 | 44.5 | 24.6 | 8.9 |
| | 17° 28′ 2 W | | | *18.1* | | | | | | |
| B7 | 20° 02′ 2 N | 54 | 35.7 | 15.1 | 3.8 | 6.2 | 12.6 | 31.1 | 14.5 | 31.8 |
| | 17° 25′ 9 W | | | *16.2* | | | | | | |
| B8 | 20° 04′ 4 N | 31 | 35.6 | 16.9 | 1.4 | 3.1 | 4.8 | 41.9 | 22.3 | 26.7 |
| | 17° 21′ 4 W | | | *17.6* | | | | | | |
| B6 | 20° 13′ 3 N | 18 | 35.9 | 16.5 | 77.7 | 11.7 | 6.1 | 3.4 | 0.52 | 0.64 |
| | 17° 08′ 5 W | | | *16.9* | | | | | | |
| Southern transect | | | | | | | | | | |
| B2B | 19° 20′ 1 N | 1374 | 35.7 | 7.5 | 2.2 | 2.9 | 2.7 | 8.6 | 15.0 | 68.7 |
| | 17° 08′ 9 W | | | *18.8* | | | | | | |
| B2A | 19° 24′ 1 N | 896 | 35.6 | 7.2 | 2.1 | 5.8 | 10.8 | 28.2 | 27.7 | 25.4 |
| | 17° 05′ 3 W | | | *18.7* | | | | | | |
| B2 | 19° 33′ 1 N | 121 | 35.7 | 13.7 | 23.4 | 8.8 | 9.6 | 27.8 | 17.0 | 13.5 |
| | 17° 00′ 6 W | | | *18.6* | | | | | | |
| B1 | 19° 33′ 7 N | 68 | 35.7 | 14.5 | 11.8 | 10.3 | 11.6 | 16.3 | 9.0 | 41.1 |
| | 16° 54′ 6 W | | | *21.4* | | | | | | |
| B3 | 19° 36′ 5 N | 27 | 35.9 | 15.4 | 0.19 | 0.83 | 0.96 | 7.9 | 18.1 | 72.0 |
| | 16° 52′ 5 W | | | *20.0* | | | | | | |
| B4A | 19° 44′ 1 N | 10 | 35.5 | 17.0 | 58.4 | 15.8 | 16.8 | 7.6 | 0.27 | 1.13 |
| | 16° 58′ 0 W | | | *17.3* | | | | | | |
| B4 | 19° 45′ 2 N | 5 | 35.4 | 17.1 | 44.6 | 31.4 | 20.4 | 3.4 | 0.14 | 0.06 |
| | 16° 53′ 9 W | | | *17.8* | | | | | | |

## Results

### *Sediment*

The geographical position of the sampling stations, some hydrographical parameters (salinity, temperature and depth) and the grain-size distribution of the sediments are summarized in Table 1. The grain-size distribution and the percentage of the fine fraction (grain size $<63\ \mu m$) of the bottom surface layer (upper 5 mm) differed considerably among various sampling sites along both transects.

The concentrations of copper, zinc, lead and cadmium analysed in the bulk sediment are presented in Fig. 1. Although a tendency for a decrease in the concentration of copper, zinc and lead in bulk sediment from the deeper off-shore

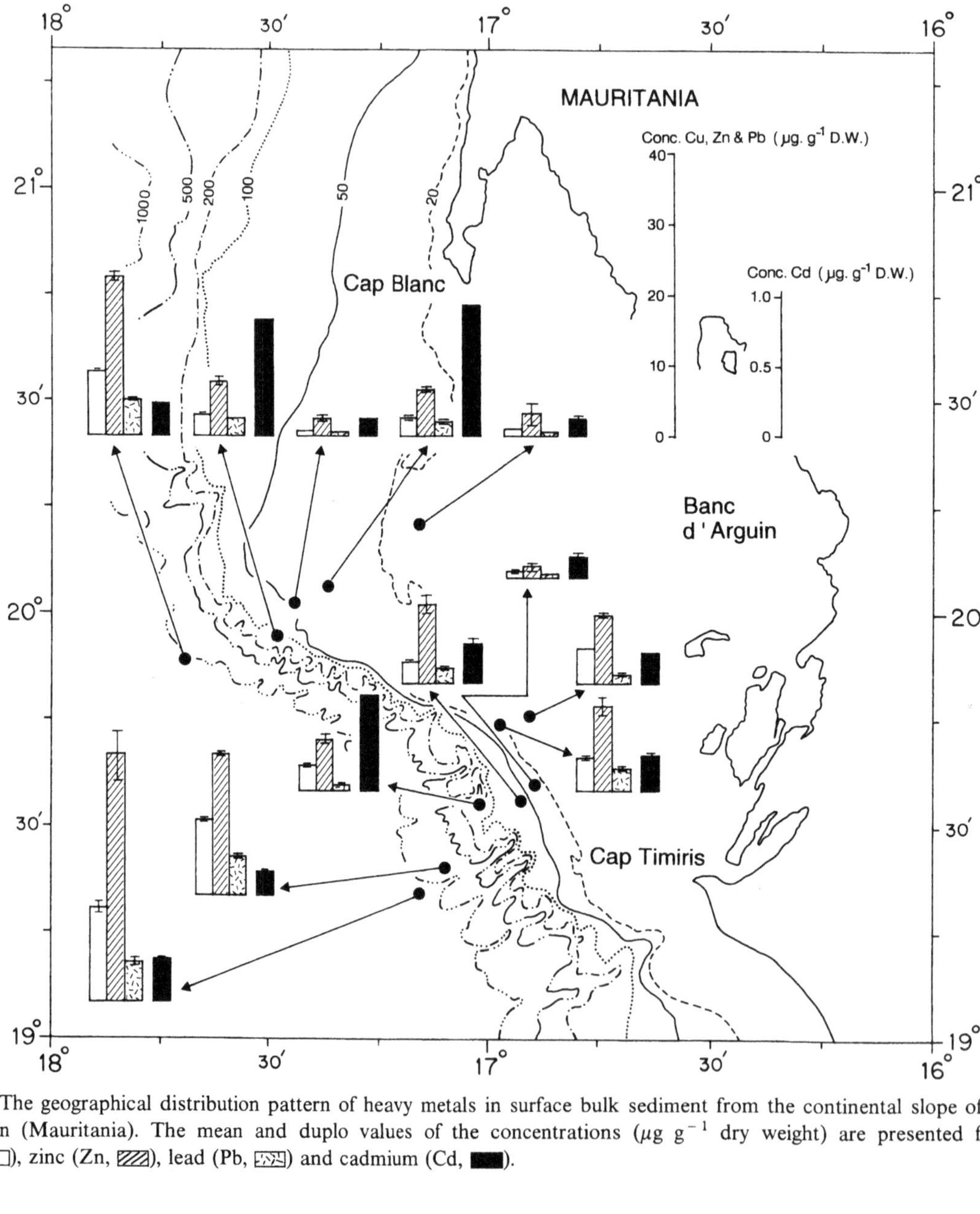


*Fig. 1.* The geographical distribution pattern of heavy metals in surface bulk sediment from the continental slope of the Banc d'Arguin (Mauritania). The mean and duplo values of the concentrations ($\mu g\ g^{-1}$ dry weight) are presented for copper (Cu, ▭), zinc (Zn, ▨), lead (Pb, ▧) and cadmium (Cd, ■).

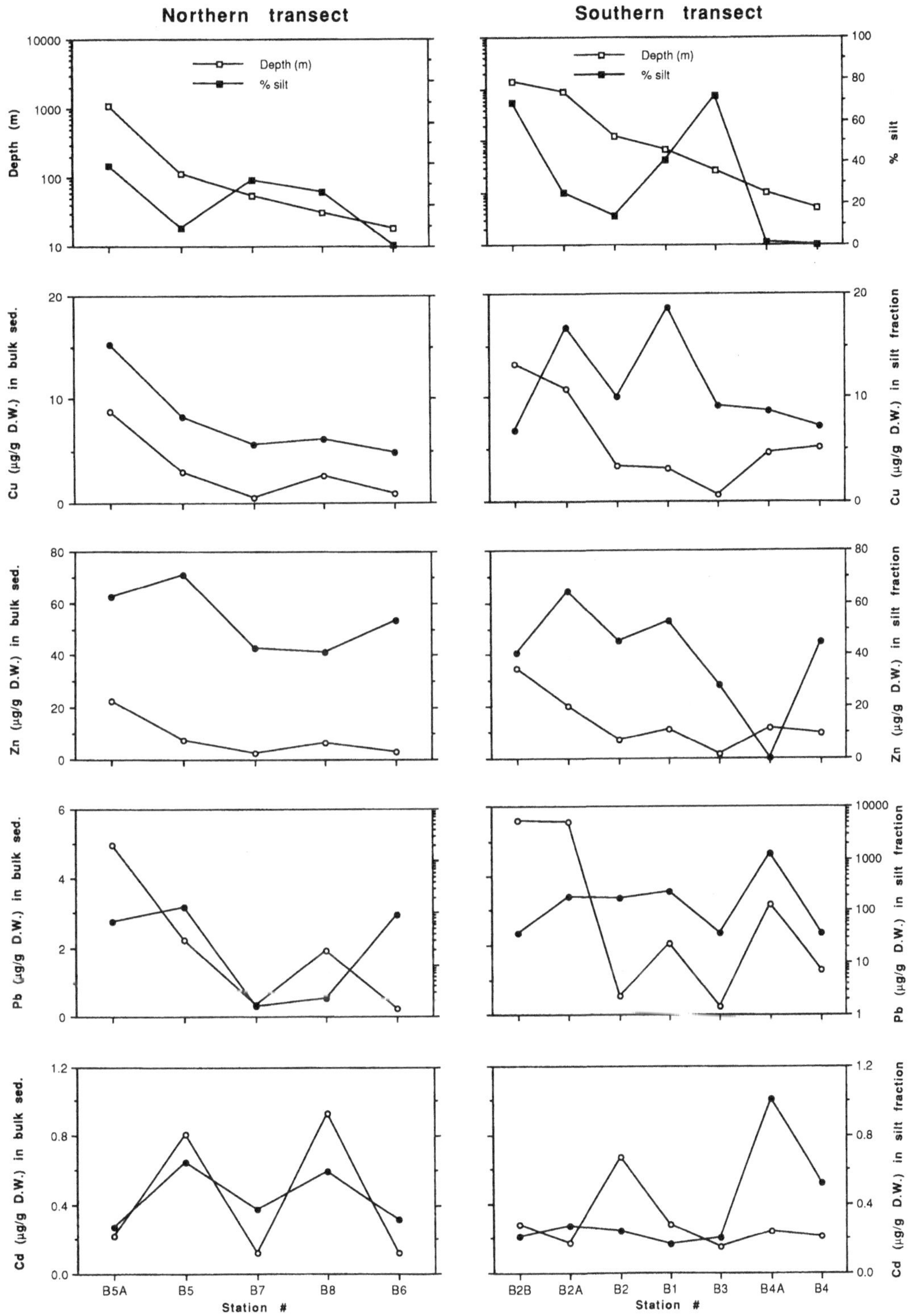


*Fig. 2.* Depth (m, logarithmic scale, —□—) of the sampling stations, the silt percentage of the surface sediment (—■—) and the concentration ($\mu g\ g^{-1}$ dry weight) of copper (Cu), zinc (Zn), lead (Pb) and cadmium (Cd) in both bulk sediment (○——○) and the silt fraction (●——●). Notice the logarithmic scale for Pb in silt.

stations towards the shallow area of the Banc d'Arguin could be described (e.g. Stations B5A-B5-B7 and B2B-B2A-B2), no distinct trends in the concentrations were measured along the two transects. The mean concentrations of copper, zinc, lead and cadmium varied between 0.5 to 13 $\mu g\,g^{-1}$, 1.5 to 34 $\mu g\,g^{-1}$, 0.2 to 5.6 $\mu g\,g^{-1}$ and 0.1 to 0.9 $\mu g\,g^{-1}$ dry weight (D.W.), respectively. In the silt fraction (grain-size <63 $\mu$m), for copper and zinc the concentrations of the heavy metals were higher than in the bulk sediment, but for cadmium and zinc at the underlimits in the same order of magnitude (Fig. 2). The concentrations of copper and zinc in the silt varied between 5 to 20 $\mu g\ g^{-1}$ and 28 to 72 $\mu g\ g^{-1}$ D.W., respectively. Cadmium concentrations in the silt frac-

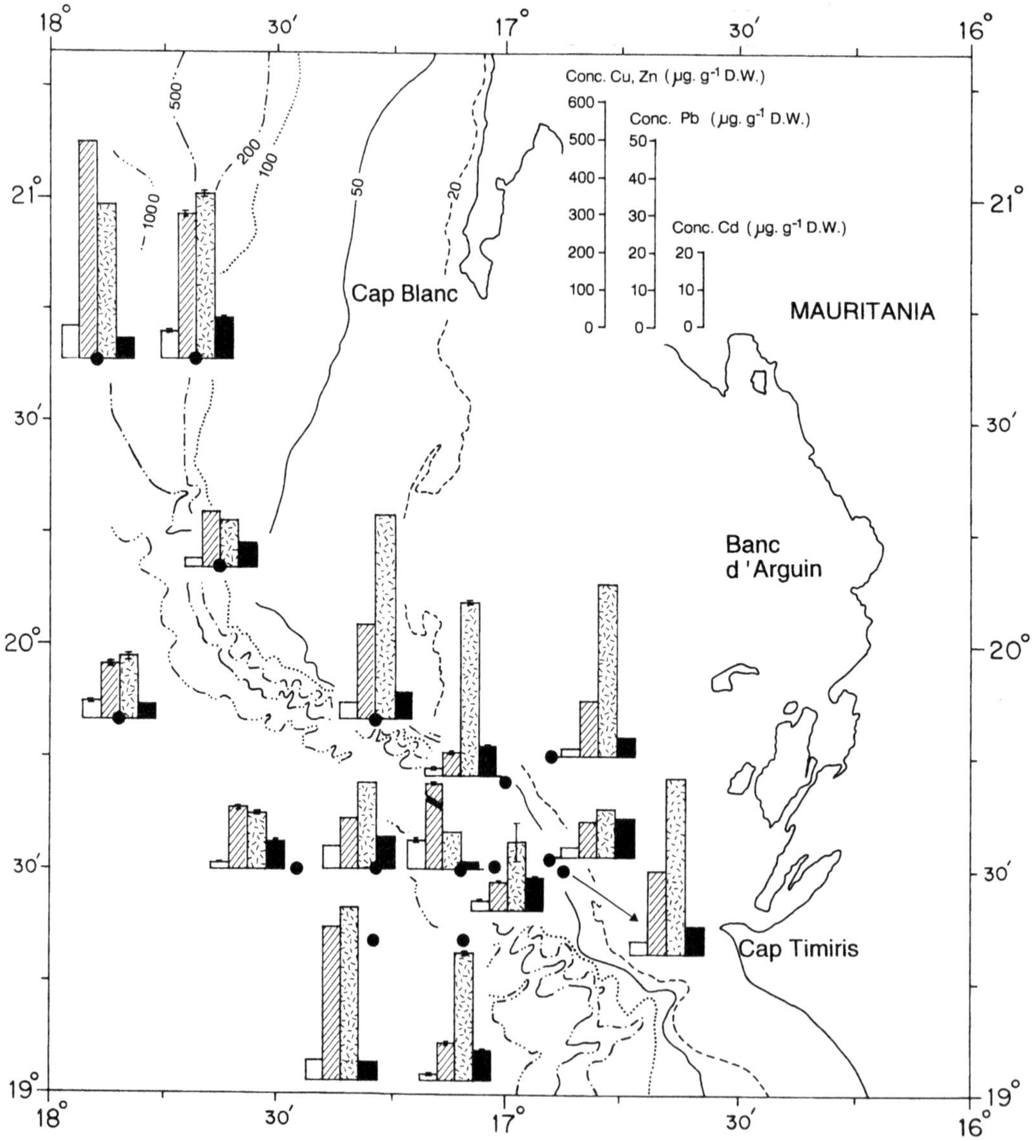


*Fig. 3.* The geographical distribution pattern of heavy metals in zooplankton from the upwelling area off-shore Mauritania. The mean and duplo values of the concentrations ($\mu g\ g^{-1}$ dry weight) are presented for copper (Cu, ▭), zinc (Zn, ▨), lead (Pb, ▦) and cadmium (Cd, ■).

tion were the same as in the bulk sediment (0.16–1.01 μg $g^{-1}$ D.W.), whereas the lead levels were a factor 10 to 100 higher, with a minimum concentration of about 2 μg $g^{-1}$ D.W. and an exceptional high concentration of about 1200 μg $g^{-1}$ D.W. in a sample from an area along the Northern transect with a low percentage of silt.

*Zooplankton*

The concentrations of copper (17–92 μg $g^{-1}$ D.W.), zinc (70–580 μg $g^{-1}$ D.W.), lead (12–54 μg $g^{-1}$ D.W.) and cadmium (4–11 μg $g^{-1}$ D.W.) in zooplankton (mainly copepods) did differ markedly for the various geographic locations

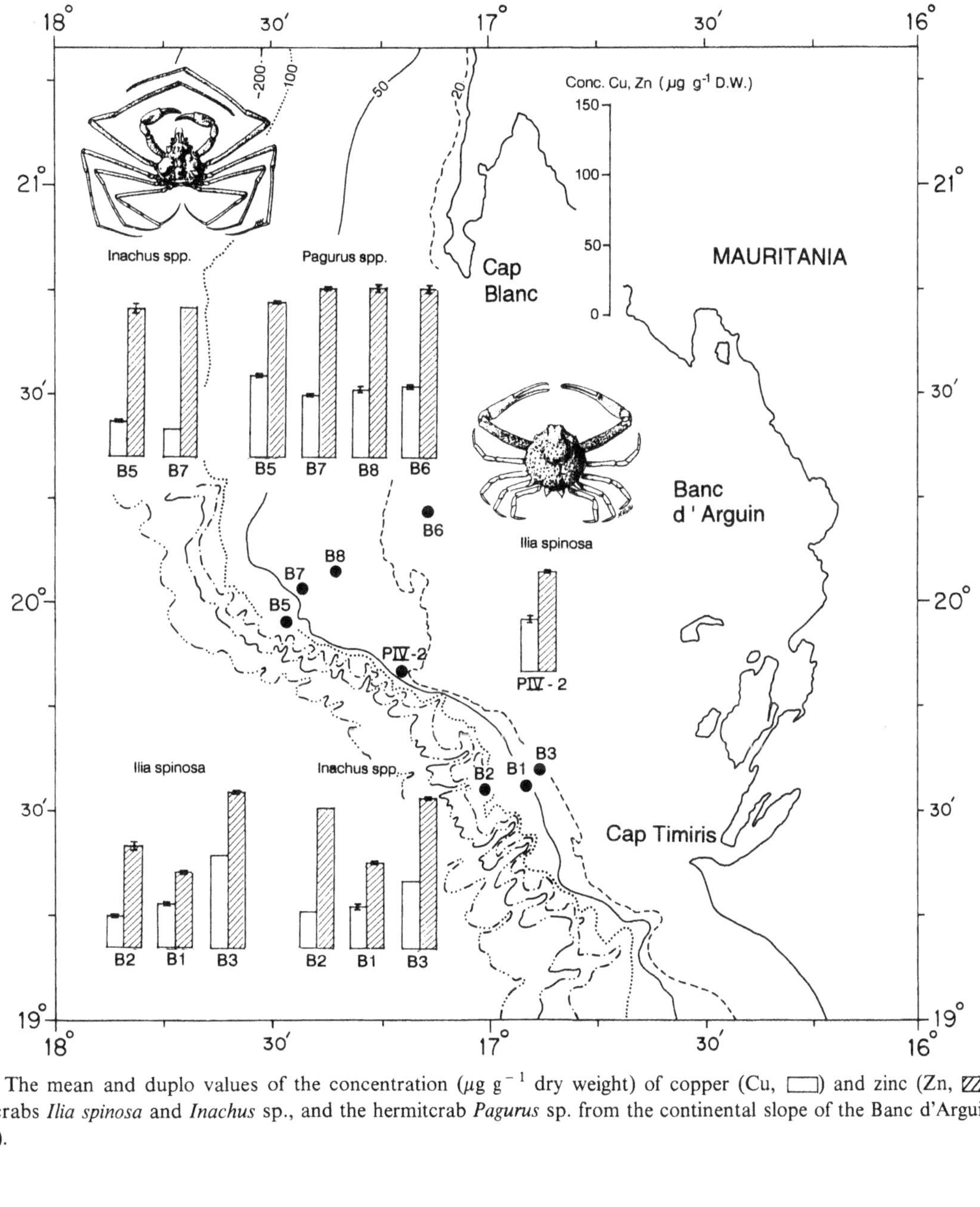


*Fig. 4.* The mean and duplo values of the concentration (μg $g^{-1}$ dry weight) of copper (Cu, □) and zinc (Zn, ▨) in the spidercrabs *Ilia spinosa* and *Inachus* sp., and the hermitcrab *Pagurus* sp. from the continental slope of the Banc d'Arguin (Mauritania).

from both the deep-sea area and the continental slope of the Banc d'Arguin (Fig. 3). In particular, the concentrations of zinc and lead showed considerable differences. Highest contents were observed in plankton from the area off the coast of Cap Blanc and in samples obtained along the edge of the Banc d'Arguin proper, down southwards to Cap Timiris.

*Epibenthic invertebrates*

The concentration of copper in whole-body samples of the crab species *Ilia spinosa* and *Inachus* sp. increased from deeper stations towards the shallow area only along the Southern transect of the Banc d'Arguin (Stations B2-B1-B3; Fig. 4). The levels of zinc did not vary among specimen

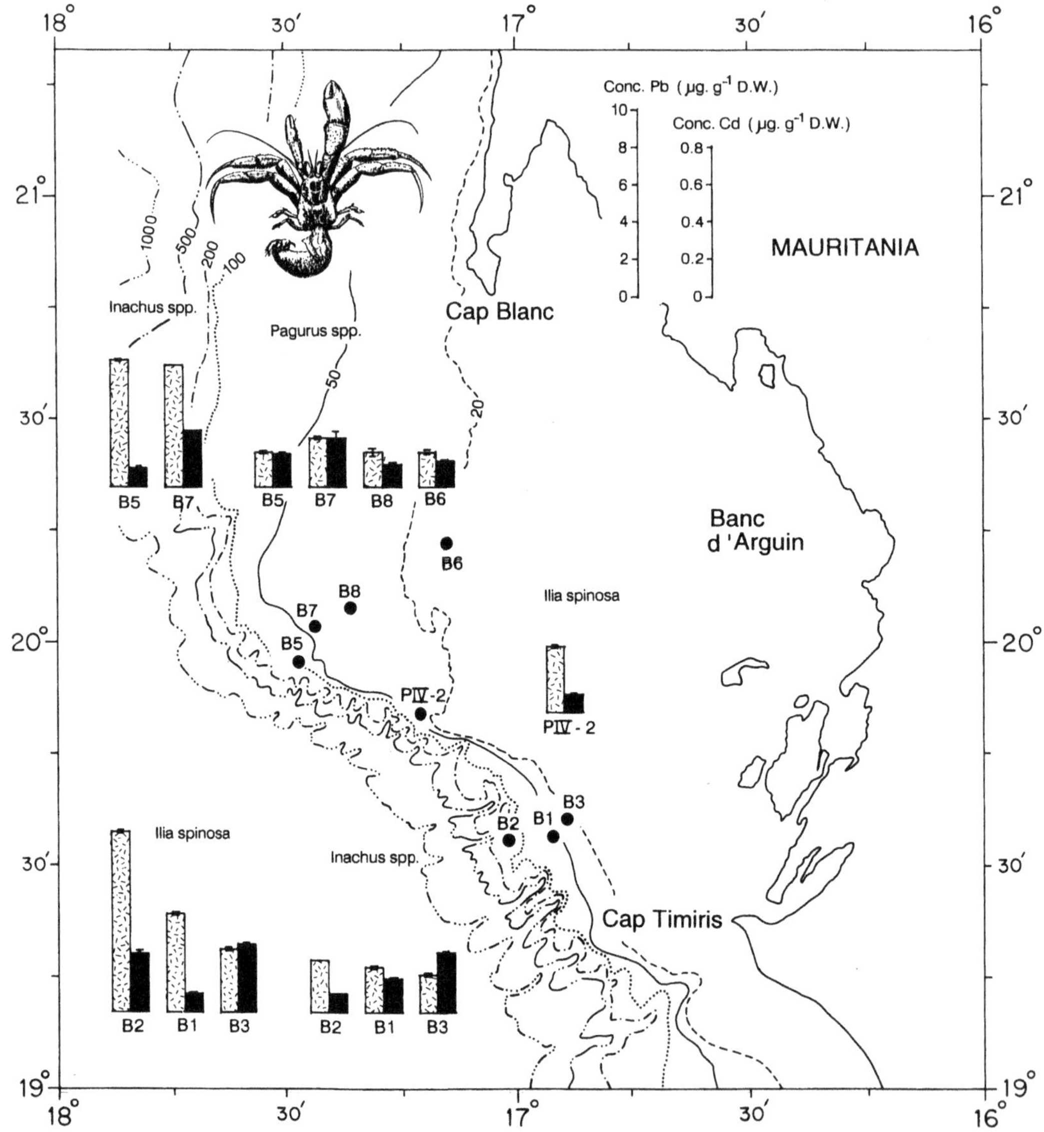


*Fig. 5.* The mean and duplo values of the concentration ($\mu g\ g^{-1}$ dry weight) of lead (Pb, ▨) and cadmium (Cd, ■) in the spidercrabs *Ilia spinosa* and *Inachus* sp., and the hermitcrab *Pagurus* sp. from the continental slope of the Banc d'Arguin (Mauritania).

from the various sampling sites. In another crustacean, the hermitcrab *Pagurus* sp., both the copper and zinc concentrations did not show any gradient (Fig. 4). The concentration of lead in *Ilia spinosa* and *Inachus* sp. from the Southern transect decreased towards the shallow area of the Banc d'Arguin, whereas the cadmium concentration showed a slight increase, in particular in *Inachus* sp. (Fig. 5). With respect to the data on the metal levels in *Inachus* sp. and *Pagurus* sp. from sampling sites along the Northern transect, no concentration gradients could be observed (Figs 4 & 5).

The concentrations of copper, zinc, lead and cadmium in whole-body of the shrimp *Processa elegantula* are presented in Fig. 6. The levels of the various metals did not differ markedly for the various geographic locations. The copper and

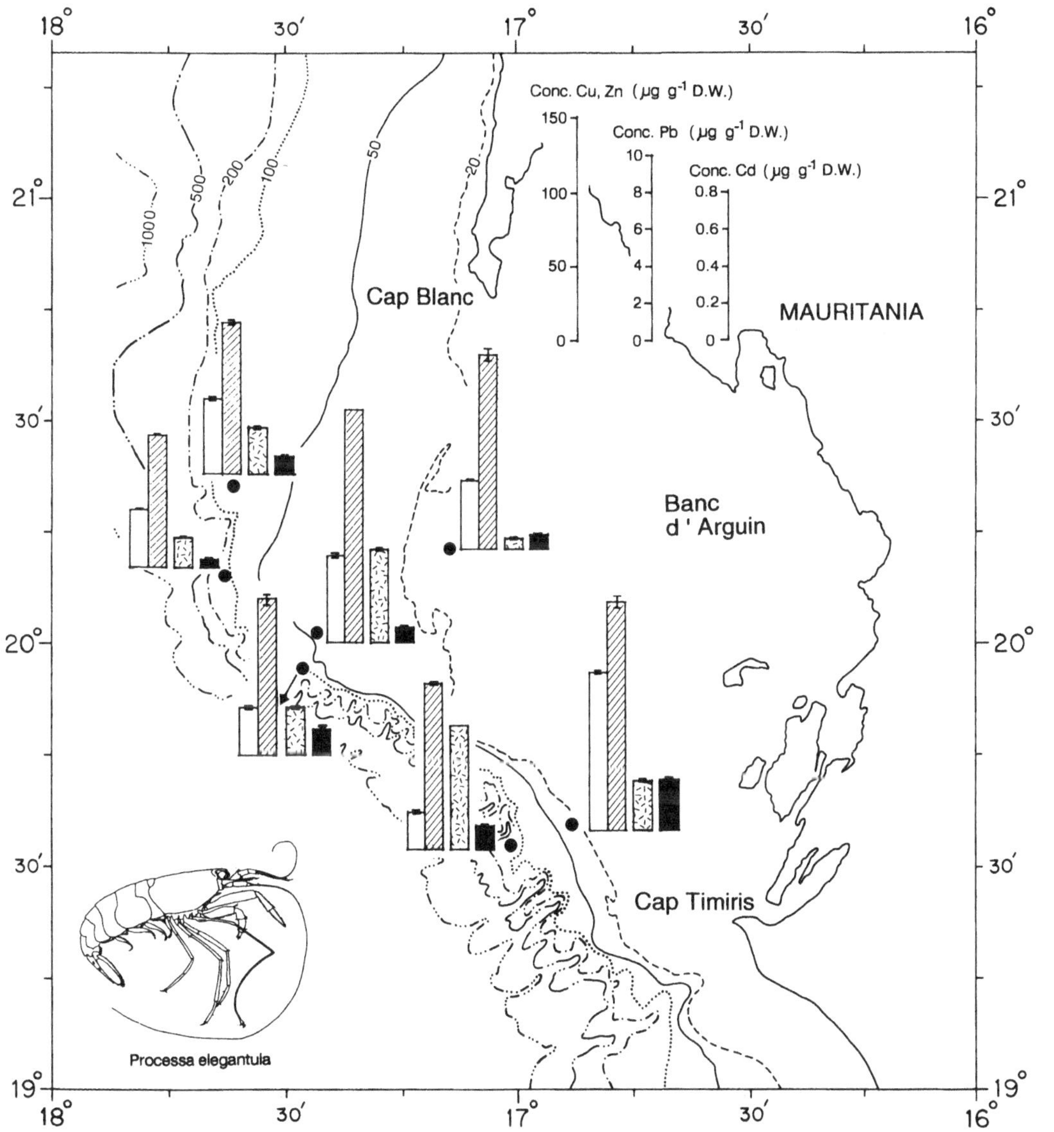


*Fig. 6.* The mean and duplo values of the concentration ($\mu g\ g^{-1}$ dry weight) of copper (Cu, ▭), zinc (Zn, ▨), lead (Pb, ▤) and cadmium (Cd, ■) in the shrimp *Processa elegantula* from the continental slope of the Banc d'Arguin (Mauritania).

zinc concentrations ranged from 20 to 100 $\mu g\,g^{-1}$ D.W. and 90 to 150 $\mu g\,g^{-1}$ D.W., respectively. The concentration of lead ranged from 0.5 to 9.7 $\mu g\,g^{-1}$ D.W. and the cadmium levels were low, varying between 0.04 and 0.27 $\mu g\,g^{-1}$ D.W.

The bivalve mollusc *Pitaria tumens* showed relatively low levels of copper, zinc and lead ranging from 9–24, 80–140 and 1.6–3.6 $\mu g\,g^{-1}$ D.W., respectively. The whole-body cadmium concentration varied between 0.13 and 0.42 $\mu g\,g^{-1}$ D.W. (Fig. 7). No significant differences in the geographical distribution pattern could be established. However, some evidence was found for a gradient of increasing concentrations of, in particular zinc and cadmium, along a northwest-southeast transect of the continental slope of the Banc d'Arguin.

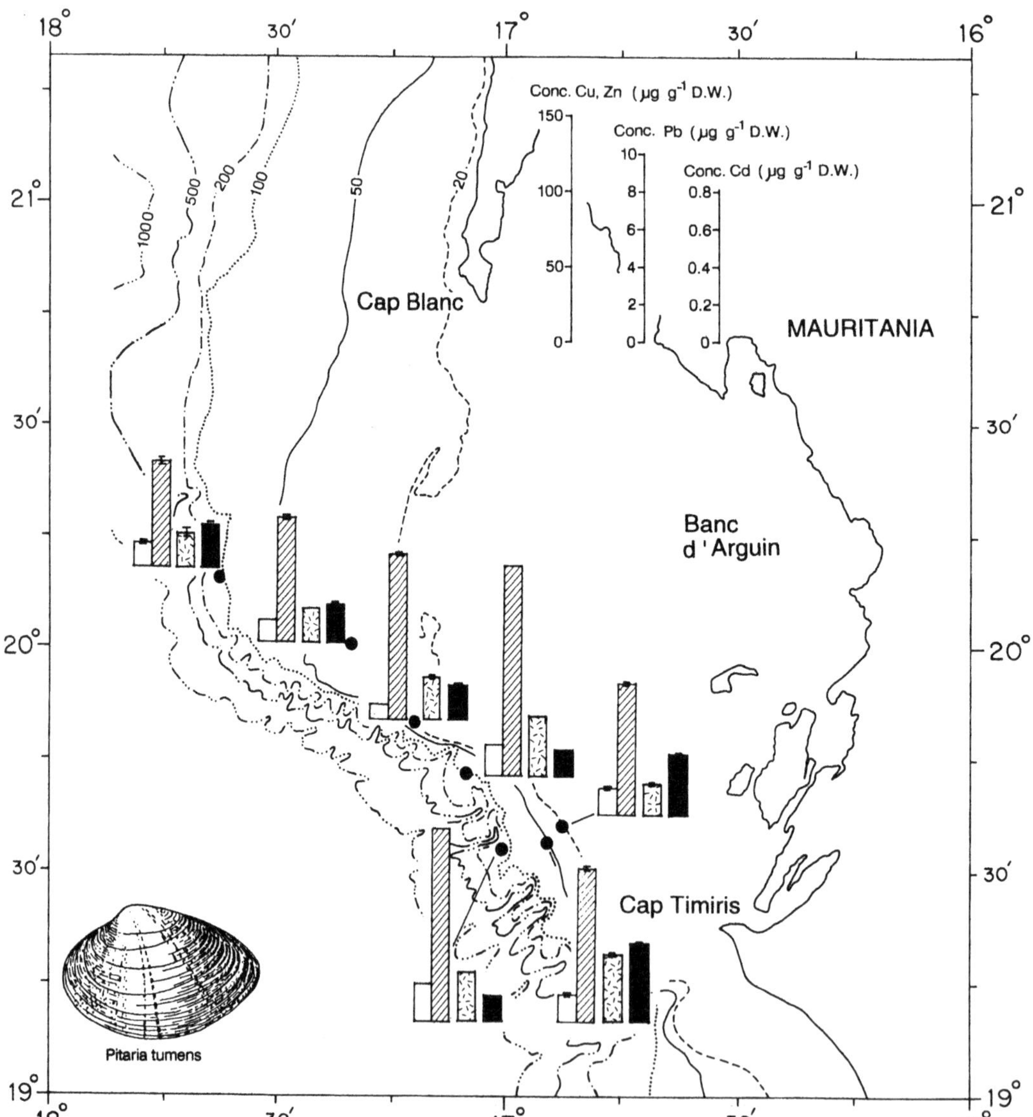


*Fig. 7.* The mean and duplo values of the concentration ($\mu g\,g^{-1}$ dry weight) of copper (Cu, □), zinc (Zn, ▨), lead (Pb, ▨) and cadmium (Cd, ■) in the bivalve mollusc *Pitaria tumens* from the continental slope of the Banc d'Arguin (Mauritania).

## Discussion

In earlier studies (Everaarts & Swennen, 1987; Everaarts & Fischer, 1992) it has been discussed already that particularly in sediments with a very low percentage of silt, the heavy metal concentrations in the silt fraction do not reflect the degree of contamination of the bulk sediment. When correlating data on the concentration of the metals in bulk sediment, the concentration in the silt fraction and the percentage silt with each other, actually no significant correlation could be described in the present study. For example, the surface sediment of station B3 (Southern transect) and station B7 (Northern transect) had high silt contents of 72% at 27 m depth and 32% at 54 m depth, respectively, but showed the lowest metal concentrations in the bulk sediment (Fig. 2). In contrast, sediment from stations B4A, B4 (Southern transect) and B6 (Northern transect) with low silt contents of 1.1% at 10 m, 0.05% at 5 m and 0.6% at 18 m depth, respectively, had metal concentrations in the bulk sediment ranging at the highest levels. In some cases this can be explained by a very high concentration of a particular metal in the silt fraction (e.g. Zn at B4, Cd at B4A and Pb at B6; Fig. 2), in other sediments the metal concentration can not be explained in terms of the silt percentage and its metal content. This indicates that specific geochemical properties (such as sediment strength, bulk density, texture and water content), together with a number of physicochemical processes predominantly determine the metal concentrations in these sediments. A comparison of the data of sampling stations either in the northern and southern region of the shelf area of the Banc d'Arguin supports this assertion. When comparing the average levels of the various metals separately in the surface bulk sediment and the silt fraction (Table 2), no statistical differences (Student's t-test) were found. However, the mean concentrations differed (e.g. copper and zinc in bulk sediment higher and lead and cadmium lower in the south than in the north), but minimum and maximum values differed extremely and the standard deviations were very large. The same applied to the metal concentrations in the silt fraction (e.g. the copper and lead concentrations in the south were higher and the zinc and cadmium levels lower than in the north). Thus, statistically no differences in the concentrations of the heavy metals in the sediment could be described, even though large dissimilarities between individual samples were obtained.

Comparison of data on the metal concentrations in marine sediments from various climatological latitudes and the data of the present study (Table 3) leads to the following conclusions: The concentrations of copper were at the lowest levels but in the same order of magnitude as the copper levels from different and similar climato-

*Table 2.* The heavy metal concentrations in surface sediment, both bulk and silt, sampled along two transects perpendicular to the continental slope of the Banc d'Arguin (Mauritania). N is the total number of samples obtained along the transect taken into account to calculate the average concentration and standard deviation.

| Sediment | Transect | N | Concentration in $\mu g\, g^{-1}$ dry weight ($\pm$ S.D.) | | | |
|---|---|---|---|---|---|---|
| | | | Cu | Zn | Pb | Cd |
| Surface bulk sediment | Northern | 5 | 1.75 | 4.94 | 1.03 | 0.49 |
| | | | (1.17) | (2.52) | (0.95) | (0.40) |
| | Southern | 7 | 2.40 | 6.64 | 0.92 | 0.37 |
| | | | (1.39) | (4.48) | (0.86) | (0.24) |
| Surface silt fraction | Northern | 5 | 8.1 | 54.2 | 58.1 | 0.44 |
| | | | (4.2) | (12.7) | (55.7) | (0.17) |
| | Southern | 7 | 11.0 | 46.0 | 273.4 | 0.38 |
| | | | (4.7) | (12.1) | (426.7) | (0.31) |

*Table 3.* The concentration of heavy metals in the surface layer of marine sediments from various geographical regions.

| Geographical area | Sediment | Concentration in $\mu g\ g^{-1}$ dry weight | | | |
|---|---|---|---|---|---|
| | | Cu | Zn | Pb | Cd |
| Dutch Wadden Sea, 1985 (1) | Silt | 10–40 | 120–225 | 40–180 | 0.78–1.7 |
| North Sea, 1989 (2) | Silt | 10–150 | 100–2500 | 30–250 | 0.17–16.1 |
| North Sea, Oyster Grounds, 1990 (3) | Silt | 15–62 | 95–220 | 32–84 | 0.05–0.42 |
| Malay Peninsula, coastal area 1987 (4) | Silt | 10–38 | 30–90 | 23–180 | 0.03–0.09 |
| Java Sea, 1989 (5) | Silt | 10–65 | 50–130 | 10–50 | 0.05–0.53 |
| | Bulk | 10–54 | 30–125 | 6–23 | 0.03–0.45 |
| Mediterranean Sea, 1991 (6) | | | | | |
| Rhône delta | Bulk | 15–50 | 250–290 | 20–26 | 1.29–2.21 |
| Continental slope | Bulk | 10–15 | 53–105 | 5–10 | 0.09–0.12 |
| Gulfe du Lion | Bulk | 30–41 | 77–87 | 11–12 | 0.07–0.26 |
| Continental slope of the | Bulk | 0.5–14 | 1.5–35 | 0.2–5.6 | 0.12–0.93 |
| Banc d'Arguin, 1991 (7) | Silt | 4.9–23 | 28–72 | 1.6–1223 | 0.16–1.01 |

*References:* (1) Kramer *et al.*, 1985; (2) Everaarts & Fischer, 1989; (3) Klamer *et al.*, 1990; (4) Everaarts & Swennen, 1987; (5) Everaarts, 1989; (6) van Hoogstraten & Nolting, 1991; (7) Present study.

logical regions. The minimum and maximum value of the zinc concentrations in both bulk sediment and silt were considerably lower than in comparable marine environments from other latitudes. The lowest concentrations of lead measured both in bulk sediment and silt were about one order of a magnitude lower than in sediment from other geographical regions. However, at a number of stations the lead levels in the silt fraction were similar to those measured in the silt fraction of sediment from areas considered to be highly contaminated, such as the North Sea. These high contents can be explained by a high atmospheric input and precipitation of lead-rich desert dust and ore-dust originating from the iron-ore loading site in Nouadhibou, which phenomenon was quite obviously visible in a certain corridor-like area southwest of the Cap Blanc area. The concentrations of cadmium were not the lowest measured in marine sediments from tropical or subtropical regions, but they varied in between the low values measured in the Java Sea (Everaarts, 1989), along the Malay Peninsula (Everaarts & Swennen, 1987) and at the continental slope of the northwestern basin of the Mediterranean Sea (Van Hoogstraten & Nolting, 1991) and higher values, such as measured in silt from the Dutch Wadden Sea (Kramer *et al.*, 1985), the Dutch part of the continental shelf of the North Sea in general (Everaarts & Fischer, 1989, 1992) and the Oyster Grounds, North Sea (Klamer *et al.*, 1990) in particular.

The concentrations of copper and zinc in zooplankton are in the same range as the levels in microplankton from a great number of open ocean areas, as was reviewed by Fowler (1986). The levels of copper found in neustonic copepods from the Mediterranean Sea (Polikarpov *et al.*, 1979) agree with the levels found in the mixture of zooplankton (the plankton composition of the samples was about 75% copepods; Kuipers *et al.*, 1993) found in Mauritanian waters. On the contrary, lead (12–54 $\mu g\ g^{-1}$ D.W.) and cadmium (4–11 $\mu g\ g^{-1}$ D.W.) levels were at least twice as high in zooplankton from the upwelling area of the continental slope of the Banc d'Arguin. Such high concentrations were also determined already in the early seventies in mixtures of copepod species from the Northwest Atlantic Ocean (Windom, 1972). The heavy metal concentrations established in zooplankton from the coastal upwelling area, off-shore Mauritania, thus agree in general with concentrations found in zooplankton in other marine areas. This holds even when these areas are considered to be highly contaminated, such as the New York Bight and Long Island

Sound (Greig *et al.*, 1977) or in other plankton species as e.g. in oceanic decapods from the Northeast Atlantic or crustaceans from highly polluted coastal sites, which show extremely high Cd values (Ridout *et al.*, 1989).

From a comparison of the average concentrations of the various metals in the selected species of epibenthic invertebrates obtained from all sampling stations, either in the northern or the southern region of the continental slope of the Banc d'Arguin (Table 4), it is seen that higher concentrations occur in animals from the southern region, except for zinc and lead in the spidercrab *Inachus* sp. Some of the concentrations were significantly higher (Student's t-test; $p<0.005$) in specimens from the southern region, such as copper in the bivalve mollusc *Pitaria tumens* and the spidercrab *Inachus* sp., or copper, zinc and lead in the hermitcrab *Pagurus*, and cadmium in the shrimp *Processa elegantula*. A comparison of the average metal concentrations in the organism with the sediment concentration (cf. Table 2) showed that there exists no or only a minor relation. When correlating the concentration of a particular metal in whole-body of a particular organism with the corresponding metal concentration in both the bulk sediment and the silt fraction, it could be concluded that there was no consistent correlation. Of the 40 relationships (linear regression) possible in the present study, 25 did not show any correlation, whereas in 9 cases a decrease-increase relationship and in 6 cases an increase-increase relationship could be described. Most of these correlations, however, were not significant. Significant linear regressions ($p<0.05$) were found for the zinc concentration in the crustacean species, negatively correlated with the zinc levels in both the bulk sediment and the silt and the cadmium concentration in the shrimp *Processa elegantula* positively correlated with the silt Cd concentration. Also Pb in the bivalve mollusc *Pitaria tumens* was significantly ($p<0.05$) correlated

*Table 4.* The heavy metal concentrations in epibenthic invertebrates sampled along two transect perpendicular to the continental slope of the Banc d'Arguin (Mauritania). N is the total number of samples obtained along the transects taken into account to calculate the average concentration and standard deviation (S.D.).

| Species | Transect | N | Concentration in $\mu g\ g^{-1}$ dry weight ($\pm$ S.D.) | | | |
|---|---|---|---|---|---|---|
| | | | Cu | Zn | Pb | Cd |
| *Ilia spinosa* | Northern | – | – | – | – | – |
| | Southern | 8 | 39.1 | 76.6 | 5.46 | 0.21 |
| | | | (18.0) | (22.9) | (2.70) | (0.13) |
| *Inachus* spp. | Northern | 5 | 23.0 | 121.5 | 5.34 | 0.21 |
| | | | (2.8) | (22.9) | (1.75) | (0.11) |
| | Southern | 5 | 36.3 | 87.6 | 2.29 | 0.21 |
| | | | (10.5) | (24.6) | (0.28) | (0.10) |
| *Pagurus* spp. | Northern | 6 | 50.7 | 117.0 | 2.02 | 0.18 |
| | | | (6.2) | (5.7) | (0.33) | (0.06) |
| | Southern | 2 | 72.2 | 161.8 | 2.81 | 0.17 |
| | | | (2.4) | (3.1) | (0.04) | (0.01) |
| *Processa elegantula* | Northern | 8 | 43.5 | 119.9 | 2.40 | 0.08 |
| | | | (10.4) | (28.1) | (1.74) | (0.04) |
| | Southern | 4 | 65.0 | 130.9 | 6.23 | 0.20 |
| | | | (46.7) | (25.0) | (4.08) | (0.08) |
| *Pitaria tumens* | Northern | 4 | 12.1 | 96.0 | 1.99 | 0.19 |
| | | | (2.5) | (15.7) | (0.25) | (0.01) |
| | Southern | 6 | 19.4 | 108.2 | 2.69 | 0.29 |
| | | | (2.6) | (21.9) | (0.94) | (0.13) |

with the Pb levels both in bulk sediment and in the silt fraction. Thus, in general it can be concluded that occasionally sediment related metals may affect the metal concentrations in the organism but the uptake and accumulation of heavy metals depend mainly on environmental factors other than the sediment concentration. Because of the complexities involved in the physicochemical processes in the sediment and in the availability of sediment-bound metals to biota, no definitive conclusion could be drawn about the role of the sediment in the route of uptake. This is in agreement with the results of a study of Samant *et al.* (1990), who concluded from both experimental and field studies that it was unclear which sediment metal fractions were responsible for the bioavailability. Besides, in the present study it was also found that different species were able to accumulate different metals to different extents. This may reflect the geochemical and physicochemical status of the sediments. However, probably this merely reflects feeding habits and habitat differences as well as homeostatic regulation of metals.

The generally higher metal levels in organisms from the southern region of the shelf area of the Banc d'Arguin may be due to differences of water quality (explained in terms of dissolved metal concentrations and the load of particulate suspended matter), of the water masses influencing the northern and southern region of the continental slope of the Banc d'Arguin. This hypothesis would agree with the mechanism described for significantly enhanced cadmium concentrations in organisms from the area north and northeast of the Dogger Bank (North Sea), correlated with augmented dissolved cadmium levels in deep water from the North Atlantic (Everaarts *et al.*, 1990). The bottom water of the southern region and the deeper parts of the continental slope of the Banc d'Arguin originates mainly from the nutrient enriched South Atlantic Central Water and depends mainly upon a poleward undercurrent (Minas *et al.*, 1982). The bottom water of the

*Table 5.* The concentration of heavy metals in whole-body of shrimp from marine environments of different climatological latitudes.

| Geographical area | Concentration in $\mu g\ g^{-1}$ dry weight ($\pm$ S.D. or range) | | | |
|---|---|---|---|---|
| | Cu | Zn | Pb | Cd |
| Dutch Wadden Sea (1) | | | | |
| *Crangon crangon* | 54.3 (25.2) | 69.2 (5.4) | – | 0.17 (0.14) |
| North Sea, coast (1) | | | | |
| *Crangon crangon* | 57.3 (8.2) | 83.1 (11.4) | – | 0.40 (0.09) |
| *Crangon allmanni* | 56.1 (12.7) | 82.8 (11.9) | – | 0.99 (0.24) |
| North Sea, central (1) | | | | |
| *Crangon allmanni* | 37.2 (7.0) | 73.7 (5.6) | – | 3.79 (1.19) |
| Malay Peninsula (2) | | | | |
| *Palaemon semmelinkii* | 60–70 | 60–70 | 0.7–1.6 | 0.2–0.4 |
| *Metapenaeus lysianassa* | 85–130 | 60–85 | 0.8–3.4 | 0.4–0.9 |
| Java Sea (3) | | | | |
| Open sea, *Penaeidae* | 26–78 | 28–82 | – | 3.3–10.3 |
| Coastal area, *Penaeidae* | 5–118 | 26–109 | – | 2.4–13.9 |
| Estuaries, *Penaeidae* | 35–120 | 28–86 | – | 0.6–3.5 |
| Shelf Banc d'Arguin (4) | | | | |
| *Processa elegantula* | 20–100 | 90–150 | 0.5–9.7 | 0.04–0.27 |

*References:* (1) Everaarts *et al.*, 1990; (2) Everaarts & Swennen, 1987; (3) Everaarts *et al.*, 1989; (4) Present study.

coastal area and the Banc d'Arguin water is mainly determined by the North Atlantic Central Water of the Canary Current. In the transition area of the shallow area and the canyons of the continental slope of the Banc d'Arguin, this water mass cascades off the Banc d'Arguin (Peters, 1976). The very complicated hydrographic situation at the Banc d'Arguin, its continental slope and the upwelling area off the Mauritania coast, makes it difficult to interpret the distribution patterns of heavy metals in the various compartments of the marine environment.

A comparison of the data on reptantian crustaceans from the literature and the present study shows that the concentrations of copper were in the same range or slightly lower than in corresponding crab species from other tropical marine environments, such as Indonesian waters (Everaarts *et al.*, 1989) and coastal waters around the Malay Peninsula (Everaarts & Swennen, 1987). However, copper concentrations were considerably (a factor 2 to 8) lower than in crab species from contaminated coastal environments from temperate (Stenner & Nickless, 1974; Lande, 1977; Karbe *et al.*, 1988) and subtropical latitudes (Stenner & Nickless, 1975; Cuadras *et al.*, 1981; Shiber, 1981). Zinc and lead concentrations in hermitcrabs, *Pagurus* sp., from the shelf area of the Banc d'Arguin equaled and doubled, respectively, the levels in *Pagurus bernhardus* from the

*Table 6.* The concentration of heavy metals in bivalve molluscs from marine environments of different geographical regions.

| Geographical area | Concentration in $\mu g\ g^{-1}$ dry weight ($\pm$ S.D. or range) | | | |
|---|---|---|---|---|
| | Cu | Zn | Pb | Cd |
| Dutch Wadden Sea (1) | | | | |
| *Mytilus edulis* | 6.4–18.0 | 61–205 | – | 1.1–4.4 |
| *Cerastoderma edule* | 3.2–6.2 | 66–101 | – | 0.1–1.1 |
| Dutch Wadden Sea (2) | | | | |
| *Mytilus edulis* | 5.5–8.9 | 55–78 | – | 0.22–1.60 |
| *Cerastoderma edule* | 2.9–4.3 | 56–71 | – | 0.13–0.25 |
| North Sea (3) | | | | |
| *Venus striatula* | – | – | 0.3–4.6 | 0.8–3.3 |
| North Sea (4) | | | | |
| *Mytilius edulis* | 7.0 (0.4) | 112 (7.0) | 1.8 (0.2) | 1.1 (0.1) |
| Malay Peninsula (5) | | | | |
| *Perna viridis* | 6.0–8.2 | 42–49 | 1.11–1.50 | 0.51–2.22 |
| *Anadara granosa* | 4.7–4.8 | 103–132 | 0.43–1.77 | 3.29–5.72 |
| *Glauconome virens* | 11.0–17.2 | 65–80 | 12.0 18.4 | 1.40–2.81 |
| Hong Kong coast (6) | | | | |
| *Perna viridis* | 11–162 | 55–138 | 3.1–46.7 | 0.12–1.51 |
| Java Sea, coast (7) | | | | |
| *Corbula* sp + *P. viridis* | 14–19 | 61–89 | – | 1.6–4.5 |
| Mauritanian coast (8) | | | | |
| *Venus verrucosa* | 4.1 (0.8) | 58 (8) | – | 2.2 (0.9) |
| *Donax trunculus* | 11.8 (1.5) | 105 (15) | – | 0.6 (0.1) |
| Shelf Banc d'Arguin (9) | | | | |
| *Pitaria tumens* | 9–24 | 80–140 | 1.6–3.6 | 0.13–0.42 |

*References:* (1) Kramer *et al.*, 1985; (2) Everaarts, unpublished data of surveys on benthic invertebrates sampled in the Dutch Wadden Sea in 1985, 1986 and 1987; (3) Kröncke, 1987; (4) Karbe *et al.*, 1988; (5) Everaarts & Swennen, 1987; (6) Chan, 1988; (7) Everaarts *et al.*, 1989; (8) Roméo & Gnassia-Barelli, 1988; (9) Present study.

North Sea (Karbe *et al.*, 1988). In spidercrabs, the zinc levels are higher than the concentrations in corresponding species from tropical and subtropical regions. The concentration of cadmium, however, is one to two orders of magnitude lower in crab species from Mauritanian waters. A comparison of the whole-body concentrations in the shrimp *Processa elegantula* with levels in other, closely related shrimp species (Table 5) shows essentially the same results: comparable Cu-, higher Zn-, Pb- and significantly lower Cd-concentrations in specimens from the continental shelf of the Banc d'Arguin. These cadmium levels are similar to the cadmium contents of shrimp species collected in uncontaminated coastal and offshore, continental shelf areas in the Northeast Pacific, which are considered to represent baseline concentration levels (Harding & Goyette, 1989).

Very few data are published on concentration levels of metals in organisms in general and in macro epibenthic organisms in particular, from the Mauritanian coastal and off-shore area. The only data available to the authors concern trace metal concentrations in two species of filter feeding molluscs, the wedge shell *Donax trunculus* and the clam *Venus verrucosa*, collected in intertidal sands near Nouakchott and in the shallow coastal region south of Cap Blanc, respectively (Roméo & Gnassia-Barelli, 1988). The concentrations of copper and zinc in whole soft parts (whole body without shells) of *Pitaria* is the same as in *Donax*, whereas the levels in *Venus* are lower (Table 6). However, cadmium levels are significantly lower in *Pitaria* (0.13–0.42 $\mu g\ g^{-1}$) than in both *Donax* ($0.6 \pm 0.1$) and *Venus* ($2.2 \pm 0.9$). Lead was not measured.

A comparison of data on bivalve molluscs from the literature and the present study (Table 6) shows that copper and zinc concentrations are similar or slightly higher than in bivalves from tropical regions and temperate latitudes, respectively. Lead concentrations are in the lower range and agree with the levels in e.g. *Venus striatula* and *Mytilus edulis* from the North Sea and the edible species *Perna viridis* and *Anadara granosa* from coastal areas of the Malay Peninsula. In contrast, the lead levels in *Pitaria* are significantly lower than in the bivalves *Glauconome virens* and *Perna viridis* from contaminated tropical coastal areas, such as Pattani Bay (Thailand) and coastal waters of Hong Kong (cf. Table 6). Cadmium concentrations in *Pitaria* from the continental slope of the Banc d'Arguin are in the lowest range determined in bivalve molluscs from estuarine and coastal areas in other geographical region.

## Conclusions

Heavy metal concentrations in surface sediments, both bulk and the silt fraction, did not differ between sampling stations in the region of the continental slope of the Banc d'Arguin. The metal levels established in the sediment did not show gradients or geographical distribution patterns on the basis of which their anthropogenic origin and input could be explained. Locally very different geochemical and physicochemical processes and hydrological inhomogeneity as well as atmospheric input were probably the main causes for different concentrations in the surface sediment.

In the bulk sediment as well as in the silt fraction, generally, the concentrations of all heavy metals analysed were considerably lower than in sediments of comparable marine environments from other latitudes. Only a few sites showed higher zinc and lead contents.

The high metal levels measured in zooplankton from the coastal upwelling area off-shore Mauritania matched with concentrations found in zooplankton from both contaminated coastal and open ocean systems in other climatological regions. In particular lead and cadmium levels were high.

The geographical distribution pattern of the heavy metals in epibenthic invertebrates from the continental slope of the Banc d'Arguin could not be explained in terms of metal levels in the bulk sediment and the silt fraction, nor in terms of the silt percentage of the surface sediment. Also, no relation was found between the metal concentrations in the zooplankton and the metal levels in the invertebrates. A species-specific uptake pattern for the various metals could be described. The concentrations of copper and zinc in the invertebrate species analysed were in the same order of magnitude, as compared to corresponding species from marine environments from other latitudes. Lead concentrations and, in particular, cadmium concentrations were low in the epibenthic species, and in fact represent baseline concentrations for comparison with similar data from contaminated marine environments.

## Acknowledgements

Many thanks are due to S. Gonzalez and H. Witte for collecting and species determination of the zooplankton samples and G. Duineveld and G. van Noort for the species determination of the crustaceans and the bivalve molluscs. Special thanks are due to Captain J. de Jong and his crew of the R. V. 'Tyro' and P. A. W. J. de Wilde for his stimulating interest as chief-scientist during the cruise. The research has been carried out as a part of the Mauritania I Expedition, co-organised and funded by the Netherlands Marine Research Foundation (SOZ).

## References

Chan, H. M., 1988. A survey of trace metals in *Perna viridis* (L.) (Bivalvia: Mytilacea) from the coastal waters of Hong Kong. Asian Mar. Biol. 5: 89–102.

Cuadras, J., A. Gimeno, R. Flos & S. Crespo, 1981. Levels of copper and zinc in tissues of the hermit crab *Dardanus arrosor* (Herbst) from the Barcelona coast (Decapoda, Anomura). Crustaceana 40: 79–87.

Engel, D. W., W. G. Sunda & B. A. Fowler, 1981. Factors affecting trace metal uptake and toxicity to estuarine organisms. I. Environmental parameters. In F. J. Vernberg, A. Calabrese, F. P. Thurberg & W. B. Vernberg. Biological Monitoring of Marine Pollutants. Academic Press, New York: 127–144.

Engel, D. W., M. Brouwer & F. P. Thurberg, 1985. Comparison of metal metabolism and metal-binding protein in blue crab and the American lobster. In F. J. Vernberg, F. P. Thurberg, A. Calabrese & W. B. Vernberg (eds), Marine Pollution and Physiology: Recent Advances. University of South Carolina Press: 229–245.

Everaarts, J. M., 1989. Heavy metals (Cu, Zn, Cd, Pb) in sediment of the Java Sea, estuarine and coastal areas of East Java and some deep-sea areas. Neth. J. Sea Res. 23: 403–413.

Everaarts, J. M. & C. V. Fischer, 1989. Micro contaminants in surface sediments and macrobenthic invertebrates of the North Sea. Neth. Inst. Sea Res., Texel. NIOZ-Rapport 1989-6: 1–44.

Everaarts, J. M. & C. V. Fischer, 1992. The distribution of heavy metals (Cu, Zn, Cd, Pb) in the fine fraction of surface sediment of the North Sea. Neth. J. Sea Res. 29: 323–331.

Everaarts, J. M. & C. Swennen, 1987. Heavy metals (Zn, Cu, Cd, Pb) in some benthic invertebrate species and in sediment from three coastal areas in Thailand and Malaysia. J. Sci. Soc. Thailand 13: 189–203.

Everaarts, J. M., J. P. Boon, W. Kastoro, C. V. Fischer, H. Razak & I. Sumanta, 1989. Copper, zinc and cadmium in benthic organisms from the Java Sea and estuarine and coastal areas around East Java. Neth. J. Sea Res. 23: 415–426.

Everaarts, J. M., E. Otter & C. V. Fischer, 1990. Cadmium and polychlorinated biphenyls different distribution pattern in North Sea benthic biota. Neth. J. Sea Res. 26: 75–82.

Fowler, S. W., 1986. Trace metal monitoring of pelagic organisms from the open Mediterranean Sea. Envir. Monit. Ass. 7: 59–78.

Greig, R. A., A. Adams & D. R. Wenzloff, 1977. Trace metal content of plankton and zooplankton collected from the New York Bight and Long Island Sound. Bull. envir. Contam. Toxicol. 18: 3–8.

Harding, L. & D. Goyette, 1989. Metals in Northeast Pacific coastal sediments and fish, shrimp and prawn tissues. Mar. Pollut. Bull. 20: 187–189.

Karbe, L., J. Gonzalez-Valero, T. Borchardt, M. Dembinski, A. Duch, H. Hablizel & R. Zeitner, 1988. Heavy metals in fish and benthic organisms from the northwestern, central and southern North Sea: regional patterns comparing dab, blue mussel and hermit crab. ICES paper C.M. 1988/E:22: 18 pp.

Klamer, J. C., R. N. Hull, R. W. P. M. Laane & D. Eisma, 1990. The distribution of heavy metals and polycyclic aromatic hydrocarbons in the sediments of the Oyster Grounds (North Sea). Neth. J. Sea Res. 26: 83–87.

Kramer, C. J. M., J. P. Boon, J. C. Duinker, J. M. Everaarts, M. Th. J. Hillebrand, R. F. Nolting & E. M. van der Vlies, 1985. Contaminants in the Dutch Wadden Sea and a comparison with surrounding areas. Report no. M201 of the Ministry of Public Health and Environment: 1–127.

Kröncke, I., 1987. Lead and cadmium contents in selected macrofauna species from the Dogger Bank and the eastern North Sea. Helgoländer Meeresunters. 41: 465–475.

Kuipers, B. R., H. J. Witte & S. R. Gonzalez, 1993. Zooplankton distribution in the coastal upwelling system along the Banc d'Arguin (Mauritania). Hydrobiologia 258: 133–149.

Lande, E., 1977. Heavy metal pollution in the Trondheimsfjorden, Norway and the recorded effects on the fauna and flora. Envir. Pollut. 12: 187–198.

Loring, D. H. & R. T. T. Rantala, 1977. Geochemical analysis of marine sediments and suspended particulate matter. Fish. mar. Serv. Canada, Techn. report no. 700.

Minas, H. J., L. A. Codispoti & R. C. Dugdale, 1982. Nutrients and primary production in the upwelling region off Northwest Africa. Rapp. P.-v. Réun. Cons. int. Explor. Mer. 180: 148–183.

Moore, J. W. & S. Ramamoorthy, 1984. Heavy metals in Natural Waters. Applied Monitoring and Impact assessment. Springer Verlag, New York, 268 pp.

Paus, P. E., 1972. Bomb decomposition of biological materials. Atomic Absorption Newsletter 11: 129–130.

Peters, H., 1976. The spreading of the water masses of the Banc d'Arguin in the upwelling area off the northern Mauritania coast. 'Meteor' Forsch Ergebn. A. 18: 78–100.

Phillips, D. J. H., 1977. The use of biological indicator organisms to monitor trace metal pollution in marine and estuarine environments – a review. Envir. Pollut. 13: 281–317.

Polikarpov, G. G., B. Oregioni, S. Parchevskaya & G. Benayoun, 1979. Body burden of chromium, copper, cadmium and lead in the neustonic Copepod *Anomalocera patersoni* (Pontellidae) collected from the Mediterranean Sea. Mar. Biol. 53: 79–82.

Ridout, P. S., P. S. Rainbow, H. S. J. Roe & H. R. Jones, 1989. Concentrations of V, Cr, Mn, Fe, Ni, Co, Cu, Zn, As and Cd in mesopelagic crustaceans from the North East Atlantic Ocean. Mar. Biol. 100: 465–471.

Roméo, M. & M. Gnassia-Barelli, 1988. *Donax trunculus* and *Venus verrucosa* as bioindicators in Mauritania coastal waters. Mar. Biol. 99: 223–227.

Samant, H. S., K. G. Doe & O. C. Vaidya, 1990. An integrated chemical and biological study of the bioavailability of metals in sediments from two contaminated harbours in New Brunswick, Canada. The Science of the Total Environment 96: 253–268.

Sanders, B. M., K. D. Jenkins, W. G. Sunda & J. D. Costlow, 1983. Free cupric ion activity in sea-water: Effects on metallothionein and growth in crab larvae. Science 222: 53–54.

Shiber, J. G., 1981. Metal concentration in certain coastal organisms from Beirut. Hydrobiologia 83: 181–195.

Stenner, R. D. & G. Nickless, 1974. Distribution of some heavy metals in organisms in Hardangerfjord and Skjerstadfjord, Norway. Wat. Air Soil Pollut. 3: 279–291.

Stenner, R. D. & G. Nickless, 1975. Heavy metals in organisms of the Atlantic coast of S.W. Spain and Portugal. Mar. Pollut. Bull. 6: 89–92.

Stumm, W. & H. Bilinski, 1972. Trace metals in natural waters: difficulties of interpretation arising from ignorance on their speciation. In S. H. Jenkins. Advances on Water Pollution Research. Pergamon Press, New York: 39–52.

Van Hoogstraten, R. J. & R. F. Nolting, 1991. Trace and major elements in sediments and in porewaters of the Northwestern basin of the Mediterranean Sea. Neth. Inst. Sea Res., Texel. NIOZ-Rapport 1991-10: 1–72.

Windom, H. L., 1972. Arsenic, cadmium, copper, lead, mercury and zinc in marine biota – North Atlantic Ocean. IDOE Workshop on Baseline studies of pollutants in the Marine Environment. Publs Brookshaven Nats. Lab. (BNL), N.Y. 24–26 May 1972.

*Hydrobiologia* **258**: 59–72, 1993.
*W. J. Wolff, J. van der Land, P. H. Nienhuis & P. A. W. J. de Wilde (eds), Ecological Studies in the Coastal Waters of Mauritania.*




# Leaf dynamics and standing stocks of intertidal *Zostera noltii* Hornem. and *Cymodocea nodosa* (Ucria) Ascherson on the Banc d'Arguin (Mauritania)

J. E. Vermaat[1], J. A. J. Beijer, R. Gijlstra, M. J. M. Hootsmans[1], C. J. M. Philippart, N. W. van den Brink & W. van Vierssen[1]
*Dept. of Nature Conservation, Agricultural University, Ritzema Bosweg 32a, 6703 AZ Wageningen, The Netherlands;* [1] *Present address: International Institute for Hydraulic and Environmental Engineering, P.O.Box 3015, 2601 DA Delft, The Netherlands*

*Key words: Zostera noltii, Cymodocea nodosa*, growth, seagrass

## Abstract

Leaf dynamics and standing stocks of intertidal seagrasses were studied in the Baie d'Aouatif (Parc National du Banc d'Arguin, Mauritania) in April and September 1988. Standing stocks of *Zostera noltii* Hornem. suggest a unimodal seasonal curve similar to what is found for populations at higher latitudes. Also, leaf growth rates (0.03 $cm^2\ cm^{-2}\ day^{-1}$ on average) were similar to those found at higher latitudes in these months. Variation in leaf loss over tidal depth, time and different locations in the Baie d'Aouatif was larger and more often significant than variation in leaf growth. In general, *Z. noltii* beds in the Baie d'Aouatif had comparable leaf growth rates and standing stocks. In both months losses were almost always higher than or equal to growth.

Variation in leaf loss over time was much higher in the plots that were situated high in the intertidal than in lower plots. This is explained by differences in susceptibility to sloughing, which is presumably higher in periods with low tide around noon for shallow depths.

In an experiment using artificial shading nets, *in situ* leaf growth was affected negatively from 94% shading onwards. This shading was observed to reduce the light intensity reaching the seagrass bed to a level below the reported range of light compensation points for *Z. noltii*.

*Cymodocea nodosa* (Ucria) Ascherson on average had higher leaf area and relative growth rates than *Z. noltii* and much lower loss rates, resulting in a positive net increase in September. Standing stocks were also higher than for *Z. noltii*. A mixed seagrass bed containing the above two species and *Halodule wrightii* Ascherson had the highest observed total biomass: 335 $g\ m^{-2}$ ash-free dry weight.

## Introduction

Seagrass beds cover vast areas in the intertidal and subtidal zones of the Banc d'Arguin and presumably are the most important primary producers in this ecosystem (Wolff & Smit, 1990). Within the scope of an analysis of the functioning of the ecosystem of the Banc d'Arguin, which was the aim of the 1988 expedition, a study of the growth dynamics of these seagrass beds was considered worthwhile. In the present paper we report on this topic for the seagrasses in the intertidal zone. The subtidal zone was studied by the teams of the Delta Institute for Hydrobiological Research (Yerseke).

The Banc d'Arguin was visited by two teams

from Wageningen Agricultural University: one in April and the other in September 1988. Logistic problems limited the stay of the April group at the Biological Station of Iouik to 10 days. In both months we collected data on leaf growth, net changes and standing stocks of *Zostera noltii* Hornem. as a function of tidal depth at a station on the shoreline of the Baie d'Aouatif. The station was located near the expedition camp at the Biological Station of Iouik. The natural light regime as well as shading by water and periphyton was measured. Also, a shading experiment was done to estimate the sensitivity of the seagrass vegetation to increased light attenuation. The September team had the opportunity to measure leaf growth at two more stations on the mudflats in the Baie d'Aouatif. Here, also data on *Cymodocea nodosa* (Ucria) Ascherson and *Halodule wrightii* Ascherson were collected. Hootsmans *et al.* (1993) report on periphyton dynamics and periphyton removal by fiddler crabs (*Uca tangeri* Eydoux) that were also studied in September.

The extensive data set that was collected allowed us to evaluate some methodic aspects of the measurement of seagrass leaf growth. In this paper we followed the method described by Vermaat *et al.* (1987), where the relative growth rate is expressed in $cm^2\ cm^{-2}\ d^{-1}$ ($rgr_{cm}$), and, consequently, area of individual leaves is measured. To test whether the more laborious measuring of leaf areas could be replaced by counting of leaves we compared $rgr_{cm}$ with $rgr^{leaves}$. The latter measures the number of leaves emerging on a plant per average number of leaves and per unit time (leaf $leaf^{-1}\ d^{-1}$). We also compared the plastochrone method of Jacobs (1979) with the 'rgr method' as used in this paper.

## Materials and methods

In April the sampling location was chosen on the shore of the Baie d'Aouatif. Along a transect perpendicular to the shore three tidal depths were selected (depths relative to mean sea level −16, −32 and −46 cm, respectively, distance between depths was 4 m). This transect will be referred to hereafter as the 'tidal depth' transect. At every depth three plots (4 $m^2$ area, distance between plots was 1 m) were set out parallel to the shore, to estimate spatial variation.

Due to higher low tide levels and short emergence times, the lowest group of plots from April could not be sampled in September. Leaf dynamics in the tidal transect were measured three times in September (cf. Fig. 3). In the last September period plants were also marked at a higher and lower plot of two stations in the Baie d'Aouatif (Ten Vaddir and Timiye, respectively, Fig. 1).

With spectrally non-selective shade nets (Agroflor), three shading levels were created in April (0%, 62% and 73% shade, each with three replicate plots). In September, this set-up was changed on the basis of the results from April. Two additional shading levels were introduced (94% and 99%) and the new 10 plots (2.25 $m^2$) were arranged in two blocks. A block had one replicate plot of each treatment. The blocks were located to the left and right of the tidal depth transect, respectively, at a depth of −32 cm relative to mean sea level.

A Campbell 21X datalogger was placed in a waterproof case on a gas-tube scaffold (2 m

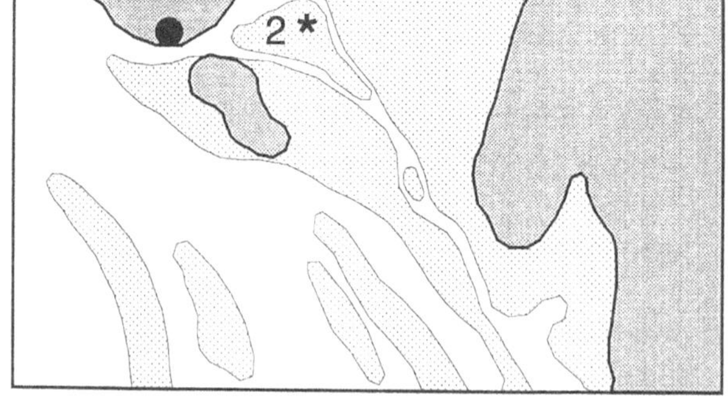

*Fig. 1.* Location of the different stations in the Baie d'Aouatif. tr = the tidal depth transect, 1 = plots at the tidal flat Ten Vaddir, 2 = plots at the Timiye flat. Map adopted from Wolff & Smit (1990).

height) that was erected above the high water line on the shore. Bottemanne submersible light sensors (type RA 200Q, measuring photosynthetically available radiation, PAR) and thermocouples were read by the datalogger every minute and an average over 15 minutes was stored. One light sensor was installed on the scaffold to measure incoming PAR, one sensor was placed just below the highest tidal depths plots (effective depth of the sensor +51 cm above mean sea level), and one was positioned just below the intermediate tidal depth plots (−24 cm). This difference between the two sensors (75 cm) allowed us to calculate attenuation of the water layer during every day-time high tide. The 4 thermocouples were placed in the seagrass vegetation in one of the higher and lower tidal depth plots each and under the shading nets (62% and 99% in September).

Following the method of Vermaat *et al.* (1987), all individual leaves of a plant were marked at a fixed distance from the sheath with a felt pen and their length was measured. After exposure of about 5 days, individual plants were re-collected, brought to the laboratory and the length (precision 1 mm) and width (precision 0.1 mm using binoculars) of all leaves was measured. All leaf and sheath material from one plot was pooled per sampling period and stored to allow for later biomass determination and the calculation of leaf area per biomass ratios.

Plant biomass samples were pre-dried at 105 °C in a stove at the Biological Station at Iouik and stored in aluminium foil in plastic airtight containers with silica gel until transport and final analysis in Wageningen. Biomass was measured as ash-free dry weight (afdw), which is dry weight (dw, 24 h at 105 °C) minus ash weight (aw, 3 h at 520 °C).

Both in April and September the plots were sampled once for above- and belowground biomass (g $m^{-2}$ afdw) and shoot density (number $m^{-2}$). In September, chlorophyll content was determined for the plants from the shading experiment. We used the method of Vernon (1960), with the acidification step modified according to Moed & Hallegraef (1978).

Statistical analysis was done with the SPSS package (Norusis, 1986). Data were tested for normality and homogeneity of variance prior to analysis of variance (ANOVA). Differences between treatments were tested with contrast tests, Tukey's HSD test or, if unequal replication occurred, with a modified lsd test maintaining an experimental error rate (EER) of 0.05. For non-orthogonal contrast tests, also an EER of 0.05 was used (Steel & Torrie, 1980).

## Results

### *Temperature and light measurements*

Daily insolation totals (PAR) and maximum temperatures at the tidal depth transect for the September period are given in Fig. 2 (upper part). Comparable total insolations were measured in April (57 E $m^{-2}$ $d^{-1}$ on average at 2 m above sea level). For a ten days' period in September the daily insolation curves are also given in Fig. 2 (lower part). The graph illustrates the progression of the high tide period over the day: On September 4th the low tide exposure covered the whole daytime between sunrise and sunset, on September 9th daytime coincided with high tide. Hootsmans *et al.* (1993) report on exponential light extinction coefficients of the high tide water.

### *Leaf dynamics*

For the tidal transect the data are summarized in Fig. 3; the results from twoway ANOVAs can be found in Table 1. In September significant differences in leaf loss existed between the different weeks, whilst shoot leaf area was significantly different over weeks and tidal depth. When comparing the April and September data, a significant interaction existed between tidal depth and 'months' for leaf growth as well as an effect of 'months' on shoot leaf area. This latter difference is also reflected in the subsequent multiple comparisons. More effects are significant when a less conservative EER of $p = 0.1$ is applied (Table 1).

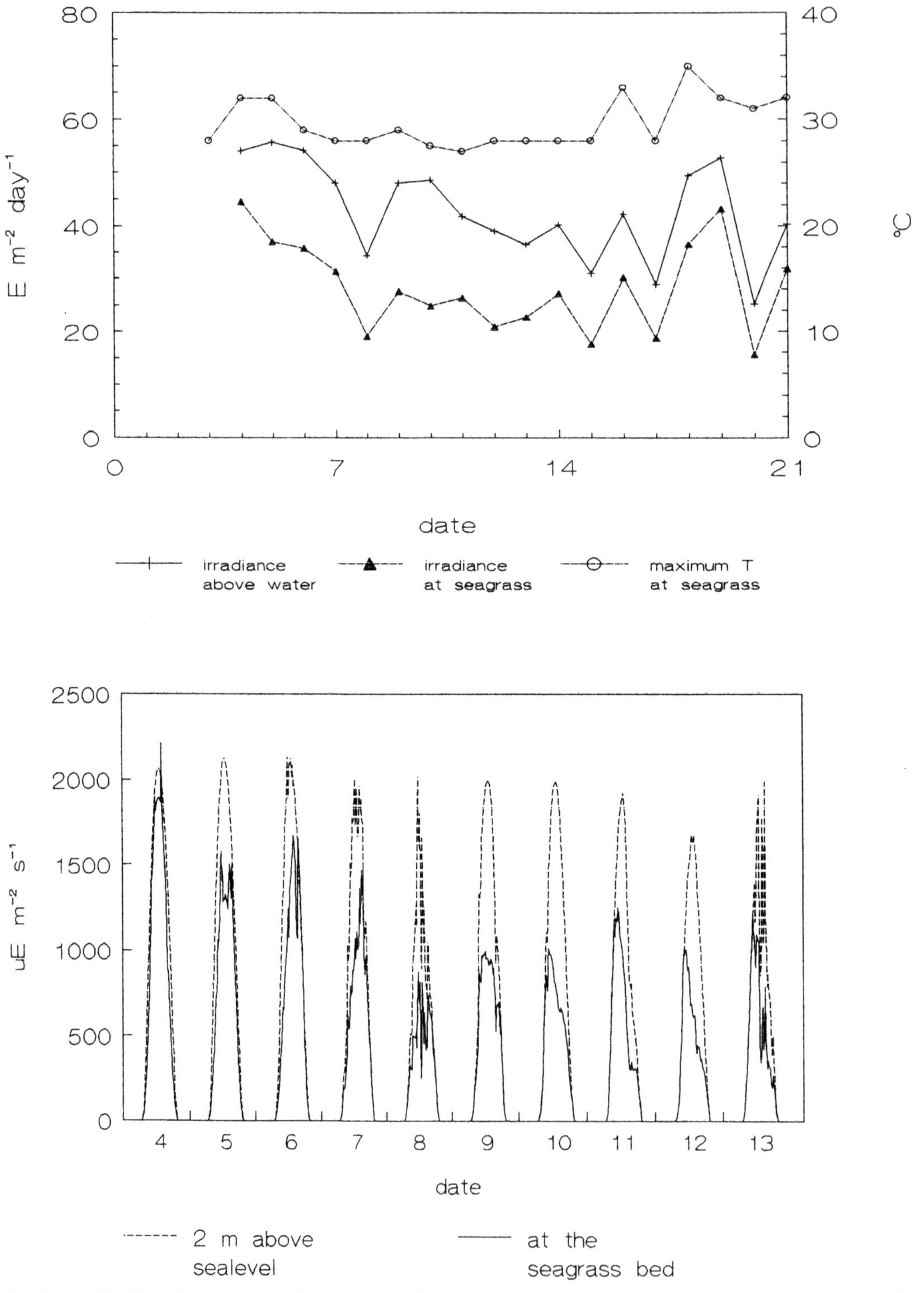


*Fig. 2.* Insolation (PAR) and temperature observations in September. The upper graph gives total daily insolation and maximum temperature on the mudflat for the whole period. The lower graph gives daily insolation curves for the first ten days.

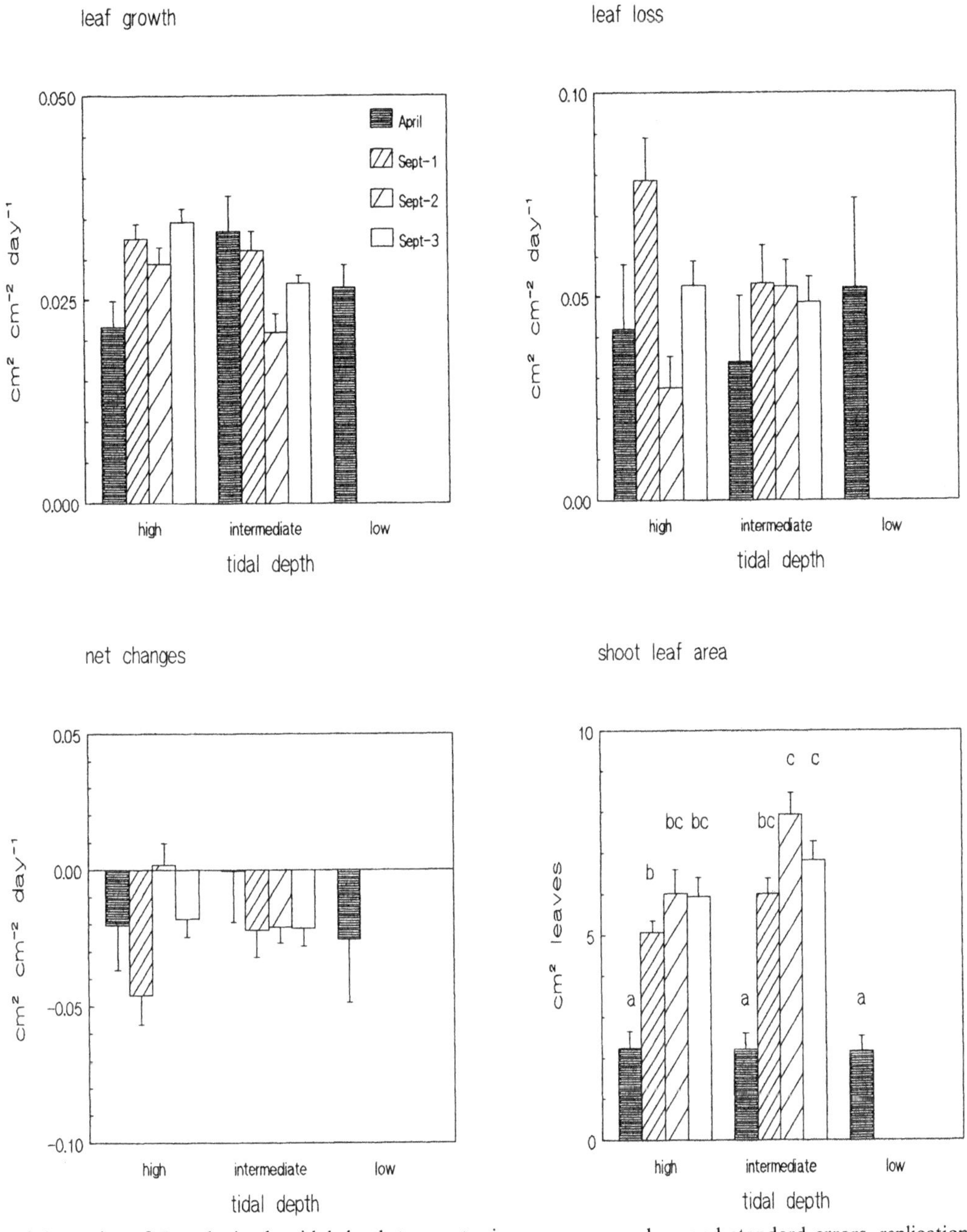


*Fig. 3.* Leaf dynamics of *Z. noltii* in the tidal depth transect; given are mean values and standard errors, replication was 14 to 44, Sept-1, Sept-2 and Sept-3 indicate the periods 3–8, 8–13 and 13–18 September. The April period lasted from 22–26 April. Shoot leaf area is the average of the area at marking and the area at harvesting. Different letters indicate significantly different means (lsd, EER = 0.05).

In that case, tidal depth had a significant effect on leaf growth, while loss was different over weeks as well as over months and differences in net changes in September are influenced significantly by leaf loss only.

The difference in variation between growth and loss is remarkable: the overall coefficient of variation (standard deviation/mean) for leaf growth was 40% while the value for loss was 106%. A comparison of *Z. noltii* leaf dynamics between

*Table 1.* Twoway ANOVAs for leaf dynamics data of the tidal depth transect. Significant probabilities ( $p<0.05$) are indicated with an asterisc.

| Factor | Growth | Loss | Net changes | Leaf area |
|---|---|---|---|---|
| *September data* | | | | |
| Weeks | 0.794 | 0.032* | 0.055 | 0.006* |
| Tidal depth | 0.096 | 0.219 | 0.404 | 0.002* |
| Interaction | 0.111 | 0.062 | 0.078 | 0.543 |
| *April versus September* | | | | |
| Months | 0.124 | 0.093 | 0.196 | 0.000* |
| Tidal depth | 0.062 | 0.422 | 0.242 | 0.246 |
| Interaction | 0.002* | 0.955 | 0.550 | 0.246 |

three sites in the Baie d'Aouatif and between *Z. noltii* and *C. nodosa* is made in Fig. 4 and Table 2. On average, the significantly larger species *C. nodosa* had respectively higher growth and lower loss values than *Z. noltii* (contrast test). The multiple comparison tests (Fig. 4), however, show that on some locations growth of *Z. noltii* may be as high as and loss as low as that of *C. nodosa*. Averaged over locations, tidal depth significantly affected growth of *Z. noltii*. No significant differences existed between the two stations Ten Vaddir and Timiye in the Baie but there were such differences between these stations and the transect: loss was significantly lower on Ten Vaddir whilst growth was lower on Timiye.

Experimental shading affected leaf growth, but also loss and, consequently, average leaf area over a marking period (Fig. 5). Only the two highest shading levels significantly reduced leaf growth. Net changes were positive in April, and significantly negative for the two highest shading in September. Similar to the findings in the tidal depth transect, shoot area in April was significantly less than in September. Chlorophyll contents are given in Table 3, here also a significant effect of a high shading level was found.

*Standing stocks*

The biomass data are presented in Table 4, the results from the statistical analyses in Table 5. In the shading experiment as well as in the tidal transect significant differences were apparent between April and September in four of the five characteristics. Only aboveground biomass was not significantly different.

Clearly, significant differences existed between species, *C. nodosa* having lower densities, higher shoot biomasses and higher standing stocks than *Z. noltii*. *H. wrightii* was only sampled in two stations, its total biomass was comparatively low. The mixed seagrass bed in the lower Timiye station had the highest total biomass: 335 g $m^{-2}$ afdw.

When entered in an ANOVA, shading had no significant effect on aboveground, belowground or total biomass. In a regression against the shading percentages however, a significant effect did exist for aboveground biomass ($p = 0.045$, $r^2 = 0.41$). This is most probably due to the different way in which the treatments are entered in the analysis: in ANOVA an equal distance is assumed between treatments.

When a comparison is made between the different locations in the Baie d'Aouatif, the pattern is different from that observed in Table 2 for the leaf dynamics data. The significant difference in net changes between tidal depths and between the transect and location 1 (Ten Vaddir) is not reflected in differences in aboveground biomass. Similarly, significant differences in shoot leaf area are not parallelled by differences in individual shoot biomass.

To allow for comparisons of our leaf production data in units of leaf area with literature data in units of biomass, we measured the proportion of non-leafy material (*i.e.* sheaths) of aboveground biomass. Comparing all plots and experimental treatments and both *Z. noltii* and *C. nodosa*, we found no significant differences (oneway ANOVA, $p = 0.885$). As an average for *Z. noltii* we used 22% non-leafy material, for *C. nodosa* we found 19%. Data on leaf dynamics per unit biomass and per unit area of seagrass bed are summarized in Table 6. Average values per plot and period were used as replicates in the analysis. For *Z. noltii* season nor tidal depth were significant, probably due to the lower number of

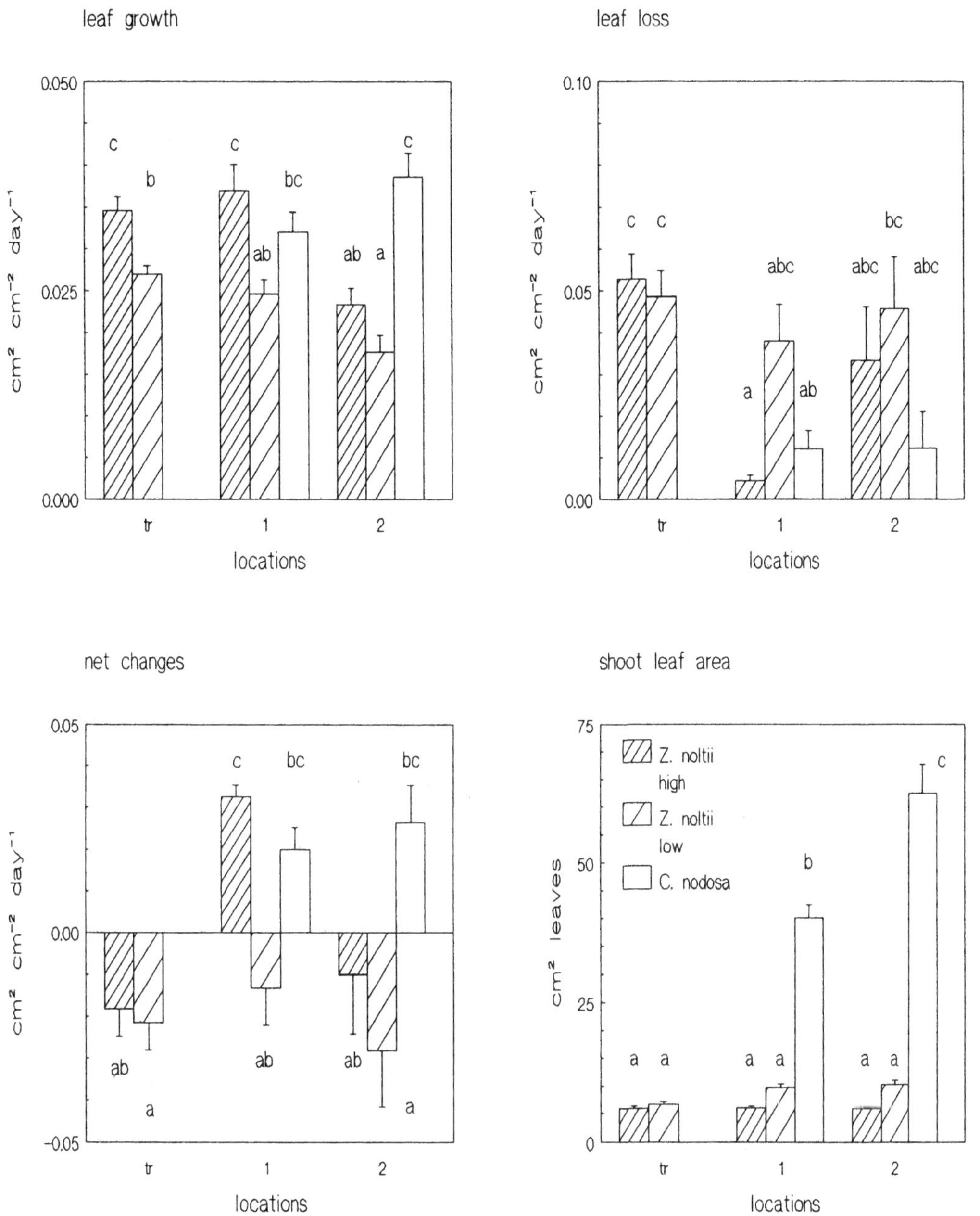


*Fig. 4.* Leaf dynamics of *Z. noltii* and *C. nodosa* at three locations and two tidal depths in the Baie d'Aouatif in the period 13–18 September; location: tr = tidal depth transect, 1 = on the mudflat of Ten Vaddir, 2 = smaller mudflat Timiye between Zira and Ten Vaddir. Replication was 8–30. The *Z. noltii* data from the tidal depth transect indicated with low are indicated as intermediate in Fig. 3, further as Fig. 3.

replicates available for these compound data. Some differences between *Z. noltii* and *C. nodosa* apparently were large enough to remain significant. Leaf area indices for *Z. noltii* ranged from 1.7 $m^2 m^{-2}$ (April) to 2.6 (September); for *C. nodosa* we measured an average value of 3.2.

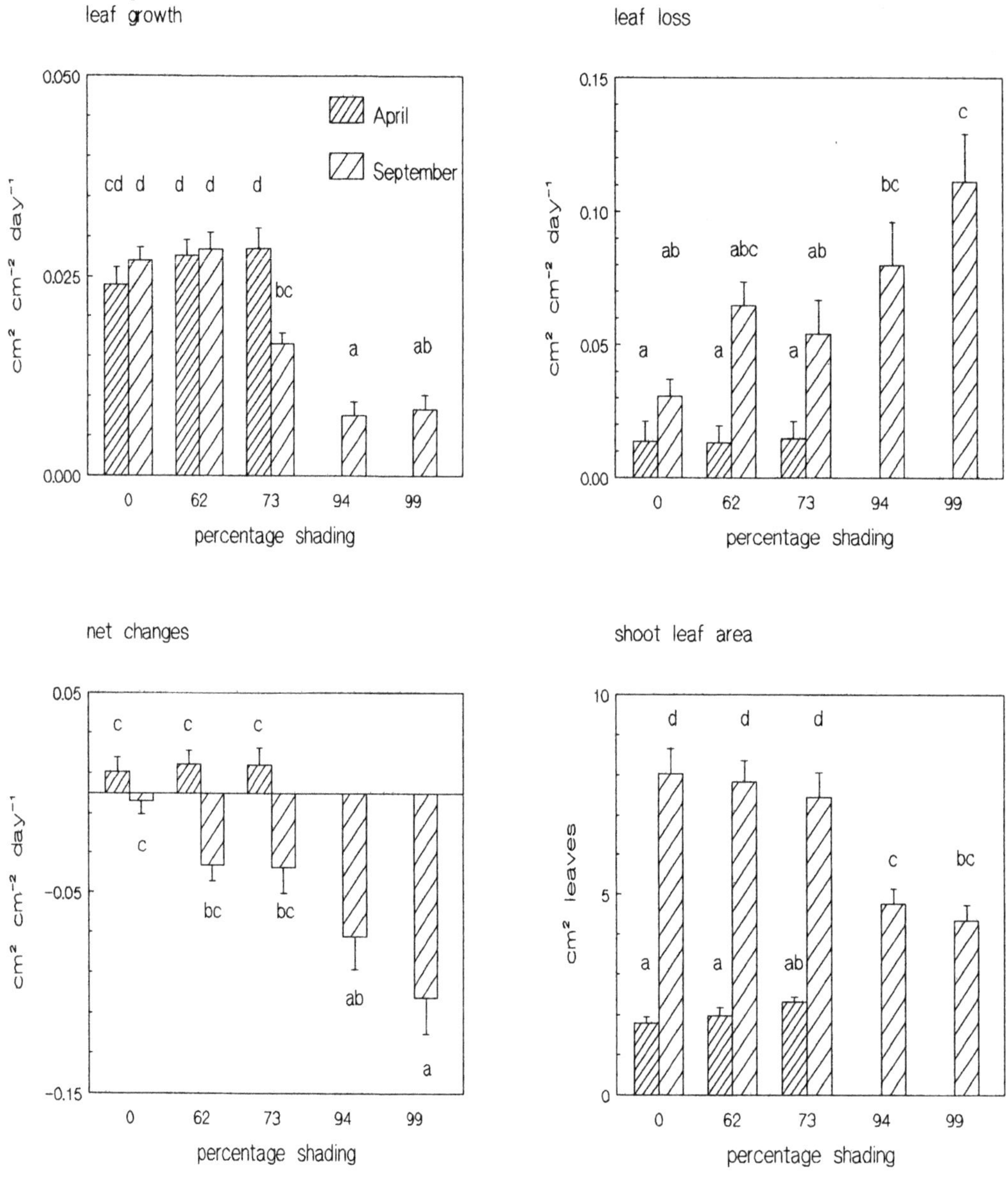


*Fig. 5.* Leaf dynamics of *Z. noltii* in the shading experiment. Replication was 15–20, further as Fig. 3. In September, total insolation at the unshaded seagrass bed was 96 E $m^{-2}$ over the experimental period (Sept. 10–14). For April 25–30 no complete insolation data set is available.

*Leaf growth measurements, methodological aspects*

The dataset of the tidal depth plots of September was used to make a methodological evaluation of (a) the plastochrone method of Jacobs (1979), and (b) relative growth and loss rates measured in leaves per leaf instead of area per area. For the evaluation of the plastochrone method we will use leaf area production per shoot as a basis of comparison.

Jacobs (1979) defined a plastochrone interval (PI) as

$$\mathrm{PI} = \frac{(\text{number of leaves marked} \times \text{length observation period})}{(\text{number of new leaves on marked shoots})}.$$

*Table 2.* Non-orthogonal contrast tests comparing leaf dynamics fo *Z. noltii* at three sites in the Baie d'Aouatif and between *Z. noltii* and *C. nodosa*. Locations are coded as explained under Fig. 4. The non-orthogonal contrasts are significant at a comparisonwise error rate of 0.009 (EER = 0.05).

| | Growth | Loss | Net changes | Leaf area |
|---|---|---|---|---|
| *Z. noltii* vs *C. nodosa* | 0.001* | 0.000* | 0.000* | 0.000* |
| Tidal depth (high vs low) | 0.000* | 0.063 | 0.005* | 0.000* |
| Location 1 vs 2 | 0.300 | 0.188 | 0.138 | 0.541 |
| Location tr vs 1 | 1.000 | 0.000* | 0.000* | 0.004* |
| Location tr vs 2 | 0.000* | 0.272 | 0.110 | 0.002* |

Number of leaves marked is meant as number marked in a plot. The inverse of this PI is a measure of the turnover rate of leaves ($d^{-1}$). The leaf blade production on a shoot (LPS) was calculated as the average biomass of a fully formed leaf blade divided by the PI. For *Zostera marina* L., Jacobs (1979) decided to use the third leaf on a shoot as the typically 'fully developed' leaf. As we use leaf area to compare the methods, a measure of the area of a fully formed leaf had to be found. We used both average leaf width × maximal leaf length and average leaf width × average leaf length, respectively, as an upper and lower limit of this fully developed area. PI values were derived from leaf counts made during the leaf marking procedure. From the leaf marking data LPS was derived by multiplication of $rgr_{cm}$ with the average area of a shoot. We also incorporated $rgr_{leaves}$ in the comparison. This resulted in four estimates of leaf production on a shoot: LPS according to Jacobs (1979) using the maximum and average leaf length respectively (methods 1 and 2), and LPS calculated with the two relative growth rates (methods 3 and 4).

Leaf morphometric data are given in Table 7. Clearly, quite some variation is present. Tidal depth had a significant effect on maximum leaf length and was almost significant ($p = 0.08$) for average leaf length. In the calculation of LPS we therefore used different values for high and low plots. Calculated mean leaf production per shoot according to the four different methods is given in Table 8. The two relative growth rates result in shoot leaf area productions that are not significantly different. The method that applies the PI and maximum leaf length, however, results in a significantly higher leaf production estimate than the other three methods.

Finally, we will evaluate the possibility to measure rgr in terms of leaf numbers instead of leaf area. Since all data were available per individual plant, it was possible to use the individual plants as replicates. Hence we had a larger data set available than the one used in Table 8. We also compared losses and net changes. Again, as in Table 8, no significant difference existed for leaf growth (paired t-test, 179 cases, $p = 0.534$) but the methods were different for loss and net changes ($p < 0.001$). Where the mean relative loss rate calculated per unit area was 0.056, calculated per leaf it was 0.041. While the averages were not significantly different for growth, variation was much higher in $rgr_{leaves}$ than in $rgr_{cm}$: coefficients of variation were 112% and 37%, respectively.

*Table 3.* Chlorophyl contents of *Z. noltii* in the shading experiment of September. Given are chlorophyll *a* + *b* (chl(*a* + *b*), mg $cm^{-2}$ leaf area and mg $g^{-1}$ afdw) and the fraction of chlorophyll *b* in the total chlorophyll (frac-b). Replication was 2, different letters indicate significant differences between means (Tukey test, EER = 0.05).

| Percentage shading | chl(*a* + *b*) mg $cm^{-2}$ | | mg $g^{-1}$ | frac-b | |
|---|---|---|---|---|---|
| | Mean | SE | Mean | Mean | SE |
| 0 | 0.023 b | 0.003 | 19.4 | 0.224 | 0.006 |
| 62 | 0.021 b | 0.001 | 14.6 | 0.227 | 0.006 |
| 73 | 0.022 b | 0.001 | 18.2 | 0.238 | 0.005 |
| 94 | 0.017 ba | 0.002 | 19.9 | 0.244 | 0.005 |
| 99 | 0.011 a | 0.002 | 10.8 | 0.246 | 0.005 |

*Table 4*. Seagrass biomass and shoot densities of the different plots sampled in April and September. Replication was 2, except for the tidal transect and all April data, where it was 3. Aboveground, belowground and total biomass are in g $m^{-2}$ afdw, shoot density is in numbers $m^{-2}$, shoot biomass is individual shoot biomass in mg $shoot^{-1}$. Ratios of afdw to dw were 0.39 and 0.53 for below- and aboveground *Z. noltii* and 0.63 and 0.40, respectively, for *C. nodosa*.

| | | Biomass | | | | | | Shoot | | | |
|---|---|---|---|---|---|---|---|---|---|---|---|
| | | Abovegr. | | Belowgr. | | Total | | Density | | Biomass | |
| | | Mean | SE | Mean | SE | Mean | SE | Mean | SE | Mean | SE |
| Tidal depth transect | | | | | | | | | | | |
| April 23 | High | 46 | 1 | 63 | 10 | 110 | 9 | 5583 | 476 | 8.5 | 0.9 |
| | Interm. | 59 | 3 | 97 | 10 | 157 | 11 | 9025 | 416 | 6.6 | 0.4 |
| | Low | 55 | 2 | 79 | 2 | 134 | 2 | 8383 | 705 | 6.7 | 0.7 |
| Sept. 18 | High | 41 | 4 | 42 | 4 | 83 | 8 | 4266 | 367 | 9.7 | 1.1 |
| | Interm. | 54 | 10 | 37 | 7 | 91 | 15 | 4383 | 479 | 12.2 | 1.6 |
| Shading experiment | | | | | | | | | | | |
| April 27 | 0% | 61 | 3 | 93 | 6 | 154 | 7 | 8467 | 535 | 7.3 | 0.7 |
| | 62% | 59 | 4 | 91 | 12 | 150 | 17 | 6658 | 516 | 9.0 | 0.8 |
| | 73% | 62 | 3 | 86 | 6 | 149 | 4 | 8417 | 603 | 7.5 | 0.7 |
| Sept. 14 | 0% | 61 | 14 | 44 | 3 | 104 | 11 | 4860 | 67 | 12.4 | 2.7 |
| | 62% | 49 | 4 | 38 | 6 | 86 | 10 | 3568 | 932 | 14.3 | 2.5 |
| | 73% | 52 | 24 | 70 | 41 | 122 | 64 | 4996 | 2837 | 11.4 | 1.7 |
| | 94% | 24 | 5 | 43 | 3 | 67 | 2 | 4527 | 327 | 5.4 | 1.5 |
| | 99% | 23 | 8 | 45 | 4 | 68 | 13 | 4114 | 193 | 5.6 | 1.8 |
| Other locations in the Baie d'Aouatif (September 16 & 17) | | | | | | | | | | | |
| *Z. noltii* | | | | | | | | | | | |
| Ten Vaddir | High | 38 | 9 | 38 | 6 | 77 | 15 | 2428 | 220 | 15.5 | 2.2 |
| | Low | 50 | 43 | 48 | 6 | 99 | 36 | 2854 | 20 | 17.8 | 15.1 |
| Timiye | High | 69 | 9 | 102 | 14 | 171 | 23 | 6148 | 1198 | 11.3 | 0.7 |
| | Low | 76 | 6 | 59 | 17 | 135 | 23 | 2158 | 55 | 35.1 | 3.9 |
| *C. nodosa* | | | | | | | | | | | |
| Ten Vaddir | High | – | – | – | – | 22 | 1 | – | – | – | – |
| | Low | 91 | 6 | 103 | 3 | 193 | 2 | 836 | 40 | 108.3 | 1.4 |
| Timiye | Low | 100 | 8 | 76 | 5 | 176 | 2 | 478 | 50 | 209.5 | 6.1 |
| *H. wrightii* | | | | | | | | | | | |
| Ten Vaddir | High | – | – | – | – | 31 | 3 | – | – | – | – |
| Timiye | Low | – | – | – | – | 24 | 14 | – | – | – | – |

## Discussion

### *Variation over season, locations and tidal depth*

Distinct differences in biomass data as well as in shoot densities of *Z. noltii* were apparent between April and September. Higher individual shoot biomasses in September were parallelled by a higher shoot leaf area for that month. Data gathered in May at Ten Vaddir (Van Lent *et al.*, 1991) were higher than our April data, suggesting an increase in biomass in summer as in other *Z. noltii* vegetations (Vermaat *et al.*, 1987; Philippart, in prep.). Pergent *et al.* (1988) presented data collected in October of 1987 at a.o. Iouik. Leaf morphometric data were comparable to those col-

*Table 5.* Oneway and twoway ANOVAs for biomass data. The non-orthogonal contrasts are significant at a comparisonwise error rate of 0.0127 (EER = 0.05). For the shading experiment only the shading levels were used that have been applied in both months.

| | Biomass | | | Shoot | |
|---|---|---|---|---|---|
| | Abovegr. | Belowgr. | Total | Density | Biomass |
| Twoway, tidal depth *Z. noltii* | | | | | |
| April versus September | 0.477 | 0.001* | 0.004* | 0.001* | 0.005* |
| Tidal depth | 0.387 | 0.675 | 0.485 | 0.141 | 0.033* |
| Interaction | 0.495 | 0.863 | 0.607 | 0.678 | 0.865 |
| Twoway, shading expt. *Z. noltii* | | | | | |
| April versus September | 0.367 | 0.000* | 0.001* | 0.000* | 0.001* |
| Shade | 0.103 | 0.186 | 0.057 | 0.636 | 0.009* |
| Interaction | 0.987 | 0.023* | 0.073 | 0.017* | 0.008* |
| Oneways, September | | | | | |
| Different species | 0.004* | 0.013* | 0.010* | 0.001* | 0.000* |
| Shading *Z. noltii* | 0.935 | 0.296 | 0.751 | 0.666 | 0.048* |
| Non-orthogonal contrasts *Z. noltii*, September | | | | | |
| Tidal depth | 0.450 | 0.123 | 0.900 | 0.030 | 0.061 |
| Location 1 vs 2 | 0.146 | 0.005* | 0.014 | 0.029 | 0.280 |
| Location tr vs 1 | 0.845 | 0.645 | 0.960 | 0.012* | 0.306 |
| Location tr vs 2 | 0.156 | 0.002* | 0.008* | 0.750 | 0.046 |

lected by us in September 1988. Average shoot density was 2932 (SE = 708), *i.e.* less than in September, similarly suggesting a seasonal trend. The significantly lower biomass in September as compared to April suggests that the timing of a peak biomass is not similar on the Banc d'Arguin and

*Table 6.* Leaf dynamics per unit aboveground biomass (mg $g^{-1}$ $day^{-1}$) and area of seagrass bed (g $m^{-2}$ $day^{-1}$). Three comparisons were made: (a) between *Z. noltii* and *C. nodosa* in September (replication 6 and 2), (b) between April and September for *Z. noltii* (replication 3 and 6) and (c) between tidal depths for *Z. noltii* in September (replication 3 and 3).

| | Per unit biomass | | | | *p* | Per unit area | | | | *p* |
|---|---|---|---|---|---|---|---|---|---|---|
| | Mean | SE | Mean | SE | | Mean | SE | Mean | SE | |
| (a) | *Z. noltii* | | *C. nodosa* | | | *Z. noltii* | | *C. nodosa* | | |
| Growth | 22 | 2 | 29 | 3 | 0.15 | 1.11 | 0.04 | 2.75 | 0.38 | 0.00* |
| Loss | − 29 | 6 | − 10 | 0 | 0.00* | − 1.65 | 0.35 | − 0.94 | 0.05 | 0.31 |
| Net changes | − 8 | 7 | 19 | 3 | 0.08 | − 0.54 | 0.35 | 1.81 | 0.33 | 0.01* |
| (b) | April | | September | | | April | | September | | |
| Growth | 21 | 3 | 22 | 2 | 0.97 | 1.16 | 0.22 | 1.11 | 0.04 | 0.77 |
| Loss | − 33 | 4 | − 29 | 6 | 0.63 | − 1.78 | 0.24 | − 1.65 | 0.35 | 0.82 |
| Net changes | − 12 | 6 | − 8 | 7 | 0.70 | − 0.62 | 0.32 | − 0.54 | 0.35 | 0.90 |
| (c) | High | | Low | | | High | | Low | | |
| Growth | 25 | 3 | 18 | 2 | 0.16 | 1.16 | 0.05 | 1.06 | 0.05 | 0.20 |
| Loss | − 24 | 11 | − 35 | 3 | 0.92 | − 1.21 | 0.54 | − 2.09 | 0.35 | 0.25 |
| Net changes | 1 | 12 | − 16 | 3 | 0.24 | − 0.05 | 0.51 | − 1.03 | 0.34 | 0.18 |

*Table 7.* Leaf morphometric data of *Z. noltii* from the tidal depth transect in September. Mean or maximum values from a plot in a period were taken as replicates, replication thus was 3 per treatment. Length is in cm, width in mm.

| Period | Tidal depth | Mean length | | Max. length | | Mean width | |
|---|---|---|---|---|---|---|---|
| | | Mean | SE | Mean | SE | Mean | SE |
| 1 | High | 11.1 | 0.9 | 19.3 | 1.3 | 1.51 | 0.07 |
| | Low | 14.1 | 0.7 | 29.0 | 1.7 | 1.41 | 0.07 |
| 2 | High | 12.8 | 0.7 | 20.6 | 1.2 | 1.43 | 0.07 |
| | Low | 14.6 | 0.8 | 24.8 | 1.5 | 1.55 | 0.12 |
| 3 | High | 12.8 | 2.1 | 21.4 | 3.6 | 1.45 | 0.06 |
| | Low | 13.7 | 1.5 | 24.7 | 2.5 | 1.46 | 0.04 |
| Twoway ANOVA | | | | | | | |
| Period | | 0.677 | | 0.782 | | 0.587 | |
| Tidal depth | | 0.077 | | 0.006* | | 0.901 | |
| Interaction | | 0.690 | | 0.295 | | 0.372 | |

*Table 8.* Comparison of four different methods to calculate the leaf production on a shoot ($cm^2$ $shoot^{-1}$ $day^{-1}$, see text). Replication was 6.

| Method | Mean | SE | Tukey |
|---|---|---|---|
| (1) (max. length × mean width)/PI | 0.32 | 0.03 | b |
| (2) (mean length × mean width)/PI | 0.18 | 0.02 | a |
| (3) $rgr_{cm}$ × mean shoot leaf area | 0.20 | 0.01 | a |
| (4) $rgr_{leaves}$ × mean shoot leaf area | 0.18 | 0.01 | a |

in The Netherlands. Also, the seasonal differences are presumably less distinct on the Banc d'Arguin, since the biomass measured in the Baie d'Aouatif in April was higher than the value observed by Vermaat *et al.* (1987). This is in agreement with the findings of Duarte (1989) who concluded in a review that seasonal variation in seagrass biomass is generally less at lower latitudes.

Leaf growth rates were remarkably similar to those measured by Vermaat *et al.* (1987) in September 1983 in the Zandkreek (The Netherlands) as well as to values observed by Philippart (in prep.). In the Zandkreek as well as in the Banc d'Arguin variation in leaf loss was distinctly higher than in leaf growth. This is probably due to the different character of the two processes. Growth is essentially continuous, whilst loss is not: fragments of a leaf are sloughed time and again.

Significant differences existed in standing stocks (Table 5) as well as in growth rates (Fig. 4) between *Z. noltii* beds at the three sampled locations in the Baie d'Aouatif. Still, on a comparative basis these beds can be broadly characterized as fairly homogeneous with maximum differences in e.g. total biomass and leaf growth hardly exceeding a factor two. The most considerable differences were observed in the leaf loss data, which is likely to be due to the discontinuity of the process to a considerable extent.

Tidal depth as well as the interaction between 'weeks' and tidal depth had a significant effect on leaf growth when a probability of 0.1 is used. This points to the phenomenon that loss was much the same over the weeks at intermediate depth, while clear differences in loss were apparent in the higher plot (Fig. 3). A hypothesis to explain this difference in loss rate variation between the higher and lower plots may be the following:

(a) During low tide exposure the chance of leaf loss through wave action is virtually nil.
(b) Low tide exposure during daytime causes desiccation stress. Severe desiccation increases the vulnerability and thereby the probability of leaf sloughing.
(c) The timing of the tidal cycle relative to the day/night cycle determines the relative importance of (a) and (b): low tide during dawn or dusk results in a low desiccation stress, and thus in less sloughing.
(d) This shifting balance is less prevalent at deeper stations due to longer submergence during a tidal cycle.

Differences in leaf length were more pronounced than differences in width. Again, the main reason probably must be sought in differences in leaf sloughing.

As an overall conclusion it can be stated that differences between locations and over depth in the Baie d'Aouatif did not affect growth of *Z. noltii* as much as they did loss of leaf material.

Data on *C. nodosa* were collected in September only. Applying our conversion factors from Table 4, Pergent *et al.* (1988) found lower biomasses, shoot densities and leaf area indices in October 1987 than we did. This may have been due to the fact that different stations were sampled. Van Lent *et al.* (1991) found comparable values for *C. nodosa* at Ten Vaddir in May. From 5.5 m depth in the Mediterranean Drew (1978) also reported lower aboveground biomass and leaf area indices than the values we found. Caye & Meinesz (1985), however, found much higher shoot densities at 0.5 m depth in the Mediterranean (1650–2060 $m^{-2}$).

*Experimental shading*

Only the two highest shading levels had affected leaf growth of *Z. noltii* after an exposure of 5 days. As pointed out by Hootsmans *et al.* (1993), the 94% shading level brought the light intensity at the seagrass bed close to or below measured light compensation points for photosynthesis of *Z. noltii*. Our experimental results from a field population thus were in agreement with laboratory data (Philippart, in prep.). Shading also influenced leaf loss (Fig. 5). Apparently, under low light the condition of the leaf material had deteriorated to such an extent that susceptibility to sloughing had increased. This was probably also reflected in lower chlorophyll contents. Our chlorophyll contents were similar to those reported by Jiménez *et al.* (1987).

*Leaf growth measurements, methodological aspects*

In various textbooks on terrestrial plant growth analysis (e.g. Hunt, 1978; Causton & Venus, 1981), it is argued that growth rates should be expressed on a relative basis to make comparisons meaningful. Vermaat *et al.* (1987) add to this that for an estimation of production per unit area of macrophyte bed, the use of relative growth rates most likely will correct for possible non-random sampling of marked shoots, *i.e.* for over-representation of larger shoots in the sample. Vermaat *et al.* (1987) validated the use of such net relative growth rates for *Z. noltii* with an independently sampled biomass curve. Our present test is in line with this: application of the plastochrone method resulted in a production estimate similar to those of the rgr methods only when an average leaf area was used. The use of the maximum leaf area, a measure close to the area of a full grown leaf, resulted in an overestimation. Whether relative growth rates necessarily are to be calculated on a log-scale depends on the time interval. For the presently used interval of 5 days we found the differences between logarithmically or linearly calculated rgr's to be 10% at most.

The use of leaf numbers instead of leaf area as a unit of measure for relative growth rates resulted in a seriously underestimated loss and much higher variation in leaf growth. The latter can be overcome by marking larger numbers of plants. For the former, however such a remedy cannot be found that easily. We thus conclude that for *Z. noltii* leaf growth measurements should be done with area as the unit of measurement. A more general point must be made here. *Z. noltii* is a species with relatively few leaves (on average 3.2 in the tidal depth samples from September). For other species with substantially more leaves, the approach using $rgr_{leaves}$ may prove to be more appropriate. An increasing number of leaves obviously increases the laboriousness of the method.

## Acknowledgements

We thank the Mauritanian authorities, in particular the Director of the Parc National du Banc d'Arguin, Mr Hadya Amadou Kane, for the permission to carry out this study and for their support. We also thank the collaborators of the Parc, Mr Abou Gueye, Mr El Hassane ould Mohammed el Abd, the late Mr Ely el Elemine and Mr Samba, and the inhabitants of Iouik for their cooperation. The other participants during April and September are thanked for their help, valuable discussions and a positive atmosphere. Tom van Spanje did an excellent job in coordinating the logistics. The project was made possible thanks to financial support by the Netherlands Marine Science Foundation. Additional support was provided by the Netherlands Research Institute for Nature Management. Koenders Instruments is gratefully acknowledged for providing us with the data-logger during the field work. Ms Frances van Lent, Prof. Dr Piet Nienhuis and Mr Ko Verschuure are thanked for allowing us to use their manuscript before publication.

## References

Causton, D. R. & J. C. Venus, 1981. The biometry of plant growth. Arnold, London, 307 pp.

Caye, G. & A. Meinesz, 1985. Observations on the vegetative development, flowering and seeding of *Cymodocea nodosa* (Ucria) Ascherson on the Mediterranean coasts of France. Aquat. Bot. 22: 277–289.

Drew, E. A., 1978. Factors affecting photosynthesis and its seasonal variation in the seagrasses *Cymodocea nodosa* (Ucria) Aschers. and *Posidonia oceanica* (L.) Delile in the Mediterranean. J exp. mar. Biol. Ecol. 31: 173–194.

Duarte, C. M., 1989. Temporal biomass variability and production/biomass relationships of seagrass communities. Mar. Ecol. Progr. Ser. 51: 269–276.

Hootsmans, M. J. M., J. E. Vermaat & J. A. J. Beijer, 1993. Periphyton density and shading in relation to tidal depth and fiddler crab activity in intertidal seagrass beds of the Banc d'Arguin (Mauritania). Hydrobiologia 258: 73–80.

Hunt, R., 1978. Plant growth analysis. Studies in Biology No. 96. Arnold, London, 64 pp.

Jacobs, R. P. W. M., 1979. Distribution and aspects of the production and biomass of eelgrass, *Zostera marina* L., at Roscoff, France. Aquat. Bot. 7: 151–172.

Jiménez, C., F. X. Niell & P. Algarra, 1987. Photosynthetic adaptation of *Zostera noltii* Hornem. Aquat. Bot. 29: 217–226.

Moed, J. R. & G. M. Hallegraef, 1978. Some problems in the estimation of chlorophyll-*a* and phaeopigments from pre- and post-acidification spectrophotometric measurements. Int. Revue ges. Hydrobiol. 63: 787–800.

Norusis, M. J., 1986. SPSS-PC$^+$ manual. SPSS Inc., Chicago, USA, 559 pp.

Pergent, G., C. Pergent & G. Vuignier, 1988. Données préliminaires sur les herbiers de phanerogames marines du Banc d'Arguin (Mauretanie). GIS Posidonie publ., Marseille, France, pp. 1–55.

Steel, R. G. D. & J. H. Torrie, 1980. Principles and procedures of statistics, a biometrical approach. Second edition. Mc Graw Hill Book Company, Singapore, 633 pp.

Van Lent, F., P. H. Nienhuis & J. M. Verschuure, 1991. Production and biomass of the seagrasses *Zostera noltii* Hornem. and *Cymodocea nodosa* (Ucria) Aschers. at the Banc d'Arguin (Mauritania, NW Africa): a preliminary approach. Aquat. Bot. 41: 353–367.

Vermaat, J. E., M. J. M. Hootsmans & P. H. Nienhuis, 1987. Seasonal dynamics and leaf growth of *Zostera noltii* Hornem., a perennial intertidal seagrass. Aquat. Bot. 28: 287–299.

Vernon, L. P., 1960. Spectrophotometric determination of chlorophylls and pheophytins in plant extracts. Analyt. Chem. 32: 1144–1150.

Wolff, W. J. & C. J. Smit, 1990. The Banc d'Arguin, Mauritania, as an environment for coastal birds. Ardea 78: 17–38.

*Hydrobiologia* **258**: 73–80, 1993.
*W. J. Wolff, J. van der Land, P. H. Nienhuis & P. A. W. J. de Wilde (eds),*
*Ecological Studies in the Coastal Waters of Mauritania.*




# Periphyton density and shading in relation to tidal depth and fiddler crab activity in intertidal seagrass beds of the Banc d'Arguin (Mauritania)

M. J. M. Hootsmans[1], J. E. Vermaat[1] & J. A. J. Beijer
*Dept. of Nature Conservation, Wageningen Agricultural University, Ritzema Bosweg 32a, 6703 AZ Wageningen, The Netherlands;* [1]*present adress: International Institute for Infrastructural, Hydraulic and Environmental Engineering, P.O. Box 3015, 2601 DA Delft, The Netherlands*

*Key words:* extinction coefficient, grazing, periphyton, seagrasses, *Uca tangeri*, *Zostera noltii*

## Abstract

Periphyton development was studied on microscopic glass slides and leaves of *Zostera noltii* Hornem. in an intertidal area in the Banc d'Arguin (Mauritania). The effects of shading, tidal depth and grazing activities by the fiddler crab *Uca tangeri* Eydoux were evaluated. For all experiments, periphyton ash content was high (52–93%) and ash-free dry weight ranged between 0.10–0.63 mg $cm^{-2}$. Slides accumulated more periphyton than leaves.

Artificial shading (62–99%) for 13 days had no effect on periphyton densities on leaves. Increased tidal depth resulted in higher ash-free dry weight on slides, but in lower ash-free dry weight on leaves. Significant variation along the coastline also existed. The effect of fiddler crabs was studied using exclosures. Presence of fiddler crabs reduced periphyton density on slides, whereas light transmittance was increased. On leaves, no significant fiddler crab effect was found. This difference between leaves and slides was probably caused by a storm at the day before the end of the experiment, and by the higher periphyton density on slides as compared with leaves. As visual inspection during the experiment showed clear differences in appearance of leaves inside and outside the exclosures, the storm probably sloughed off mainly the older leaves, *i.e.* those on which the differences in periphyton cover were the highest.

It is hypothesized that periphyton accumulation is higher with increased tidal depth, whereas fiddler crab grazing pressure also increases in this direction. The result is a decreased periphyton density with increased tidal depth.

The presently found light extinction coefficients (mean 0.8 $m^{-1}$) and periphyton light attenuance (up to 25%) in Banc d'Arguin are not likely to affect seagrass leaf growth.

## Introduction

Periphyton development on macrophytes may have negative effects on macrophyte photosynthesis and growth (Sand-Jensen & Søndergaard, 1981; Bulthuis & Woelkerling, 1983; Sand-Jensen & Borum, 1984). Both shading and competition for carbon sources are important in this respect (Sand-Jensen, 1977). Periphyton development seems to play a decisive role in the start of the decline of macrophyte vegetation during eutrophication (Phillips *et al.*, 1978; Silberstein *et al.*, 1986).

In line with these observation, periphyton grazing by herbivorous gastropods and crustaceans has been demonstrated to be beneficial for macrophyte growth (Brönmark, 1985; Hootsmans & Vermaat, 1985). Brönmark (1989) gives a review on the interaction between periphyton, macrophytes and snail grazers.

In this study we present some information on periphyton density and shading on the intertidal seagrass *Zostera noltii* Hornem. in the Banc d'Arguin, Mauritania. Also, the effect of feeding activities by fiddler crabs (*Uca tangeri* Eydoux) on periphyton development is evaluated. Almost nothing seems to be known about their possible impact and the consequences for seagrass growth. The information was collected in September 1988 as part of a large Dutch expedition to this wetland on the Sahara coast from April–October 1988. At the same time and location, data were gathered on seagrass biomass and production under various conditions. These are presented in Vermaat *et al.* (1993).

## Material and methods

The data were collected at the shore of the Baie d'Aouatif, near the Biological Station of Iouik. Exponential extinction coefficients of the water-layer were calculated using insolation data collected in April and September (for a description of the position of the light sensors and the experimental site, see Vermaat *et al.*, 1993). An extinction coefficient was calculated for each quarter hour during which two sensors were submerged. The number of replicate estimates used to calculate a daily mean was dependent on the coincidence of the tides with daylight. Light was measured during two days only in April. This was due to a delayed arrival of the first team at the start of this expedition to Banc d'Arguin.

To be able to measure the transmittance of a periphyton layer, microscopic glass slides were used. Although differences may occur in the quantity and quality of the periphyton developing on leaves of *Z. noltii* plants as compared to such slides (Eminson & Moss, 1980), with slides it is possible to assess the transmittance of an undisturbed periphyton layer. Vermaat & Hootsmans (1991) found significant systematic differences between the true transmittance of undisturbed periphyton and that of suspensions of the same samples. Microscopic glass slides were incubated with one side flat on top of the dense vegetation, not directly on the sediment to limit burying by sediment movement. The slides were fixed at one end using a large paperclip and ironwire nails. This allowed for a slight up and down movement. Thus, slides were continuously horizontal, whereas leaves generally stand more or less upright during submergence and may clean themselves by rubbing against each other. Due to their feeding behaviour (see Results section), fiddler crabs can be expected to have less impact on periphyton on slides compared to leaves. Both differences between leaves and slides may lead to higher periphyton densities on slides as compared to leaves. Thus, the data obtained from slides may be seen as an overestimation of the real periphyton situation on seagrass leaves and an underestimation of fiddler crab grazing effects. Still, when the attenuation data from slides do not suggest light limitation of seagrass photosynthesis, it can be concluded that the present periphyton densities in the Banc d'Arguin are not an important limiting factor for seagrass growth and development.

We studied the effects of three different factors on periphyton development: shading, tidal depth and feeding activities of fiddler crabs. The effect of shading was studied in a transect along the shore at a depth of −32 cm (all depths are relative to mean sea level). This transect consisted of two groups of five 2.25 $m^2$ plots, each with five different shading levels (0, 62, 73, 94 and 99%) created by using neutral density netting (Agroflor). Both groups of plots had the same spatial sequence of shade netting. From each plot, three random samples of about 10 leaves were collected on September 15th, after they had been subjected to these shading conditions for 13 days.

The effect of tidal depth on periphyton was studied on slides and on leaves. Four locations were selected, two at a depth of −16 cm and two at −71 cm. On September 13th, 10 microscopic glass slides were incubated at each location. The slides were collected again on September 20th. Leaf samples of about 10 leaves were collected on September 14th at 10 stations at 1.5 m intervals along a 14.5 m transect perpendicular to the shore line. The transect ranged from a depth of −10 cm to −71 cm. Station 2 of the transect coincides with the tidal depth of −16 cm.

The effect of fiddler crabs was studied by constructing two chicken-wire (mesh size 2 cm) exclosures of 4 $m^2$, each sided by a reference plot, in a line along the shore at −22 cm. The walls of the exclosures inclined outwards by 45° to prevent the fiddler crabs from climbing the sides. In each of the plots, 10 slides were incubated on September 9th and harvested on September 20th. On this date, from each plot 3 random samples of about 10 leaves were collected also.

Immediately after harvesting, the lower side of each slide was cleaned and its transmittance was measured relative to a clean slide using a Bottemanne quantum sensor (measuring photosynthetically active radiation). Sensor and slide were submerged in seawater and exposed to full sunlight. Subsequently, the intact periphyton of the one side remaining on two slides was combined to one sample because of the relatively low amount of periphyton present. Periphyton samples were stored in 100 ml polyethene bottles in filtered seawater with 4% formalin for transport to The Netherlands.

As young leaves were completely free of periphyton, only old leaves were scraped clean using a razor blade. Periphyton from these leaves was stored in the same way as described for the slides. The leaf samples were pre-dried at 80 °C in Mauritania and stored in aluminum foil in plastic airtight containers with silica gel until final analysis.

In The Netherlands, periphyton samples were filtered over pre-ashed and pre-weighed Whatman GF/C glass-fiber filters. Dry weights of all plant and periphyton samples were determined after drying for 24 hours at 105 °C. Ash-free dry weight and ash content were calculated after ashing for 2 hours at 520 °C.

Statistical analyses were done with the GLM option in SAS (SAS Institute Inc., 1985). Prior to analysis of variance (ANOVA), normality and homogeneity of residual variances were checked. Multiple comparisons were done with the LS-MEANS option, using an experimentwise error rate of 0.05.

Samples taken on September 20th were influenced by a storm the day before harvesting. This particularly affected the fiddler crab experiment.

## Results

### *Extinction coefficient*

The mean extinction coefficients with standard errors for April and September are shown in Fig. 1. The peak values on September 6 and 20 are due to a previous storm. The average value for the two periods together was 0.8 $m^{-1}$.

### *Shading effects on periphyton*

No significant effects of shading were found in an ANOVA analysis of all periphyton characteristics that were discerned: dry weight (dw), ash weight (aw), ash-free dry weight (afdw) and the ash fraction of dry weight (afr). The mean values of dw, aw and afdw, all expressed per $cm^2$ of old leaf surface, are shown in Fig. 2. Data of the two replicate plots per shading level are combined.

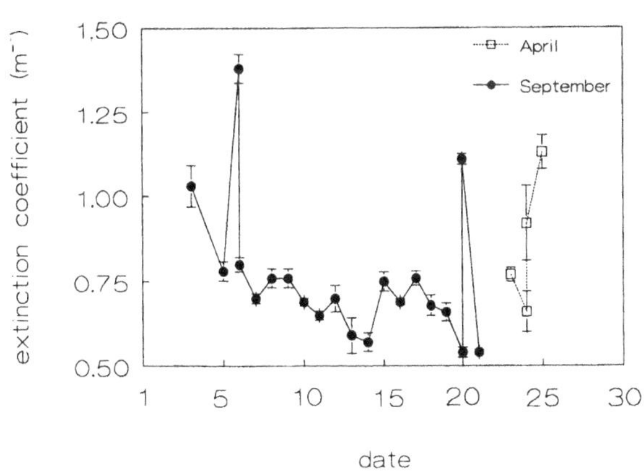


*Fig. 1.* Mean extinction coefficients and their standard errors in the Baie d'Aouatif, April and September 1988.

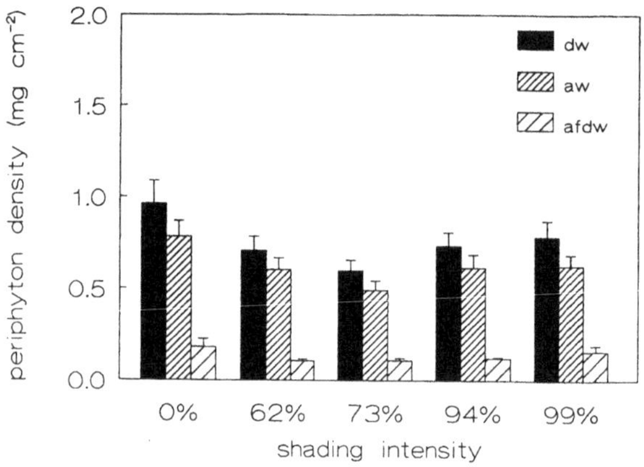


*Fig. 2.* Dry weight (dw), ash weight (aw) and ash-free dry weight (afdw) of periphyton on old leaves of *Z. noltii* (in mg $cm^{-2}$) in a shading transect at a depth of −32 cm relative to mean sea level. In this and the next figures mean values + 1 standard error are indicated.

## *Tidal depth*

The results from the glass slides and the transect are shown in Fig. 3. For the periphyton on slides, significant effects were found on afdw for tidal depth ($p < 0.0001$) and the interaction between tidal depth and replicate plots ($p < 0.0034$). Thus, a shallow tidal depth resulted in lower afdw content of the periphyton on slides, and this was more pronounced in one of the two sets of plots. Tidal depth also had a significant effect on afr ($p < 0.0003$): ash content was higher on slides at the shallow tidal depth. No effects were found on transmittance.

Along the transect, significant changes were found in afdw (oneway ANOVA, $p < 0.0006$) and afr ($p < 0.0055$). Although the variation along the transect was quite strong, it seems that with increased tidal depth, periphyton on old *Z. noltii* leaves decreased in afdw and increased in afr, *i.e.* opposite to the effects found on glass slides. In a direct comparison between the stations 2 and 10, afr on slides was significantly higher than afr on leaves. Slide results may have been influenced by the storm prior to their harvesting.

## *Effects of fiddler crab activities*

Periphyton data inside and outside the fiddler crab exclosures (situated at −22 cm) are shown in Fig. 4. Data for the leaves were averaged over replicate plots because no significant position effect was found for all characteristics in a twoway ANOVA (testing effects of replicate plots and grazing on leaf periphyton data only).

The four density characteristics from glass slides and plants were analyzed together in a threeway ANOVA. Significant position effects (*i.e.* differences between the replicate plots) were found for dw, aw and afdw. Differences between periphyton on slides and on plants were apparent

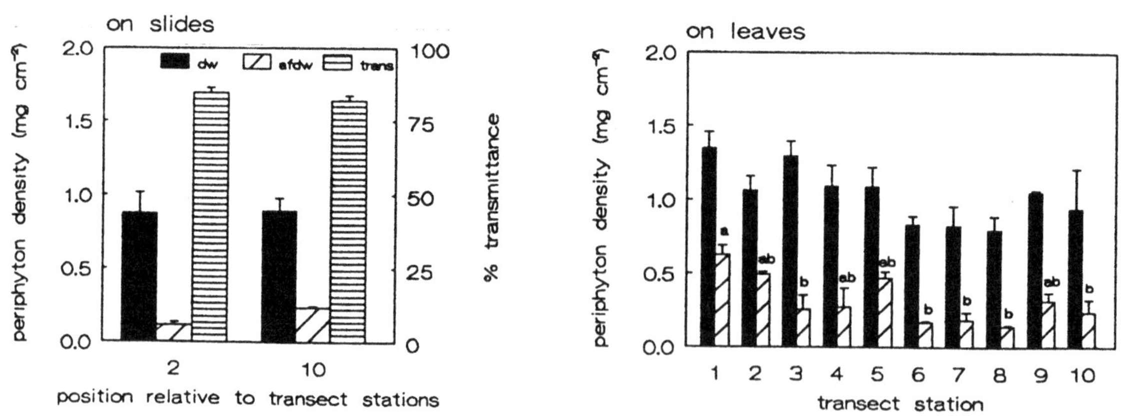


*Fig. 3.* Dry weight, ash-free dry weight and transmittance (trans) of periphyton on glass slides and dry weight and ash-free dry weight on old leaves of *Z. noltii* in a tidal depth transect (station 1 is at −10 cm, station 10 at −71 cm). Significant differences between afdw means on leaves along the transect are indicated with different letters.

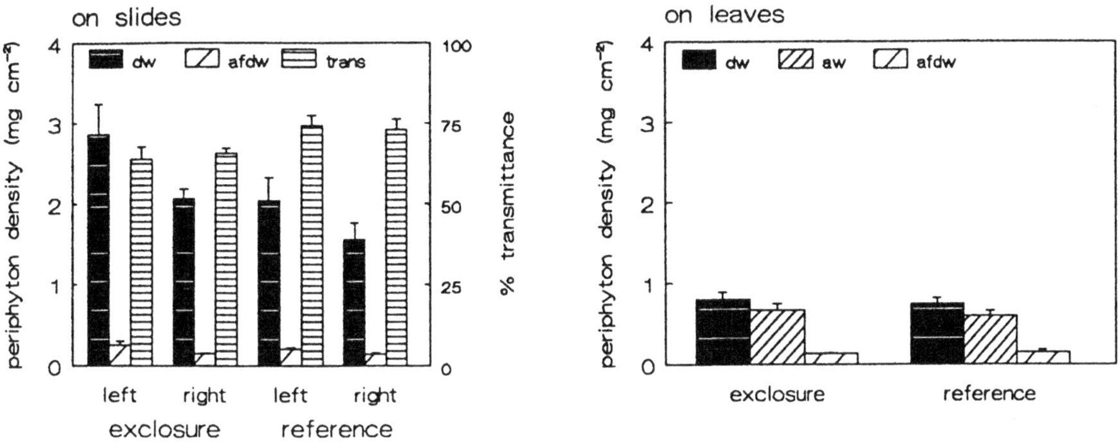


*Fig. 4.* Dry weight, ash-free dry weight and transmittance of periphyton on glass slides in the replicate plots, and dry weight, ash weight and ash-free dry weight of periphyton on *Z. noltii* leaves, both inside and outside the fiddler crab exclosures.

for all four characteristics: the slide values were significantly higher than those from plants. Significant effects of the exclusion of fiddler crab activities were found for dw, aw and afr. All three were higher in the absence of *U. tangeri*. In a twoway ANOVA with leaf periphyton data no significant effects were found apart from the effect of the interaction between *U. tangeri* presence and position on afr. In a twoway ANOVA with slide periphyton data significant effects of the exclosures were found for dw, aw and afr. Transmittance was also significantly increased by fiddler crab activity.

The virtual absence of any fiddler crab effect on periphyton on plants is remarkable, as visual inspection of the plots during the incubation revealed a clear difference in appearance between grazed and ungrazed plants. This occurred already soon after their installment. Inside the exclosures, all plants were oriented in the direction of the outgoing tide. They had a greyish look. Outside the exclosure, fiddler crab activity resulted in a haphazard orientation of leaves, which looked greenish. Unfortunately, the storm event just before the day of harvesting (see the extinction coefficient data in Fig. 1) effectively removed these obvious differences on seagrass leaves before we could quantify them.

Observation of feeding fiddler crabs showed that a leaf was picked up in one claw, the leaf tip was fixed by the mouth parts, followed by a streak along the leaf with a claw to scrape off periphyton in the direction of the mouth.

## Discussion

A general characteristic of all periphyton data collected in Mauritania was the high ash content (52–93%). Periphyton density expressed in afdw ranged between 0.10–0.63 mg $cm^{-2}$. Not many marine periphyton data, at least expressed in afdw, are available in literature. Borum & Wium-Andersen (1980) found peak periphyton values on *Zostera marina* L. of 1.9 mg $cm^{-2}$ dw. Bulthuis & Woelkerling (1983) reported periphyton densities of 1.5–4.5 mg $cm^{-2}$ dw on *Heterozostera tasmanica* (Martens ex Aschers.). Their ash content was about 35%. A comparison between data of various sources of course is obscured by effects of differences in incubation time, substrate type and grazing pressure.

The low afdw content may contribute to the explanation of the absence of a shading effect on periphyton density. The major part of the periphyton apparently is formed by trapping inorganic sediment particles, and light cannot be expected to influence this process.

The effect of periphyton grazing by fiddler crabs is not quite clear from the experimental results.

Fiddler crab presence did influence periphyton dw, aw and transmittance on glass slides, but the effect on afdw was not significant. If such an effect in fact was present, it may have been obscured because of the low afdw content. 'Grazing' on slides probably should be interpreted mainly as removal of periphyton by walking: the characteristic feeding technique on leaves of course could not be applied to slides. Based on our observations we expected the effect of grazing to be much stronger on leaves. Its absence was quite unexpected, but can be attributed to the storm event. The accumulated amount of periphyton on slides inside the exclosures as compared to leaves apparently was high enough to remain statistically detectable, even after removal of part of the periphyton due to the storm. As was mentioned already in the methods section, it can be assumed that leaves during submergence will have endured much more scouring due to wave action and thus were cleaned rather completely from loose material (sediment particles, *i.e.* the ash fraction, which forms a large part of the periphyton). Furthermore, the older leaves, which have the most periphyton, are more likely to break off due to strong currents. Thus, it can be expected that the average periphyton cover on leaves inside and outside the exclosures did not differ very much directly after the storm. On the basis of these arguments, we still conclude that fiddler crabs are able to limit the periphyton densities on leaves considerably. It seems worthwhile to try the use of artificial seagrass leaves, for example made from transparant, flexible and floating material.

Submergence clearly influenced periphyton characteristics. The contradictory results for afdw and afr in the transect as compared to the slides might simply be caused by the observed variation in periphyton density along the transect. Another explanation may be as follows. Fiddler crab activity during low tide was concentrated most on the seagrass close to the low water line (pers. obs.), and the animals probably cleaned the slides less completely than the leaves. The growth rate of periphyton afdw was lower in the plots higher in the intertidal area because of photoinhibition and/or desiccation. This would lead to a decrease in afdw on leaves along the transect towards the sublittoral when *U. tangeri* grazing pressure was high enough, while afdw on slides would increase in the same direction.

Crucial in this respect is the question whether slides were indeed cleaned less efficiently than leaves by the crabs. Accumulation on the slides used in the grazing experiment took place in 11 days, which is comparable to the turnover rate of seagrass leaves (about 10 days; Vermaat *et al.*, 1993). However, the periphyton density on these slides was much higher than on leaves. The slides in the submergence experiment were incubated for 7 days, but already accumulated as much periphyton as old seagrass leaves. These results can support the above statement on fiddler crab cleaning efficiency, but it cannot be excluded that accumulation on (horizontal) slides was more rapid than on leaves (see Fig. 4, exclosure data, and the Methods section).

The transect data indicate a higher availability of afdw close to the highwater line, and thus *U. tangeri* grazing would be expected to be most intense there. In view of bird predation, this feeding location is also comfortably close to their safe burrows. However, a lower afdw growth rate near the high water line combined with a high grazing pressure would quickly exhaust the afdw supply. This might explain why, although the risk of predation is higher, the fiddler crabs concentrate on the seagrass far from their burrows.

The mean extinction coefficient of the waterlayer in our study area ($0.8\ m^{-1}$) is lower than values known from the Dutch Waddensea (1–$4.3\ m^{-1}$, various locations in the period 1970–1980; Giesen *et al.*, 1990). Even lower values are known from Australia (Cockburn Sound, $0.13\ m^{-1}$; Cambridge *et al.*, 1986) and the USA (near Woods Hole, $0.3\ m^{-1}$; Dennison, 1987). Again, as with the periphyton values, comparisons between localities this far apart are difficult to make. The month of September in Mauritania coincided with the highest high tides. leading to flooding of the salt plains (sebhka). This appeared to have caused algal blooming (W. J. Wolff, pers. comm.). Thus, the presently reported attenuation by the waterlayer might represent an upper limit for the study area.

Measurements of light attenuation in the waterlayer and by periphyton allow an evaluation of the light climate experienced by the seagrasses during high tide. For a depth of 1 m, our mean extinction coefficient of 0.8 $m^{-1}$ results in an average light reduction of 55%. An additional shading caused by periphyton of 25% results in about 35% of incident light reaching the seagrass vegetation (reflection from the surface not included, and bearing in mind that the presently measured shading by periphyton on slides may well be an overestimation). Thus, regarding the high light intensities in Mauritania (average maximum in September 2000 $\mu E\ m^{-2}\ s^{-1}$ in air), the present effect of periphyton shading on seagrass production in the Banc d'Arguin seems to be unimportant. Indeed, Vermaat *et al.* (1993) did not find a decrease in productivity along the same shading transect as used in this study up to and including 66%. At 94%, leaf growth was negatively affected. This represents a light intensity of 2% of incident light, roughly 20 $\mu E\ m^{-2}\ s^{-1}$ assuming an average light intensity of 1000 $\mu E\ m^{-2}\ s^{-1}$ passing through the watersurface during the day (see Table 1). The light compensation point (LCP) for *Z. noltii* is 30–35 $\mu E\ m^{-2}\ s^{-1}$ according to Jiménez *et al.* (1987). Philippart (pers. comm.) measured photosynthesis of *Z. noltii* from our field location in the laboratory in The Netherlands; she found an LCP of 40 $\mu E\ m^{-2}\ s^{-1}$. Thus, a deterioration of the vegetation as observed under these conditions can be expected.

*Table 1.* Calculation of the amount of light reaching the leaf surface of *Z. noltii* in the Banc d'Arguin under heavily shaded conditions during high tide.

| | |
|---|---|
| Mean extinction coefficient | 0.8 $m^{-1}$ |
| Depth of the waterlayer | 1 m |
| Transmittance through periphyton | 75% |
| Artificial shading | 94% |
| Mean light intensity below the water surface | 1000 $\mu E\ m^{-2}\ s^{-1}$ |
| Mean light intensity reaching the periphyton | 450 $\mu E\ m^{-2}\ s^{-1}$ |
| Mean light intensity after periphyton | 340 $\mu E\ m^{-2}\ s^{-1}$ |
| Mean light intensity after artificial shading | 20 $\mu E\ m^{-2}\ s^{-1}$ |

## Acknowledgements

We thank the Mauritanian authorities, in particular the Director of the Parc National du Banc d'Arguin, Mr Hadya Amadou Kane, for the permission to carry out this study and for their support. We also thank the collaborators of the Parc, Mr Abou Gueye, Mr El Hassane ould Mohammed el Abd, the late Mr Ely el Elemine, Mr Samba and the inhabitants of Iouik for their cooperation. The other participants during the September expedition are thanked for their help and valuable discussions. The logistics were perfectly organised by Tom van Spanje. René Lobbes of IHE analysed the periphyton samples. The project was made possible thanks to financial support by the Netherlands Marine Science Foundation.

## References

Borum, J. & S. Wium-Andersen, 1980. Biomass and production of epiphytes on eelgrass (*Zostera marina* L.) in the Øresund, Denmark. Ophelia, Suppl. 1: 57–64.

Brönmark, C., 1985. Interactions between macrophytes, epiphytes and herbivores: an experimental approach. Oikos 45: 26–30.

Brönmark, C., 1989. Interactions between epiphytes, macrophytes and freshwater snails: a review. J. Moll. Stud. 55: 299–311.

Bulthuis, D. A. & W. J. Woelkerling, 1983. Biomass accumulation and shading effects of epiphytes on the leaves of *Heterozostera tasmanica*, in Victoria, Australia. Aquat. Bot. 16: 137–148.

Cambridge, M. L., A. W. Chiffings, C. Brittan, L. Moore & A. J. McComb, 1986. The loss of seagrass in Cockburn Sound, Western Australia. II. Possible causes of seagrass decline. Aquat. Bot. 24: 269–285.

Dennison, W. C., 1987. Effects of light on seagrass photosynthesis, growth and depth distribution. Aquat. Bot. 27: 15–26.

Eminson, D. & B. Moss, 1980. The composition and ecology of periphyton communities in freshwaters 1. The influence of host type and external environment on community composition. Br. Phycol. J. 15: 429–446.

Giesen, W. B. J. T., M. M. van Katwijk & C. den Hartog, 1990. Eelgrass condition and turbidity in the Dutch Wadden Sea. Aquat. Bot. 37: 71–85.

Hootsmans, M. J. M. & J. E. Vermaat, 1985. The effect of periphyton-grazing by three epifaunal species on the growth of *Zostera marina* L. under experimental conditions. Aquat. Bot. 22: 83–88.

Jiménez, C., F. X. Niell & P. Algarra, 1987. Photosynthetic adaptation of *Zostera noltii* Hornem. Aquat. Bot. 29: 217–226.

Phillips, G. L., D. Eminson & B. Moss, 1978. A mechanism to account for macrophyte decline in progressively eutrophicated waters. Aquat. Bot. 4: 103–126.

Sand-Jensen, K., 1977. Effect of epiphytes on eelgrass photosynthesis. Aquat. Bot. 3: 55–63.

Sand-Jensen, K. & J. Borum, 1984. Epiphyte shading and its effect on photosynthesis and diel metabolism of *Lobelia dortmanna* L. during the spring bloom in a Danish lake. Aquat. Bot. 20: 109–119.

Sand-Jensen, K. & M. Søndergaard, 1981. Phytoplankton and epiphytic development and their shading effect on submerged macrophytes in lakes of different nutrient status. Int. Revue ges. Hydrobiol. 66: 529–552.

SAS Institute Inc., 1985. SAS/STAT Guide for personal computers, version 6 edition. Cary, N.C., pp. 378.

Silberstein, K., A. W. Chiffings & A. McComb, 1986. The loss of seagrass in Cockburn Sound, Western Australia. III. The effect of epiphytes on productivity of *Posidonia australis* Hook. f. Aquat. Bot. 24: 355–371.

Vermaat, J. E. & M. J. M. Hootsmans, 1991. Periphyton dynamics in a temperature-light gradient. In M. J. M. Hootsmans & J. E. Vermaat, Macrophytes, a key to understanding changes caused by eutrophication in shallow freshwater ecosystems. IHE, Delft, The Netherlands, IHE Report Series 21: 157–187.

Vermaat, J. E., J. A. J. Beijer, R. Gijlstra, M. J. M. Hootsmans, C. J. M. Philippart, N. W. Van den Brink & W. Van Vierssen, 1993. Leaf dynamics and standing stocks of intertidal *Zostera noltii* Hornem. and *Cymodocea nodosa* (Ucria) Ascherson in the Banc d'Arguin (Mauritania). Hydrobiologia 258: 59–72.

*Hydrobiologia* **258**: 81–93, 1993.
*W. J. Wolff, J. van der Land, P. H. Nienhuis & P. A. W. J. de Wilde (eds),
Ecological Studies in the Coastal Waters of Mauritania.*




# Primary production and distribution of phytopigments in the water column and sediments on the upwelling shelf off the Mauritanian coast (Northwest Africa)

E. M. Berghuis, G. C. A. Duineveld & J. Hegeman
*Netherlands Institute for Sea Research, P.O. Box 59, 1790 AB Den Burg, Texel, The Netherlands*

*Key words:* primary production, plant pigments, sediment trap, Mauritanian shelf

## Abstract

In May 1988, water and sediments of the shelf area off the Banc d'Arguin were studied for plant pigments and organic carbon. Phytoplankton primary production was also measured and sedimentation of organic matter was estimated using moored sediment traps. On the basis of simultaneous hydrographic measurements, two water types could be distinguished. Upwelling water covered the southern shelf and the outer part of the northern shelf. The other water type had more coastal characteristics and covered the inshore part of the northern transect. The distribution of chlorophyll *a* and particulate organic carbon conformed with the distribution of the two water types. A similarity analysis of the pigments from the sediment traps showed a difference in the composition of the algal material depositing from the two water types. The primary production was also found to be distinctly different, *i.e.* higher in the coastal water type. The pigments in the surface sediments did not reflect the respective overlying water types due to the increasing accumulation of pigments at the offshore stations of the northern transect. This accumulation seems to correspond with the progressive reduction of the benthic respiratory activity in an offshore direction. Along the southern transect such a relation could not be established, *i.e.* pigment accumulation along the shelf edge was not apparent. Special attention was paid to the presence of specific seagrass pigments on the shelf. No evidence for a major influx of seagrass material was found.

## Introduction

The benthos of the upwelling area along the northwestern African shelf has a high level of activity (Christensen & Packard, 1977) and a high standing stock (Nicols & Rowe, 1977; Duineveld *et al.*, 1993). Since there is no input from rivers in the immediate vicinity, this benthic enrichment is attributed to the enhanced primary production resulting from the upwelling. Regional differences in upwelling intensity are accordingly reflected by the activity and biomass of the benthic fauna (Pfannkuche *et al.*, 1983). The Mauritanian shelf, however, may have an additional source of primary production in the form of dense seagrass beds on the shallow Banc d'Arguin (Fig. 1). Because nothing is known about the transport of organic matter from the offshore upwelling area onto the Banc, Wolff & Smit (1990) assumed that the ecosystem within the limits of the Banc d'Arguin thrives on the organic carbon initially produced by the seagrass. Information about the transport of seagrass from the Banc in offshore direction is not available either.

Hydrographic data, which could provide answers to these questions, show contradicting evi-

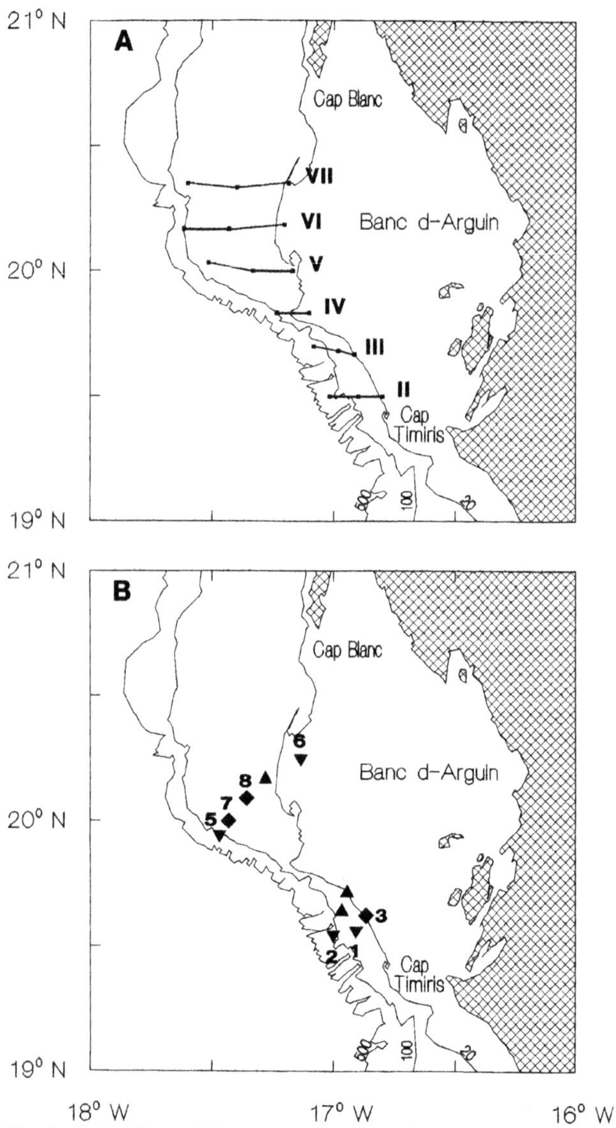


*Fig. 1.* A. Position of 6 transects along which POC and chlorophyll *a* profiles were made. B. Position of stations where either primary production was measured (▲) or pigments in the traps and the sediments were analysed (▼) or both (◆).

dence for the exchange of Banc water and shelf water. The general flow pattern across the Banc d'Arguin is described in Wolff & Smit (1990). Cold offshore water enters the shallows in the north near Cap Blanc, after which it flows southward over the Banc, thereby increasing in temperature and salinity. A front (see Minas *et al.*, 1982) along the border of the Banc prevents the immediate outflow of this water, although Peters (1976) observed isolated bodies of Banc water being released along the entire length of the front. In the south, near Cap Timiris, the dense Banc water crosses the narrow shelf and sinks down the slope. Deviations from this pattern were found by Mittelstaedt (1976) and Peters (1976) who report an outflow of Banc water in the north during times that a southern outflow was absent. Illustrative for the erratic hydrographic events on the Banc is the observation by Peters (1976) that in the beginning of May 1971, the Banc was covered with fresh oceanic water coming from the south. This condition, however, abruptly changed into a more normal regime.

It is evident that the possibilities for transport of organic matter in both directions are still open. During the Dutch TYRO expedition on the offshore Mauritanian shelf in May 1988, various aspects of the benthic and pelagic ecosystem were studied in order to shed light on the directions of and possibilities for transport of organic matter. The present paper discusses the results of primary production measurements as well as the distribution of phytopigments in the water column, sediment traps and the sediment. A detailed analysis of the different pigment types collected in sediment traps and in the top layer of the sediment was performed to distinguish areas with a different quantity and quality of settling and deposited algal material. This classification is compared with the patterns of biomass and activity of the shelf benthos that was found during the same expedition.

## Material and methods

The general distribution of chlorophyll *a*, total phaeopigments and POC (particulate organic carbon) in the water column over the shelf was assessed on the basis of 17 stations situated along 6 transects across the shelf (Fig. 1A). At each station three samples (surface, mid depth and bottom) were collected with a Rosette sampler equipped with a Guildline 8705 CTD probe. Depth profiles of temperature, conductivity and oxygen concentration at the stations were recorded. The chlorophyll *a* and phaeopigment content of the samples were determined on board with the spectrophotometric method described by Lorenzen (1967). Organic carbon was analyzed using the wet persulphate oxidation method of Menzel & Vaccaro (1964). For POC (particulate organic carbon) 1 liter samples were filtered over precombusted GF/F glassfibre filters. The filters were sealed in a precombusted ampule which had first been purged with $O_2$ to remove the $CO_2$ (see Cadée, 1982 for details). In the laboratory the amount of $CO_2$ was measured with a Total Carbon System (Oceanography International Corp., Texas, USA).

For a more detailed study of the quantity and composition of pigments in depositing material and bottom deposits, 7 sampling stations were selected (Fig. 1B). Concurrently, benthic respiration was measured and faunal samples were taken at these stations (Duineveld *et al.*, 1993). Depositing material was collected 1.5 m above the sea bed with sediment traps that were attached to the frame of the bell-jars deployed for *in situ* benthic respiration measurements. Each trap consisted of a perspex tube with a diameter of 7.3 cm and a height of 33 cm. The entrance of the trap consisted of 14 smaller tubes (1.4 cm wide and 8.2 cm long) to reduce turbulence. Before deployment, the trap was filled with filtered seawater to which NaCl was added until a salinity of 90 was reached. At each station two bell-jars, each with an attached trap, were placed less than one mile apart on the sea bed and left for a period of about 8 hours. After retrieval of the bell-jars, the content of each trap was thoroughly mixed and samples of 100 ml from this mixture were filtered through Whatman GF/F glassfibre filters. These filters were subsequently deep-frozen (−30 °C).

For the analysis of phytopigments in the sediment, one sample per station was taken from an intact boxcore using a coretube (2.6 cm diameter). The cores were sectioned in slices of 1 cm under *in situ* temperatures in a thermostatted room and the subsamples were deep-frozen. Pigments in the sediment samples (1 g) and filters were extracted in the laboratory with buffered methanol (2% ammonium acetate). The mixture was desintegrated in an ultrasonic generator for 1 minute and centrifuged. The supernatant was filtered through a GF/F glassfibre filter and an aliquot of 100 $\mu$l was immediately injected in the HPLC system. Reverse-phase HPLC was performed according to the procedures described in Veldhuis & Kraay (1990).

Primary production was measured at 6 stations (Fig. 1B) using the method of Steemann Nielsen (1952). Radioactive labeled bicarbonate (5 $\mu$Ci in 100 ml glass bottles) was added to water samples that were placed in a thermostatted incubator on the ship's deck. A similar type of incubator is described in Gieskes *et al.* (1990). The incubator consists of 7 tubes, each covered by a set of grey and blue filters transmitting respectively 90, 60, 40, 15, 7, 3.5 and 0.7% of the surface irradiance; dark values were obtained by incubating a set of samples in darkened bottles. An incubation lasted for half a daylight period. Surface irradiance was measured throughout the whole day with a Kipp Solarimeter, in order to extrapolate the production measured over the incubation time to a full daylight period. The water samples to be incubated were collected from three depths in the euphotic layer, *i.e.* surface, half-way and at the lower boundary. The depth of the euphotic layer was read from an *in situ* light profile made before sampling (light meter mounted on Rosette sampler). The water samples, each consisting of four replicates, were distributed in the incubator tubes on the basis of the *in situ* light conditions from which they were collected. The tube containing the surface samples was cooled with surface water; the remaining tubes were kept at the average temperature of the euphotic layer. Bicarbonate carbon at the three water depths was measured according to the method of Strickland & Parsons (1968). A production-depth curve was constructed by plotting the production figures from each tube at the depth where the *in situ* light level corresponded with that inside the tube. The primary production per square meter was calculated from the surface area under the production-depth curve.

Statistical analyses were performed with the

MS-DOS version 4.1 of SYSTAT (SYSTAT inc., Evanston, USA).

## Results

### *Hydrography and benthic sampling stations*

A detailed description of the hydrographical situation on and along the shelf during the time of the expedition can be found in Loktionov (1993). Therefore, in this section only a brief summary will be given of the main water types found during the collection of samples for the present study.

Figure 2A shows the temperature profiles along the 6 transects across the shelf. Because the oxygen and salinity data yielded essentially the same trends, only the temperature distribution is depicted. The T-S values of the water samples are shown in Fig. 3. Two principal water types can be discriminated in Figs 2A and 3. In the southern part of the investigated area (transects II & III) relatively cold upwelling water (with a low $O_2$ content and salinity) covered the entire (narrow) shelf up to depths of 20 m. Higher temperatures were only found in the surface layers due to surface heating. On the more northern transects, cold upwelling water only reached the outer part of the transects (up to 75 m depth). The shallower stations were covered by water of a higher temperature ($O_2$ and salinity). In the subsequent sections, this water type will be designated as 'coastal water'.

Table 1 summarizes the most important abiotic parameters of the benthic sampling stations. The depth of the stations varied from 20 to 100 m and the sediment type ranged from mud (88% particles <63 μm) to coarse sand (median grain size 1500 μm). T, S and $O_2$ data from the benthic stations clearly show the two different water types on the northern transect: 5 & 7 with upwelling water and 6 & 8 with coastal water.

### *General distribution of chlorophyll* a *and POC*

Figures 2A, B show the distribution of temperature and chlorophyll *a* along the 6 transects (cf. Fig. 1A). High chlorophyll *a* values were mainly associated with the presence of relatively warm 'coastal water' over the shallow parts of the northern transects. In contrast, the cold upwelling water covering the southern part of the shelf, had low chlorophyll *a* levels. Figure 3 shows the relation between chlorophyll *a* levels and the T-S properties of the two water types. POC (particulate organic carbon) shows a similar correspondence with temperature as chlorophyll *a*. This is illustrated by the linear relation between POC and chlorophyll *a* shown in Fig. 4. The positive intercept of the straight line in Fig. 4 suggests that when chlorophyll *a* is zero, still an amount of POC is present, probably representing the detrital fraction. When this fraction is calculated for the separate transects, zero values are found for transects II to V (T-test, $P<0.05$). Transect VI yielded a significant detrital fraction of about 80 μg C $l^{-1}$. Further north, on transect VII, no significant linear relation was found between POC

*Table 1.* Abiotic parameters of the benthic sampling stations. Positions of stations are displayed in Fig. 1b.

| | SOUTHERN off ·····> inshore | | | NORTHERN TRANSECT off ·····> inshore | | | |
|---|---|---|---|---|---|---|---|
| | 2 | 1 | 3 | 5 | 7 | 8 | 6 |
| Depth in m | 103 | 62 | 22 | 113 | 50 | 33 | 19 |
| Median Grainsize μm | 298 | 40 | 35 | 220 | 160 | 125 | 1500 |
| Perc. silt <63 μm | 13 | 88 | 86 | 14 | 16 | 21 | <1 |
| Bottom temp. °C | 13.6 | 14.5 | 14.7 | 14.0 | 14.8 | 16.7 | 16.5 |
| Bottom salinity ‰ | 35.5 | 35.5 | 35.5 | 35.5 | 35.5 | 35.9 | 35.9 |
| Bottom $O_2$ sat. % | 24 | 24 | 22 | 23 | 28 | 86 | 84 |

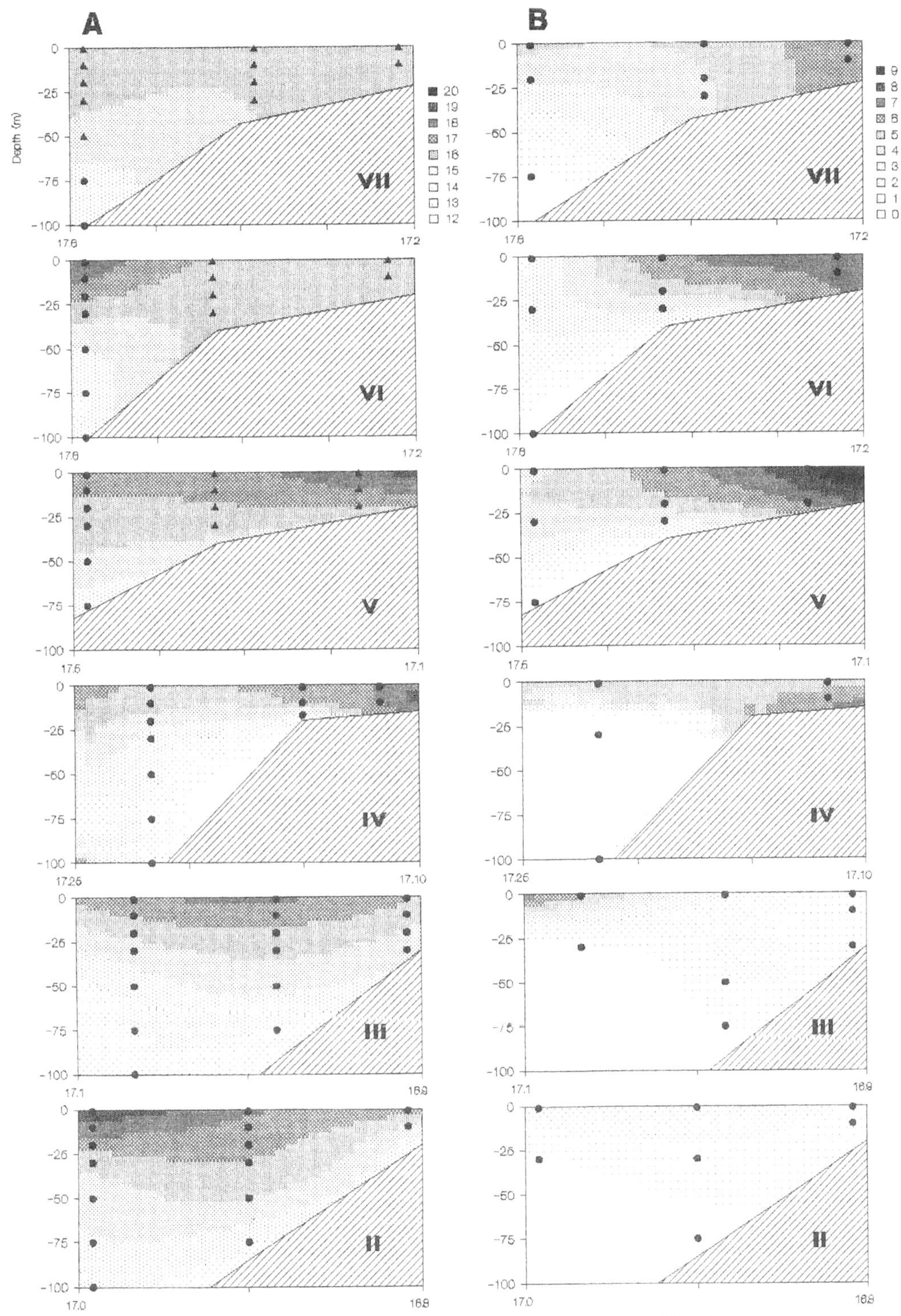


*Fig. 2.* A. Profiles of temperature (°C) and B. profiles of chlorophyll *a* ($\mu g\ l^{-1}$) along the 6 transects shown in Fig 1A. The positions of the samples have been indicated. The two types of symbols in Fig. 2A discriminate the upwelling water (●) from the 'coastal' water (▲) (see text).

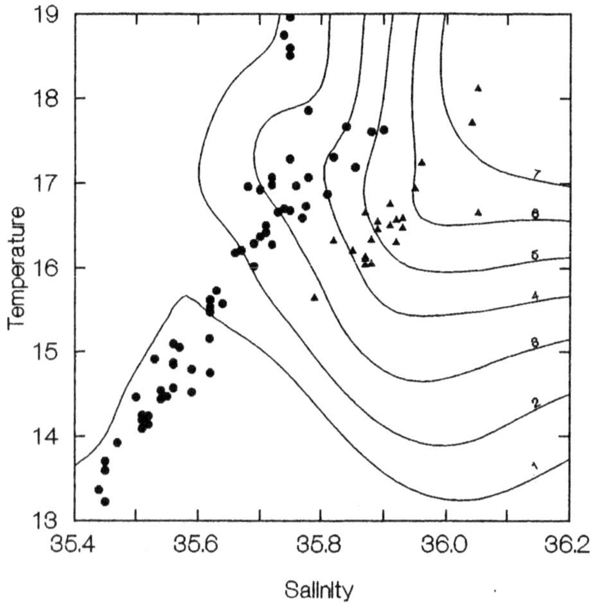


*Fig. 3.* Temperature (°C) and salinity (‰) of water samples shown in Fig. 2A. Symbols indicate upwelling and coastal water in Fig. 2A. Contours denote the associated chlorophyll *a* level (in $\mu g\ l^{-1}$).

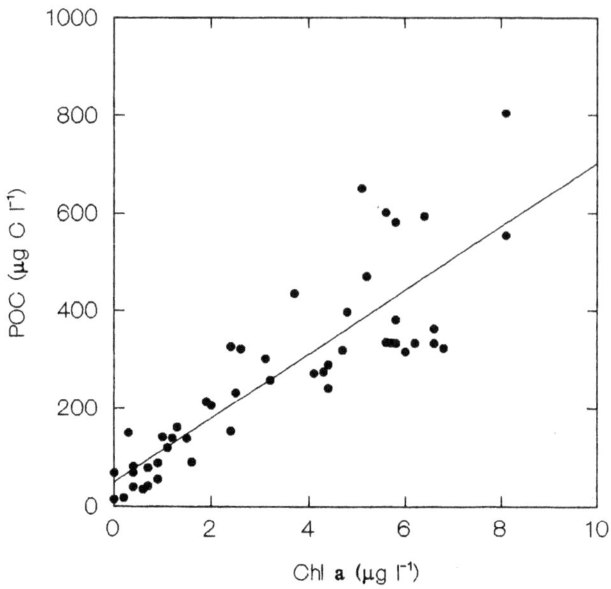


*Fig. 4.* Relation between chlorophyll *a* and POC (particulate organic carbon) data of all transects. The line was obtained with linear regression ($r^2 = 0.74$, $p < 0.05$).

and chlorophyll *a* nor between POC and total pigment. This implies an increasing dominance of the detrital fraction in the northern direction. Nevertheless, the highest amounts of POC found on the shelf (cf. Fig. 2B) seem to be due to fresh algal material.

## *Pigments in sediment traps and bottom deposits*

The concentrations of the principal pigments in the sediment and trap samples have been summarized in Tables 2 and 3. Average values are given for the traps, because the replicates turned out to have very similar contents as demonstrated by the cluster diagram in Fig. 5.

*Table 2.* Concentration of principal pigments in the 0–1 cm layer of the sediment. Quantities expressed in $\mu g\ g^{-1}$ dry sediment; b.d. means below detection limit. For the position of the stations see Fig. 1B.

| SEDIMENT 0–1 cm | SOUTHERN off ········> inshore | | | NORTHERN TRANSECT off ········> inshore | | | |
|---|---|---|---|---|---|---|---|
| Pigmenttype | 2 | 1 | 3 | 5 | 7 | 8 | 6 |
| Chlorophyll *c* | 0.13 | 0.18 | 0.19 | 3.29 | 1.65 | 0.87 | 0.08 |
| Fucoxanthin | 0.17 | 1.03 | 0.82 | 10.34 | 3.32 | 2.79 | 0.43 |
| Lutein/zeaxanthin | 0.58 | 0.94 | 0.55 | 2.47 | 1.16 | 1.08 | 0.05 |
| Chlorophyll *a* | 0.72 | 2.91 | 1.59 | 4.01 | 2.42 | 2.06 | 0.68 |
| Chlorophyll *b* | b.d. | b.d. | b.d. | b.d. | b.d. | b.d. | 0.48 |
| β-Carotene | 0.98 | 2.33 | 0.66 | 6.83 | 3.29 | 2.48 | 0.06 |
| Phaeophorbide-1 | 0.47 | 1.39 | 0.69 | 31.30 | 11.13 | 4.69 | 0.59 |
| Phaeophorbide-2 | 4.14 | 6.19 | 5.65 | 62.59 | 14.32 | 5.75 | 0.78 |
| Phaeophytin | 0.18 | 2.80 | 0.80 | 1.97 | 3.86 | 2.80 | 0.29 |
| SUM | 7.36 | 17.75 | 10.95 | 122.80 | 41.15 | 22.52 | 3.43 |

*Table 3.* Deposition rate of principal pigments expressed in $\mu g\ cm^{-2}\ d^{-1}$; b.d. means below detection limit. For the position of the stations see Fig. 1B.

| SEDIMENTTRAP | SOUTHERN off ·····> inshore | | | NORTHERN TRANSECT off ·····> inshore | | | |
|---|---|---|---|---|---|---|---|
| Pigmenttype | 2 | 1 | 3 | 5 | 7 | 8 | 6 |
| Peridinin | b.d. | b.d. | b.d. | b.d; | 0.53 | 8.16 | 1.33 |
| Chlorophyll *c* | 0.05 | 0.02 | 0.06 | 0.03 | 0.11 | 0.69 | 0.22 |
| Fucoxanthin | 0.18 | 0.07 | 0.31 | 0.16 | 0.79 | 5.13 | 1.34 |
| Lutein/zeaxanthin | 0.07 | 0.03 | 0.04 | 0.03 | 0.12 | 0.46 | 0.03 |
| Chlorophyll *a* | 0.45 | 0.21 | 0.61 | 0.32 | 1.03 | 6.21 | 1.62 |
| Chlorophyll *b* | 0.03 | b.d. | b.d. | b.d. | b.d. | b.d. | b.d. |
| $\beta$-Carotene | 0.02 | 0.02 | 0.01 | 0.01 | 0.08 | 0.50 | 0.07 |
| Phaeophorbide-1 | 0.05 | 0.04 | 0.13 | 0.09 | 0.78 | 7.69 | 1.07 |
| Phaeophorbide-2 | 0.31 | 0.34 | 0.87 | 0.50 | 1.44 | 7.93 | 1.13 |
| Phaeophytin | 0.02 | 0.04 | 0.04 | 0.02 | 0.41 | 2.26 | 0.66 |
| SUM | 1.17 | 0.77 | 2.07 | 1.14 | 5.27 | 39.02 | 7.48 |

One of the questions of this study concerned the presence of seagrass matter on the shelf. The most abundant species on the tidal flats of Banc d'Arguin is *Zostera noltii* (see Wolff & Smit, 1990). Among the main characteristic pigments of seagrass are chlorophyll *b* and lutein (Van den Hoek, 1978). Because we could not separate lutein from zeaxanthin with our methods, the two pigments are grouped. In HPLC chromatograms of *Zostera noltii* from the Dutch Wadden Sea, chlorophyll *b*: lutein/zeaxanthin occur in a 1:1 ratio (rough counts; K. Philippart, pers. comm.). As shown in Tables 2 and 3, chlorophyll *b* was nearly absent in our samples; only the sediment from the innermost station 6 of the northern transect yielded a significant amount of chlorophyll *b*. Lutein/zeaxanthine, on the other hand, was present in all chromatograms although in relatively low concentrations.

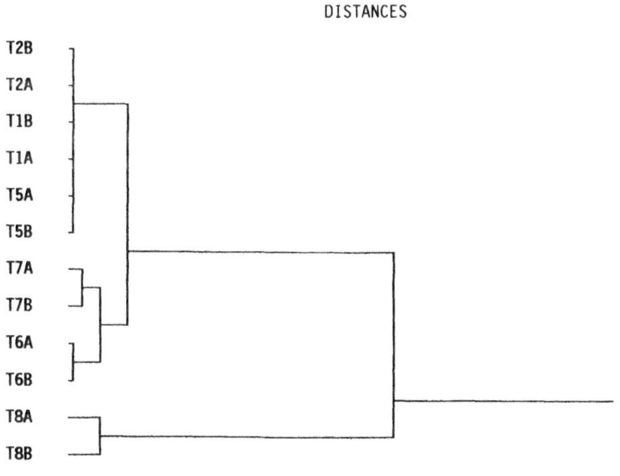


*Fig. 5.* Cluster diagram based on phytopigment types of the two replicate trap samples (a and b) from the 6 stations shown in Fig. 1B. Code T1a stands for the first trap from station 1 etc. Similarity was measured as the normalized euclidean distance between the raw data.

The total pigment content in the traps and of the corresponding surface layer of the sediment is depicted in Fig. 6. This figure with Tables 2 and 3 show that the deposition rates as well as the sediment contents on the northern transect on average were higher than those on the southern transect. Another conspicuous feature of Fig. 6 is the sharp increase of the pigment content in the sediment towards the shelf edge of the northern transect. This contrasts with the opposite trend in the deposition rates on the three outer stations. This combination could point at an accumulation of pigments in an offshore direction.

It is of interest to examine whether the differences in quantity of depositing and deposited pigments are associated with different types of algae (and water masses) and different rates of primary

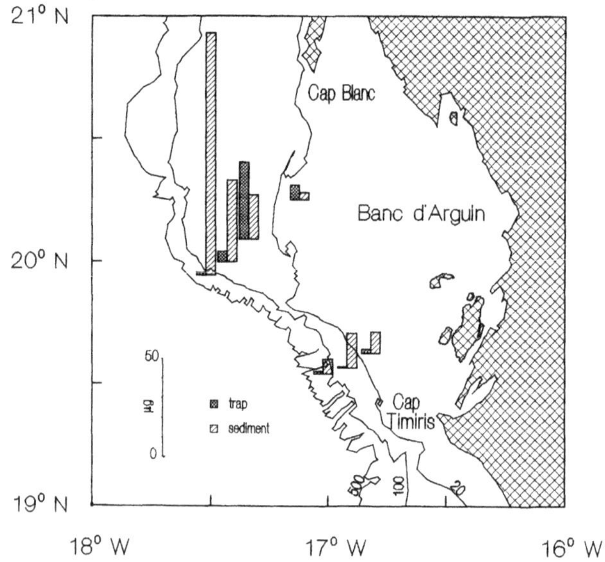


*Fig. 6.* Total amount of selected pigments (cf. Table 2, 3) in the sediment traps (in $\mu g\ cm^{-2}\ d^{-1}$) and in the upper 1 cm of the sediment (in $\mu g\ g^{-1}$ dry sediment).

productivity. In order to compare the pigment composition among traps and between traps and sediments on a normalized basis, the accessory pigments (and phaeopigments) were expressed as ratios of chlorophyll *a*. Since each phytoplankton class has a specific ratio accessory pigment: chl-*a* (Gieskes & Kraay, 1983; Everitt *et al.*, 1990), variations in these ratios among the samples are indicative for differences in the composition of the algal material.

Based on the pigment: chl-*a* ratios, the (distance) similarity between the traps was calculated and graphically depicted by means of multidimensional scaling (MDS; Kruskal, 1964). Figure 7 shows that the contents of the traps from station 6, 7 and 8 (cf. Fig. 1B) were different from the other stations. This is for most part accounted for by the presence of peridinin in the traps from stations 6, 7 and 8. Peridinin is characteristic for dinoflagellates (Van den Hoek, 1978) and was not found elsewhere. This group of stations is furthermore distinguished by relatively high proportions of fucoxanthin and of the phaeopigment

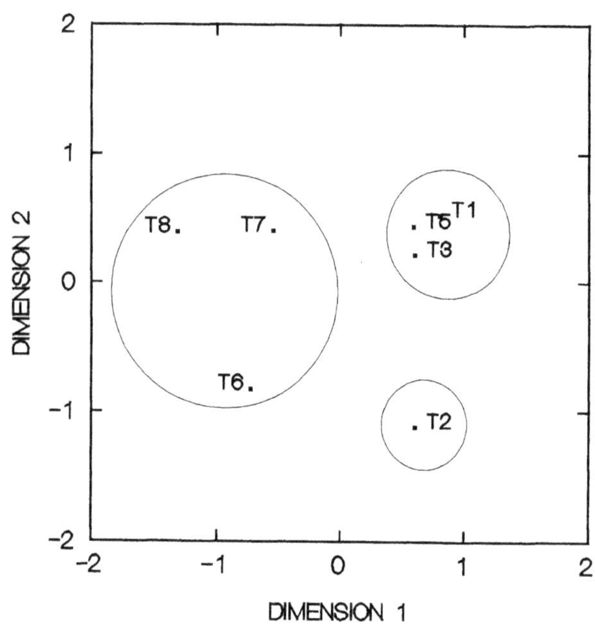


*Fig. 7.* Multidimensional scaling plot of trap phytopigment contents, showing the dissimilarity between the trap contents from the separate stations. Code T1 stands for the mean value from the traps at station 1 etc. Prior to the analysis the data were expressed as percentages of the chlorophyll *a* concentration.

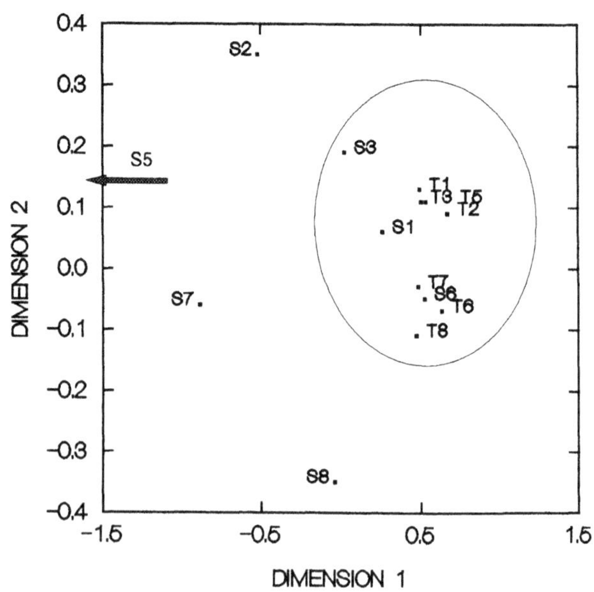


*Fig. 8.* Multidimensional scaling plot of traps (T) and sediments (S), showing the dissimilarity between the contents in the traps and the sediment. Prior to the analysis the data were expressed as percentages of the chlorophyll *a* concentration. The arrow near S(ediment) 5 denotes that this point falls beyond the scale of the diagram.

phaeophytin. The somewhat deviating position for the trap contents from station 2 (Fig. 7) is due to the comparatively low proportion of phaeopigments.

The differences in composition of depositing algae on the shelf, suggested a comparison be made between the pigment composition in the traps and in the corresponding toplayer of the sediment. The MDS diagram comparing trap and sediment data (Fig. 8), reveals only a partial resemblance between the two. The greatest similarity is found at stations 1, 3 and 6. The largest discrepancy is found at station 5, which is caused by the high proportions of fucoxanthin and phaeopigments in the sediment. The same pigments were responsible for the dissimilarity between the sediment and trap of station 7 and 8. Peridinin was excluded from this analysis, since including it would have further enlarged the dissimilarities in station 6, 7 and 8.

To be able to explain the discrepancies between the traps and the corresponding sediments, it is necessary to have some idea of the degradation rates of the various pigments. Enhanced proportions of a particular pigment in the sediment, for instance, could be due to the stability of that pigment. To obtain some insight in the stability of the various pigments, some deeper sediment layers of the stations on the southern transect were analysed. Because of differences in composition and activity of the benthic fauna at the three investigated stations (see Table 4), degradation rates can be expected to vary among the stations. Nevertheless some consistent trends can be discerned in Fig. 9. At all stations $\beta$-carotene and lutein/zeaxanthin appeared to be the most stable pigments, whereas chlorophyll *a* was still present in substantial amounts up to a depth of 5 cm. Chlorophyll *c*, fucoxanthin and phaeophorbides, on the other hand, appeared to decrease at a higher rate with depth.

*Primary production*

The results from the primary production measurements have been plotted in Fig. 10. Noticeable are the relatively low production values on the southern transect which was covered by the upwelling water. This low production coincided with depressed levels of chlorophyll *a* in the water column (Fig. 2B) and with low deposition rates of pigments (Table 3). Along the northern transect an increase of primary production in inshore direction was found, culminating in a maximum of about 9 g C $m^{-2}$ $d^{-1}$ at the shallowest station. The pattern of deposition rates at the stations near the ones where production was measured was in agreement with this increasing production.

*Table 4*. Synopsis of macro- and meiobenthic standing stock, benthic respiratory activity, concentration of selected phytopigments in the sediment and their deposition rates. For the position of stations see Fig. 1B.

| | SOUTHERN | | | NORTHERN TRANSECT | | | |
|---|---|---|---|---|---|---|---|
| | off ··················> inshore | | | off ···························> inshore | | | |
| | 2 | 1 | 3 | 5 | 7 | 8 | 6 |
| Macrofauna biomass (g AFDW $m^{-2}$) | 9.4 | 7.5 | 7.9 | 6.2 | 21.0 | 10.9 | 42.1 |
| Meiofauna abundance (no. ind * $10^6$ $m^{-2}$) | 3.0 | 5.6 | 7.6 | 8.3 | 9.5 | 3.6 | 6.9 |
| Benthic respiration ($\mu$mol $O_2$ $m^{-2}$ $h^{-1}$) | 249 | 584 | 1268 | 600 | 306 | 1005 | 2288 |
| Phytopts. sediment ($\mu$g $g^{-1}$ dry sed.) | 7.4 | 17.7 | 11.0 | 122.8 | 41.2 | 22.5 | 3.4 |
| Phytopts. deposition ($\mu$g $cm^{-2}$ $d^{-1}$) | 1.2 | 0.8 | 2.1 | 1.1 | 5.3 | 39.0 | 7.5 |

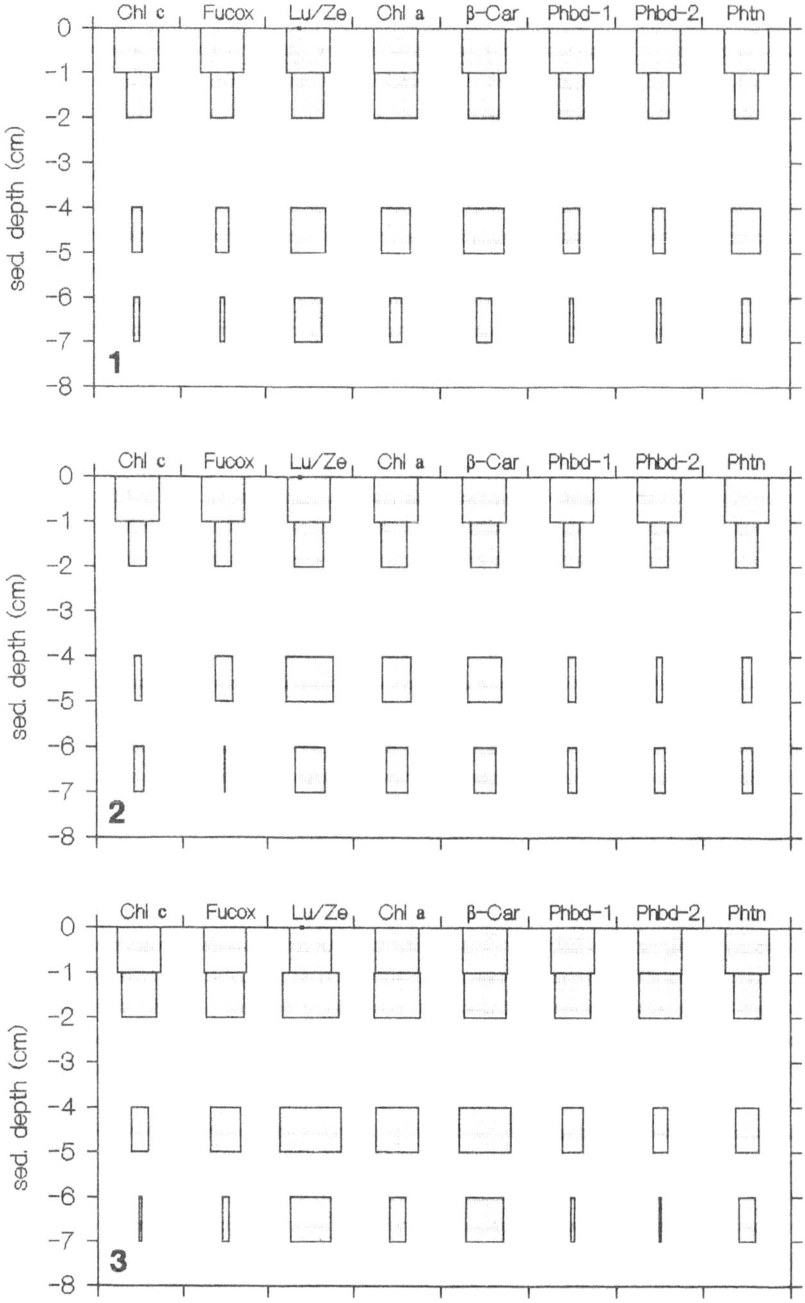


*Fig. 9.* Vertical distribution of pigments in the sediment of stations 1, 2 and 3 at the southern transect (cf. Fig. 1B). Width of the bars are scaled as percentages of the concentration in the upper 1 cm (100%, absolute values in Table 2).

## Discussion

It is evident from the near absence of chlorophyll *b* in the traps and shelf deposits that there is no substantial input of seagrass from the Banc d'Arguin at the time of our study. As a consequence, it could be inferred that significant transport of organic matter from the Banc d'Arguin towards the shelf is absent.

Our data, nevertheless, show that part of the

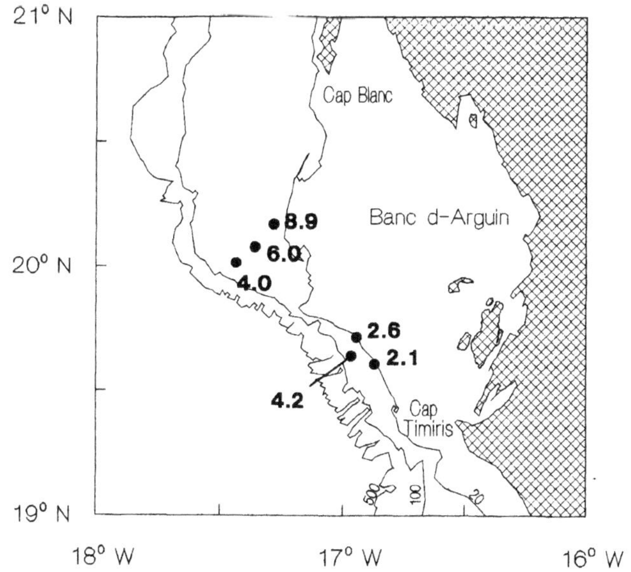


*Fig. 10.* Primary production (in g C $m^{-2}$ $d^{-1}$) at the stations displayed in Fig. 1B.

shelf was covered by water with coastal characteristics. Support for the distinction of this water type comes from the observations by Kuipers *et al.* (1993) who found that the zooplankton community in this 'coastal water' is dominated by neritic copepods whereas upwelling species were absent. According to the hydrographical data the 'coastal water' of uniformly high temperature, salinity and $O_2$ content, reaches as far offshore as station 8 (Table 1). Beyond this station the properties of the water near the bottom change abruptly: the bottom water at the adjacent station 7 closely resembles that of the upwelling water found on the shelf edge (station 5) and further offshore (see Loktionov, 1993). Our analysis of the trap pigments corroborates the distribution of the water types as far as it concerns stations 6 and 8. At station 7, where upwelling water covered the bottom, the trap pigments were still closely affiliated to those of stations 6 and 8. A possible explanation for this discrepancy could be mixing in the upper water layers due to the strong winds (13 m $s^{-1}$) at the time of the expedition.

Other characteristics of the 'coastal water' on the northern transect include the relatively high primary production and deposition rates of pigments. Reports from earlier research in the area mention the presence of a belt with enhanced chlorophyll *a* levels along the edge of Banc d'Arguin (Weichart, 1974; Peters, 1976), which could be indicative for an elevated primary production similar to what we observed. However, because these earlier cruises did not cover the shallow Banc d'Arguin, the inshore extension of the chlorophyll belt and, therefore, its relation to the Banc remains obscure. Minas *et al.* (1982) argue that the chlorophyll *a* belt could result from a mixture of upwelling water and old descended Banc water reaching the euphotic zone on the shallow shelf. Such a cross-shelf origin for our 'coastal water' seems unlikely because of its sharp boundaries with the upwelling water. There are also reasons to assume that our 'coastal water' is not derived from the inner Banc d'Arguin. First is the absence of seagrass traces. Second, previous measurements in the innermost parts of the Banc d'Arguin in April–May (Reyssac, 1977) invariably yielded much lower chlorophyll *a* and production figures (resp. 0–3 μg $l^{-1}$ and 40–110 mg C $m^{-3}$ $d^{-1}$) than the ones found on the edge of the Banc near our northern transect (resp. 7.5 μg $l^{-1}$ and 90–270 mg C $m^{-3}$ $d^{-1}$). Although we cannot resolve the source of the 'coastal water', it is evident from its hydrographical and biological characteristics (e.g. copepods; see Kuipers *et al.*, 1993) that it must have resided for some time on the shallow part of the shelf. Through its high primary production it likely exerts a positive effect on a significant part of the offshore shelf.

Another result of our pigment analyses was the lack of correspondence between the quantity and quality of pigments in the traps and in the sediment. For example, station 6 had second highest deposition rate, but the lowest sediment content. The presence of coarse sediment (1500 μm median and no silt) in combination with a macrofauna composed of thick-shelled suspension-feeders and interstitially living infauna, point at a turbulent, non-depositional environment at station 6. The extremely high benthic respiration rate of this station (see Table 4) contributes furthermore to a rapid disappearance of organic matter.

The opposite situation, *i.e.* a low deposition rate and a high amount of pigments in the sediment, was found at station 5. This station lies at the end of a gradient of increasing pigment accumulation in the sediment in spite of a reduced deposition rate. This could be explained by a decrease in the rate at which pigments are degraded to non-fluorescent substances. The first evidence for a delayed degradation is the extraordinarily high number of peaks (45) in the chromatogram of station 5 when, for instance, compared to the 16 peaks detected in sediment from station 6. Further support comes from the increasing proportions of phaeopigments and fucoxantin in the sediment despite their constant proportions in the traps. According to the depth profiles (Fig. 9), especially these pigments should be most liable to decomposition. An offshore decrease of the degradation activity also fits well with the observed reduction in benthic respiratory activity towards the shelf edge at the northern transect (Table 4).

A similar relation between benthic respiration and pigment contents is lacking for the southern transect. Because the deposition rate and primary production at station 2 were comparable with those of station 5, one would expect a similar accumulation of pigments. In fact, benthic respiration at station 2 was only half that of station 5 and neither faunal density nor biomass differed much between the stations (Table 4). Speculations about the causes for the differences between the shelf edge stations readily incorporate the spatial and temporal variation in upwelling intensity. Upwelling may negatively affect benthic respiration through the associated low oxygen levels (Duineveld *et al.*, 1993). Depending on the duration of this stress, pigments can accumulate in varying amounts. It is conceivable that in the southern area, where short-term variations in upwelling have previously been observed (Peters, 1976), pigments in the sediment became depleted during an absence of upwelling prior to our visit. An alternative explanation is that in the absence of upwelling dense Banc d'Arguin water moves downward over the shelf in the vicinity of the southern transect. This could create adverse conditions for sedimentation or even erosion, especially at the steep shelf edge.

Whatever the causes are for the observed differences in pigment content between the transects, our dataset shows that no overall unambiguous relation can be established between benthic respiration on one hand, and either pigment deposition rates or sediment content or a combination of both, on the other hand (cf. Table 4). This stands in contrast with the strong positive correlation between sediment respiration and pigment content found by Pfannkuche *et al.* (1983) across the slope (130–3000 m) off Morocco. The interpretation of our results, though admittedly speculative, shows that in light of the complicated and variable conditions on the Mauritanian shelf such a simple relationship is not to be expected.

## Acknowledgements

We are greatly indebted to G. W. Kraay for his assistance with the HPLC analyses and his invaluable information on the identification of the pigments and their concentration equivalents. In this respect we also wish to thank Anita Buma. P. A. W. J. de Wilde and G. C. Cadée gave critical comments on the manuscript which led to major improvements. We furthermore wish to thank the crew and technicians on board of the 'TYRO' for their assistance. The S.O.Z. is acknowledged for their financial support of the expedition.

## References

Cadée, G. C., 1982. Tidal and seasonal variation in particulate and dissolved organic carbon in the western Dutch Wadden Sea and Marsdiep tidal inlet. Neth. J. Sea Res. 15: 228–240.

Christensen, J. P. & T. T. Packard, 1977. Sediment metabolism from the northwest African upwelling system. Deep Sea Res. 24: 331–343.

Duineveld, G. C. A., P. A. W. J. de Wilde, E. M. Berghuis & A. Kok, 1993. The benthic infauna and benthic respiration off the Banc d'Arguin (Mauritania, Northwest Africa). Hydrobiologia 258: 107–117.

Everitt, D. A., S. W. Wright, J. K. Volkman, D. P. Thomas & E. J. Lindstrom, 1990. Phytoplankton community compositions in the western equatorial Pacific determined from chlorophyll and carotenoid pigment distributions. Deep Sea Res. 37: 975–997.

Gieskes, W. W. C. & G. W. Kraay, 1983. Dominance of Cryptophyceae during the phytoplankton spring bloom in the central North Sea detected by HPLC analysis of pigments. Mar. Biol. 75: 179–185.

Gieskes, W. W. C., G. W. Kraay, A. Nontji, D. Setiapermana & Sutomo, 1990. Monsoonal differences in primary production in the eastern Banda Sea. Neth. J. Sea Res. 25: 473–483.

Kruskal, J. B., 1964. Nonmetric multidimensional scaling: a numerical method. Psychometrika 29: 115–129.

Kuipers, B. R., H. J. Witte & S. R. Gonzalez, 1993. Zooplankton distribution in the coastal upwelling system along the Banc d'Arguin, Mauritania. Hydrobiologia 258: 133–149.

Loktionov, Y., 1993 Hydrological observations west of the Banc d'Arguin, Mauritania, in May 1988. Hydrobiologia 258: 21–32.

Lorenzen, C. J., 1967. Determination of chlorophyll and phaeopigments. Spectrophotometric equations. Limnol. Oceanogr. 12: 343–346.

Menzel, D. W. & R. F. Vaccaro, 1964. The measurement of dissolved and particulate carbon in seawater. Limnol. Oceanogr. 9: 138–142.

Minas, H. J., L. A. Codispoti & R. C. Dugdale, 1982. Nutrients and primary production in the upwelling region off Northwest Africa. Rapp. P.-v. Réun. Cons. int. Explor. Mer 180: 148–183.

Mittelstaedt, E., 1976. On the currents along the northwest African coast south of 22° North. Dt. hydrogr. Z. 29: 97–117.

Nichols, J. & G. T. Rowe, 1977. Infaunal macrobenthos off Cap Blanc, Spanish Sahara. J. mar. Res. 35: 525–536.

Peters, H., 1976. The spreading of the water masses of the Banc d'Arguin in the upwelling area off the northern Mauritanian coast. 'Meteor' Forsch Ergebn. A 18: 78–100.

Pfannkuche, O., R. Theeg & H. Thiel, 1983. Benthos activity, abundance and biomass under an area of low upwelling off Morocco, Northwest Africa. 'Meteor' Forsch Ergebn. D 36: 85–96.

Reyssac, J., 1977. Hydrologie, phytoplancton et production primaire de la baie du Lévrier et du Banc d'Arguin. Bull. Inst. Franç. Afrique Noire, Sér. A, T. 39: 487–554.

Steemann Nielsen, E., 1952. The use of radioactive carbon ($C^{14}$) for measuring organic production in the Sea. J. Cons. perm. int. Explor. Mer 18: 117–140.

Strickland, J. D. H. & T. R. Parsons, 1968. A practical handbook of seawater analysis. Bull. Fish. Res. Bd Can. 167: 1–311.

Van den Hoeck, C., 1978. Algen. Einführung in die Phykologie. Thieme Verlag, Stuttgart, 481 pp.

Veldhuis, M. & G. W. Kraay, 1990. Vertical distribution and pigment composition of a picoplanktonic prochlorophyte in the subtropical North Atlantic: a combined study of HPLC-analysis of pigments and flow cytometry. Mar. Ecol. Prog. Ser. 68: 121–127.

Weichart, G., 1974. Meereschemische Untersuchungen im nordwestafrikanischen Auftriebsgebiet 1968. 'Meteor' Forsch Ergebn. A 14: 33–70.

Wolff, W. J. & C. J. Smit, 1990. The Banc d'Arguin, Mauritania, as an environment for coastal birds. Ardea 78: 17–38.

*Hydrobiologia* **258**: 95–106, 1993.
*W. J. Wolff, J. van der Land, P. H. Nienhuis & P. A. W. J. de Wilde (eds),*
*Ecological Studies in the Coastal Waters of Mauritania.*



# Distribution of sponges on the Mauritanian continental shelf

R. W. M. van Soest
*Institute of Taxonomic Zoology, University of Amsterdam, P.O. Box 4766, 1009 AT Amsterdam, The Netherlands*

*Key words:* Biogeography, faunal barrier, Porifera

## Abstract

Transect sampling on the continental shelf off the Banc d'Arguin, Mauritania, revealed a patchy distribution of sponge species and populations. Major sponge habitats are inshore coastal sandstone flats harbouring a diverse infaunal sponge fauna, and the moving substrate of shells of live gastropods or empty shells occupied by hermit crabs which harbour a numerically important but species-poor incrusting sponge fauna. The latter forms occur over most of the depth profile. Horizontal distribution of the sponges shows a distinct gap in species and numbers associated with a muddy bottom area where the sediment-rich Banc d'Arguin run-off is found. North and south of this gap the specific composition of the infaunal sponges is considerably different and this is explained as evidence for limitation of range extensions, notably of the southern sponge fauna, by the muddy area.

## Introduction

Somewhere along the sandy continental platform off Mauritania or Senegal, the Mediterranean-Atlantic marine shallow-water biota reach their southern border to be replaced by the tropical marine biota of West Africa. Often quoted faunal borders are Cabo Verde (Senegal) and Cap Blanc (Mauritania) (e.g. Briggs, 1974; for sponges: Boury-Esnault & Lopes, 1985). Abiotic and ecological factors responsible for the faunal changes in these localities are, however, barely documented. One of the purposes of the Tyro Mauritania II Expedition (1988) (Van der Land, 1988) was to examine in detail the bottom conditions and quantitative distributions of the benthic organisms in a North-South series of transects set off at right angles to the coast of Mauritania. The results of this study, of which the part concerning the distribution of the sponges is presented here, may contribute in understanding the nature of the faunal barriers in this area.

To date, no published reports are available on sponges from Mauritanian waters. A major reason for this lack of distributional data from that area may be the sandy nature of the sea bottom of the continental shelf off Mauritania, which does not invite the study of hard substrate dwellers such as sponges. The Tyro Mauritania II Expedition nevertheless secured approximately one hundred different species of sponges between 0 and 1900 m, and the majority of these were not collected from the often hard-bottomed continental slope as might be expected, but came from shallow waters on the continental platform between 0 and 100 m. There, two major substrates are available to sponges: the mobile substrate presented by extremely common gastropod shells and their empty shells inhabited by hermit crabs, and the hard sandstone platforms and ridges covered by a sufficiently thin layer of sand or mud, which harbour stalked or infaunal species. Less importantly, sponges may be epizoic on stems of hydroids, algae and gorgonids.

**Study area and methods**

The Tyro Mauritania II Expedition was part of a large programme of multidisciplinary marine biological investigations in the period of April to October, 1988, aimed at studying the area of the Banc d'Arguin (20° 55′–18° 49′ N). The expedition used the Dutch research vessel Tyro as a platform for making eight bottom sampling transects at right angles to the coast of Mauritania from Cap Blanc in the north to just south of Cap Timiris (I–VIII in Fig. 1). The surface currents in the area are well known from previous oceanographic expeditions. In summary, there is a strong relatively cool current coming from the north (Canary Current), present all the year round, which gives off a branch into the Banc d'Arguin area south of Cape Blanco through the Passe de Lévrier. This branch leaves the Banc d'Arguin north of Cape Timiris. In the south a weak northwardly directed coastal current is present in the summer months. Eight transects were chosen at approximately equal distances of 15 geographic minutes (cf. Fig. 1: I–VIII), each from inshore coastal waters (depth limitation of the research vessel was 16 m) westward to continental slope depths (400–1900 m).

Sampling along the transects was done at regular depth intervals, using a variety of bottom gear, with a fixed routine for each gear-type: Van Veen grab (4 grabs at each station), Hamon grab (idem), Agassiz 1.2, 2.4 and 3.5 m trawls (single hauls with standard bottom time), rectangular dredge (idem). Samples were sorted immediately when on deck; sponges were isolated from the samples, characterized briefly, pre-identified, and preserved in ethanol. Of grab samples, bottom sediment was noted (categories: sandstone, sand, gravel, muddy sand, mud); distribution of these bottom sediments in the depth range of 10–100 m is presented in Fig. 2. In the area where the run-off of Banc d'Arguin water is concentrated (transects IV–VI), bottom sediment samples were almost exclusively mud. Hard bottoms with thin layers of sediment are found mostly inshore, although here and there hard bottoms occur at depths of 50–60 m. The continental slope between 100 and 1000 m may be quite steep and canyon-like; in such areas the bottom is often rocky and may carry deep-water reefs (*Lophelia*). These deep-water reefs were found to contain an extensive sponge fauna. Species composition was found to resemble that of similar Northern Atlantic deep-water reefs, with such typical representatives as *Pachastrella monilifera* (Schmidt, 1868), *Desmacidon fruticosa* (Montagu, 1818), *Myxilla incrustans* (Johnston, 1842) and *M. rosacea* (Lieberkühn, 1859). Few if any species were found both there and in the shallow-waters. In other areas the slope is gradual and the bottom muddy, and sponges are absent. Deep-water samples (> 100 m) were identified to species level, but are ignored in the present study, because they clearly appear to belong to a separate fauna with a different ecological range and geographic distribution.

Sampling efforts varied somewhat over the various transects; the distribution over the eight transects of a total of 96 grab and trawl/dredge stations, and another 4 stations (3, 4, 7 and 8) involving wading/snorkling or other means of collecting, is presented in Fig. 3. Depths covered by the sampling efforts also varied, dictated by the nature of the continental platform in the various transects. The distribution of 100 stations over the depth profile in 10 m depth intervals is given in Fig. 4. In both Figs 3 and 4 sampling by grabs and trawls/dredges are presented also separately because they are widely different in area covered and have different capture efficiency with regard to mobile vs. infaunal sponges. Graphs of depth- and horizontal distributions of sponges presented below need to be compared with Figs 3 and 4.

Drs C. Franssen identified the hermit crab associates of 'mobile sponges'; Mr R. Moolenbeek identified the gastropod associates. Mr J. J. Vermeulen assisted in collecting and preserving the sponges. Dr J. van der Land allowed me to join

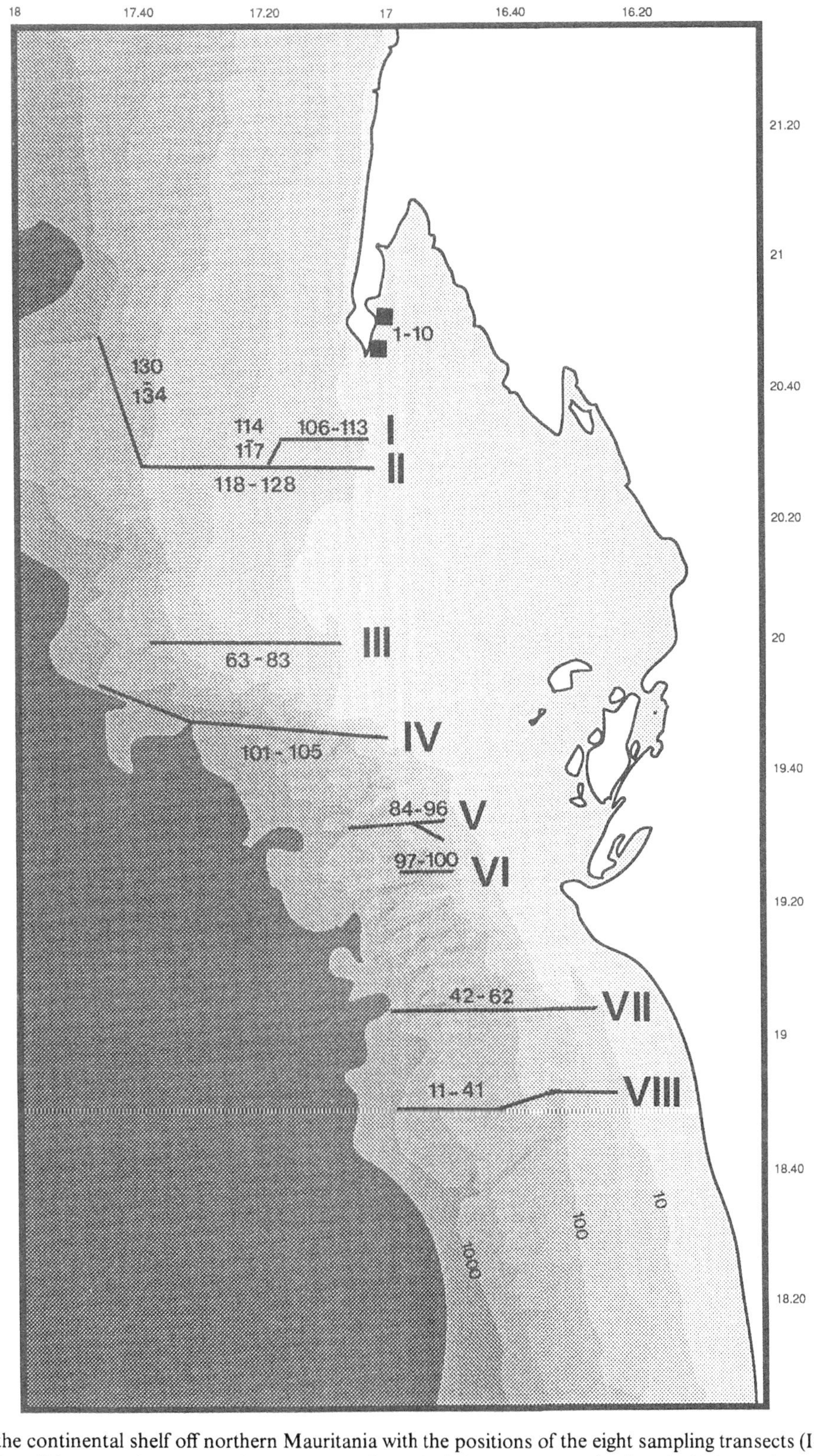


*Fig. 1.* Map of the continental shelf off northern Mauritania with the positions of the eight sampling transects (I–VIII) of the Tyro Mauritania-II Expedition; numbers along the transects refer to individual stations details of which are found in Van der Land (1988).

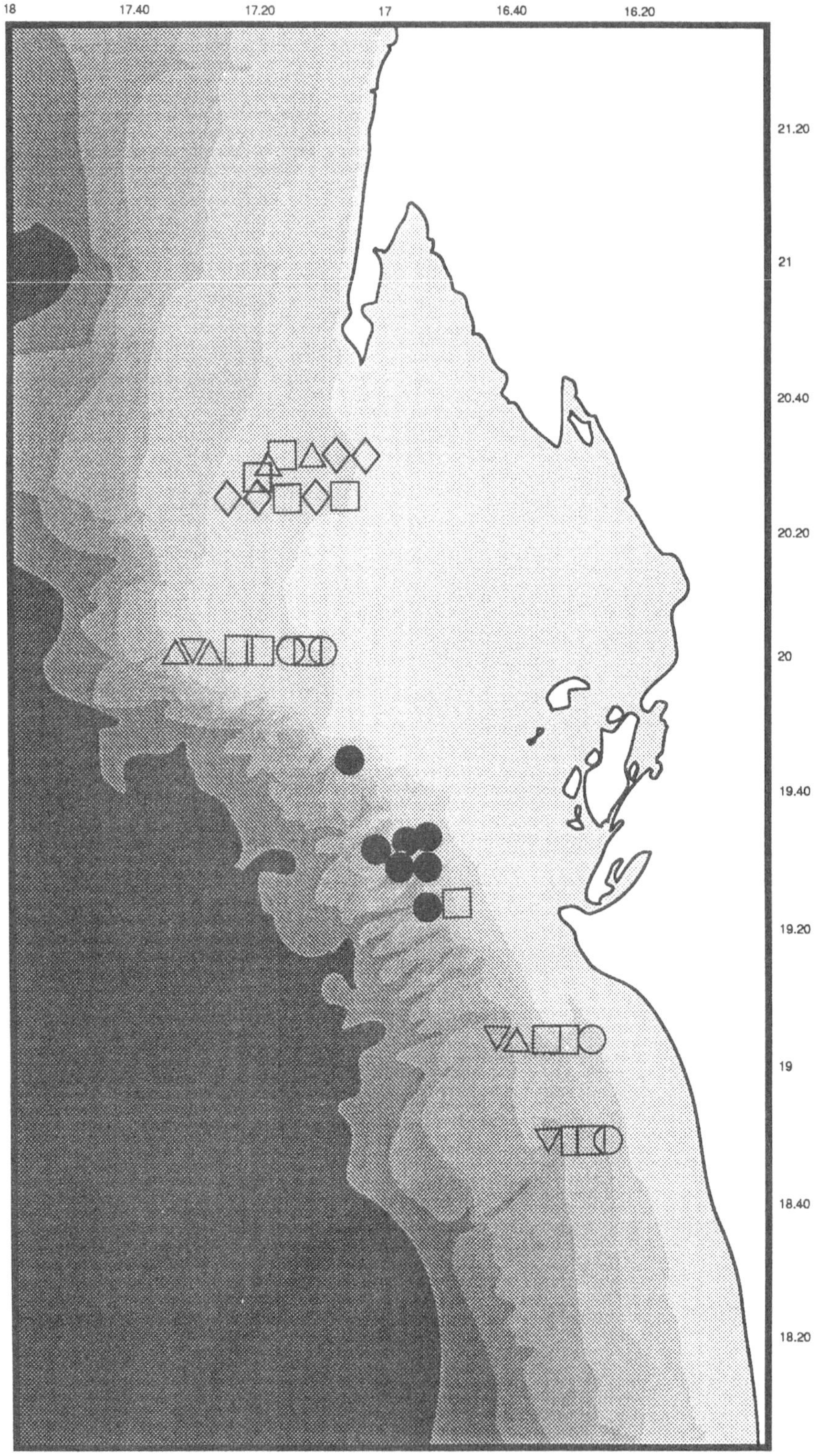


*Fig. 2.* Bottom conditions on the continental shelf off Mauritania as found in the eight sampling transects of the Tyro Mauritania-II Expedition. Symbols indicate subjective categories: black circles: mud, triangles: muddy sand, squares: sand, open circles: sandstone flats and ridges, diamonds: gravel.

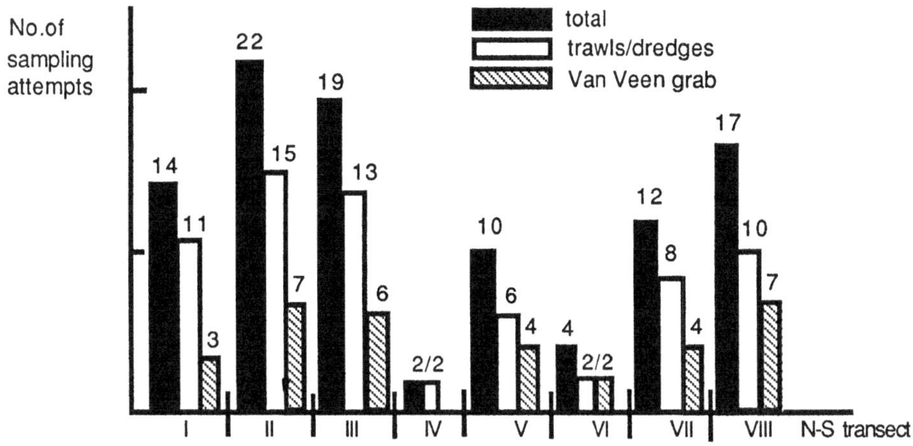


*Fig. 3.* Sampling efforts of the Tyro Mauritania-II Expedition over the eight transects; black bars: total number of samples, white bars: Agassiz-trawl and rectangular dredge samples, hatched bars: Van Veen grab samples (4 grab attempts each).

the expedition and provided the maps of Figs 1 and 2.

## Results

*Vertical distribution*

The dominant substrate of the continental shelf off Mauritania (and indeed of large areas off the Moroccan, Saharan and North Senegal coasts) is sand. This habitat presents such difficulties for sponges (almost no settlement surface for larvae, clogging of inhalant and exhalant orifices), that only specially adapted growth forms can succesfully thrive there. Depending of the thickness of the sediment layer, there are two ecological sponge groups, viz. stalked or long-papillated sponges, and those incrusting on shells of gastropods and hermit crabs ('mobile sponges'). The

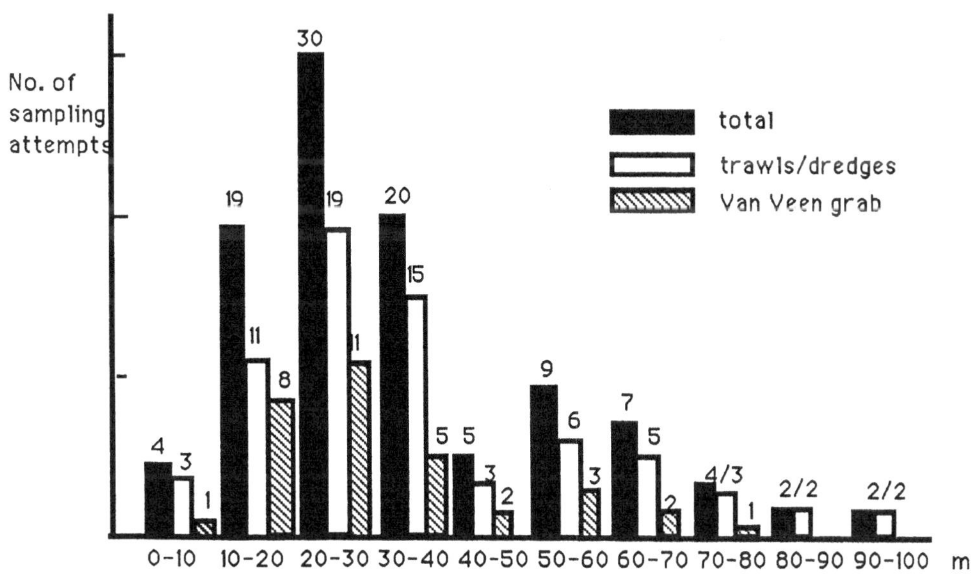


*Fig. 4.* Sampling efforts of the Tyro Mauritania-II Expedition over the depth profile; the attempts of the eight transects are here combined to number of attempts at 10 m depth intervals.

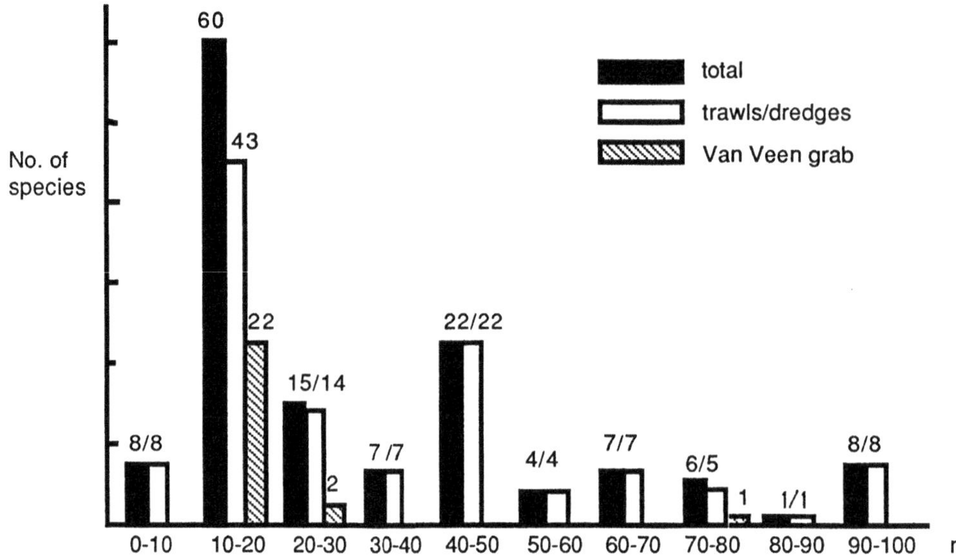


*Fig. 5.* Depth distribution of sponge diversity on the shelf off the Banc d'Arguin. The results of the eight transects are combined to species numbers found at 10 m depth intervals.

former group is only viable when the sand overlies a hard surface (bigger shells of molluscs and sandstone platforms and ridges).

Diversity of sponges over the 0–100 m depth profile (expressed as the total no. of species found in 10 m depth intervals over the entire area, see Fig. 5) shows a peak (for both grab and trawl/dredge samples) at 10–20 m. This is contrary to the sampling efforts which were highest in the 20–30 m depth interval (see Fig. 4). This inshore high diversity is caused by the high frequency and diversity of the stalked-papillated sponges at these depths, correlated with the occurrence of sandstone platforms and ridges close inshore. Fig. 5, in fact, presents a rather misleading picture of gradual decline of diversity towards greater depths. Six of the 100 stations contribute disproportionately towards the diversity counts (station 11: 13 species, 27: 9, 49: 28, 53: 11, 63: 15, 72: 22), and together account for almost half the number of collected samples (*i.e.* one or more specimens of a given species collected at a given station) and 81% of the recorded species (0–100 m). Of these six stations five were inshore and made on hard bottom covered with a thin layer of sand (the sixth station was made on a gravel bottom at 40–50 m). Growth forms of the various species were either stalked (examples: *Suberites* aff. *syringella* (Schmidt, 1862) (Plate 1c), *Adreus micraster* (Burton, 1956), and *Dictyonella plicata* (Schmidt, 1862) (Plate 1d) or massively sprawling with long papillae sticking out beyond the sediment (*Topsentia porrecta* (Topsent, 1928) (Plate 1a, b).

Most stations yielded a low number of species: average diversity was 3.5 sponge species per station; if the above mentioned six stations are omitted, average diversity of the remaining stations was only 2.1. Most of the stations other than the six mentioned above had bottoms consisting of a thick layer of sand or muddy sand, and the only viable sponge group in such habitats are the 'mo-

---

*Plate 1*

Figs. a-d. Stalked/papillated sponges found on inshore sand stone platforms off the coast near the Banc d'Arguin.

a. *In-situ* photo of *Topsentia porrecta* papillae sticking out from under the sediment at approximately 20 m depth off the coast near Cap Timiris.

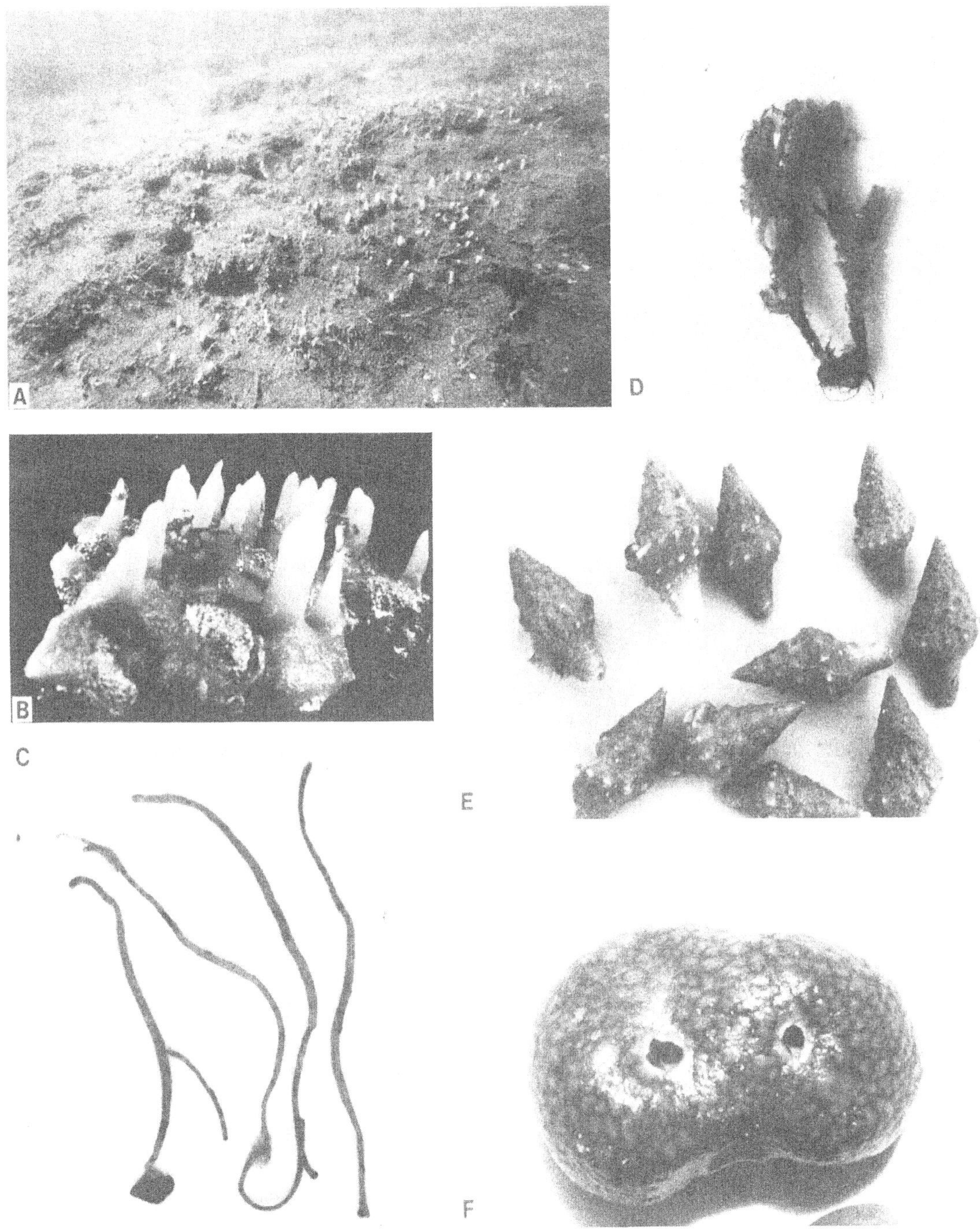


b. *Topsentia porrecta* (Topsent, 1928), specimen photographed upon collection.
c. *Suberites* aff. *syringella* (Schmidt, 1868), photographed upon collection.
d. *Dictyonella plicata* (Schmidt, 1868), photographed upon collection.
*Figs. e-f.* 'Mobile' sponges found occurring widely on the continental shelf off the Banc d'Arguin.
e. *Clathria (Microciona) atrasanguinea* (Bowerbank, 1866) incrusting on gastropod *Clavatula*, photographed upon collection.
f. *Suberites domuncula* (Olivi, 1791) photographed upon collection.

*Table 1.* Depth distribution and frequency of occurrence (% of samples containing specimens) of six species of 'mobile' sponges on the continental platform of the Banc d'Arguin, Mauritanian, between 0 and 100 m.

| Depth (m)<br>Species | 10–20 | 20–30 | 30–40 | 40–50 | 50–60 | 60–70 | 70–80 | 80–90 | 90–100 |
|---|---|---|---|---|---|---|---|---|---|
| *Suberites domuncula* | 27 | 44 | 88 | 100 | 66 | 20 | 33 | 0 | 33 |
| *Suberites ficus* | 0 | 0 | 0 | 0 | 34 | 40 | 0 | 50 | 100 |
| *Clathria atrasanguinea* | 45 | 22 | 30 | 33 | 80 | 20 | 0 | 0 | 0 |
| *Clathria* spec. (beige) | 0 | 0 | 8 | 0 | 0 | 20 | 0 | 0 | 0 |
| *Hymedesmia coriacea* | 10 | 0 | 15 | 66 | 17 | 60 | 0 | 0 | 100 |
| *Tricheurypon* spec. | 10 | 0 | 0 | 0 | 0 | 0 | 0 | 0 | 0 |
| Number of trawl attempts | 11 | 19 | 15 | 3 | 6 | 5 | 3 | 2 | 2 |

bile sponges'. These comprised the following six species (see also Table 1 for a summary of distributional data): *Suberites domuncula* (Olivi, 1871), (Plate 1f) almost exclusively the 'marmorated' form (see Topsent, 1900 for an extensive description of this species). These are by far the most numerous of the mobile sponges. They form big (up to 11 cm in greatest dimension) ellipsoid potato-shaped, mottled orange-brown-slate coloured, associations of sponges and hermit crabs. The gastropod shell (mostly either of *Turritella* or of *Natica*) originally colonized by the hermit crabs and subsequently incrusted by the sponge, is often completely 'dissolved' in the later growth stages, presumably by acidic agents secreted by the sponges. However, the gastropod cavity is still preserved by the residing hermit crab, which also keeps a wide passage open to the outside of the sponge. The sponges host at least three different hermit crab associates: *Dardanus* spec., *Pagurus* spec. 1 and spec. 2, and there seems to be a clear size partitioning (Fig. 6) with small sponges hosting small hermit crab species (*Pagurus* spec. 1 in sponges up to 3.3 cm in greatest dimension, *Pagu-*

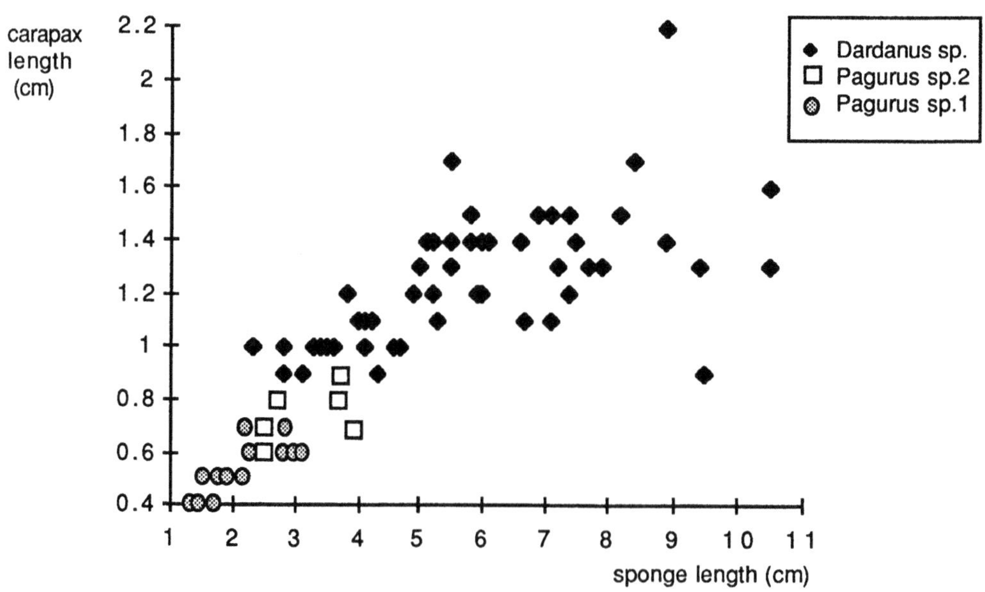


*Fig. 6.* Relationship between sponge length and carapax length of the *Suberites domuncula* (Olivi, 1791) – hermit crab association found abundantly off the Banc d'Arguin. Note that the hermit crabs belong to three different species.

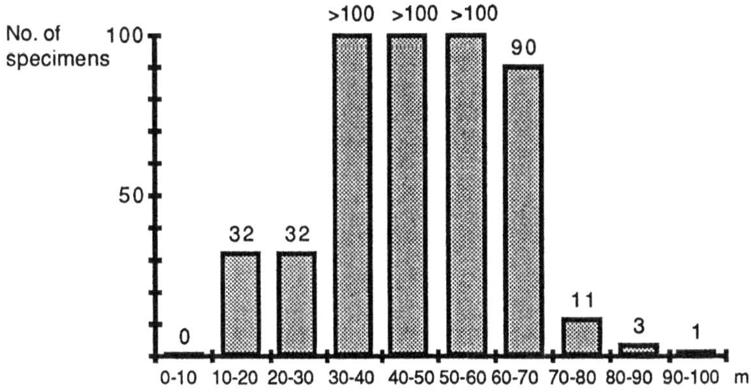


*Fig. 7.* Number of specimens of *Suberites domuncula* (Olivi, 1791) – hermit crab associations collected over the depth profile on the shelf off the Banc d'Arguin. The results of the eight transects are combined to numbers found at 10 m depth intervals. The numbers for the 30–60 m depth intervals were very much higher (up to 300 specimens or more in individual trawl samples) than the 100 limit.

*rus* spec. 2 in sponges up to 4.0 cm) and large sponges hosting large hermit crabs (*Dardanus* spec. in sponges of 4–11 cm). Depth distribution is from 18 to 95 m, with a clear optimal depth range between 30 and 70 m (Fig. 7). *S. domuncula* is a Mediterranean-Atlantic species occurring at least as far south as the Gulf of Guinea (Lévi, 1959).

At the most shallow stations (and upwards into the littoral region), this symbiotic association is replaced by a similar association of hermit crabs and large nodular calcareous bryozoa; the two associations were not collected together, so presumably there is a form of resource partitioning. *Suberites ficus* (Esper, 1794): dull light grey-brown coloured incrustations on gastropod shells (a.o. *Xenophora* and *Calliostoma*) with hermit crabs. The incrustations of this species seldom exceed 0.5 cm in thickness. Depth distribution 50–> 100 m (down to at least 200 m). This distribution may point to habitat partitioning between *S. domuncula* and *S. ficus*. The latter is known from the North Sea, Mediterranean, and the west coast of Africa as far south as Namibia (Uriz,1989).

*Clathria (Microciona) atrasanguinea* (Bowerbank, 1866): (Plate 1e) thin reddish or orangy velvety crusts on both live gastropods (often *Clavatula*) and empty shells with hermit crabs. Common between 10 and 40 m, found down to 60 m. It is a North Eastern Atlantic species.

*Clathria (Microciona)* spec.: thin velvety beige-coloured crusts on mostly live gastropods (*Clavatula*). Found at 30–40 m and 60–70 m.

*Tricheurypon* spec.: thin red, very hispid crust on *Favartia emersoni* resembling the cosmopolitan species *T. viridis* (Topsent, 1889), but red in colour instead of blue-green. Found only once at 10–20 m.

*Hymedesmia coriacea* (Fristedtt, 1887) is a smooth, slippery, dull greyish brown crust with a delicate faint blue venal pattern on gastropod shells (a.o. *Marginella*) with hermit crabs. The taxonomy of this sponge has recently been revised (Van Soest, 1987); it occurs on various substrates (shells, gravel, rocks) from Norway southwards, to reach its southern border in the studied area.

Occasionally sponges were found attached to legs and thorax of spider crabs, mostly at depths exceeding 70 m.

### *Horizontal distribution*

Although sampling efforts in the mid-transects (IV–VI, cf. Fig. 1) were considerably lower than those in other transects, both the number of

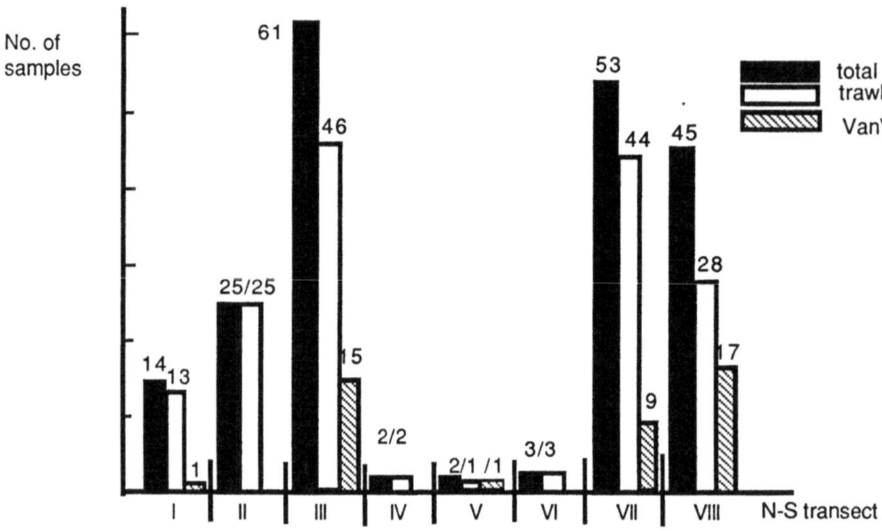


*Fig. 8.* Number of sponge samples (*i.e.* one or more specimens of a given species in a given sampling attempt) collected in the eight transects off the Banc d'Arguin. Black bars: total sample number, white bars: Agassiz- trawl and rectangular dredge samples, hatched bars: Van Veen grab samples.

sponge samples (7), and the number of species (4) caught in these transects were disproportionately low compared to those from the other transects (Figs 8 and 9). The only species found in low numbers (2–4 specimens) were *Cliona* spec. (a sponge boring in empty shells), the mobile *Clathria* spec., and the ubiquitous mobile sponges *Suberites domuncula* and *Clathria atrasanguinea.*

The overall horizontal distribution (Fig. 8) of collected sponge samples shows distinct northern (transects I–III) and southern (transects VII–VIII) maxima, with a gap in the area of transects IV–VI. If we compare Figs 1 and 2 with Fig. 8, it is obvious that the observed gap coincides with muddy bottoms, caused by the Banc d'Arguin run-off of sediment-rich waters.

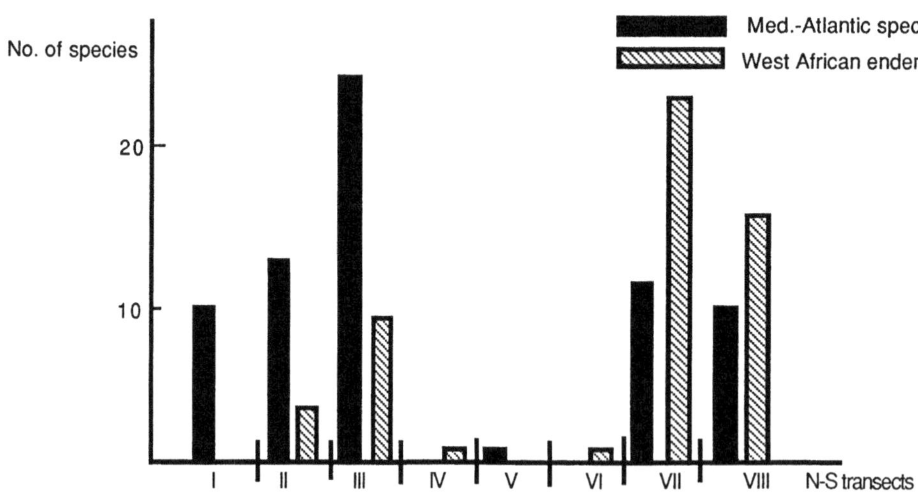


*Fig. 9.* Number of sponge species collected in the eight transects off the Banc d'Arguin subdivided in Mediterranean-Atlantic (black bars) and West African (hatched bars) species.

Although habitats and growth forms of stalked and papillated sponges in the northern and southern transects are similar (cf. above), species composition differs dramatically: only about 23% of the species (18 species) is shared between the two areas. These are for the most part (17%) species with a wide distribution to the north (Mediterranean-Atlantic), and relatively few (6%) with a southern distribution (West African endemics) (for a listing of West African sponges and their affinities cf. Van Soest, in press). If we tabulate the main distribution of the species found in the various transects (Fig. 9), we observe going southwards an expected increase of West African endemics and an expected decrease of Mediterranean-Atlantic species. However, this is not a proportional increase-decrease phenomenon: of the 53 species occurring in the southern transects 23 (43%) are known from the Mediterranean-Atlantic, whereas of the 49 species occurring in the northern transects only 10 (20%) are known only from the coasts of West Africa (most of these are undescribed). Combined these figures point to a low proportion of West African forms in the northern transects and a relatively high proportion of Mediterranean-Atlantic forms in the southern transects. Some care with the interpretation of the data is in order as about half the species were collected only once.

## Discussion

The observed horizontal and vertical distribution patterns as well as the high proportion of single records point to a patchy distribution of sponge species and populations in the area off the Banc d'Arguin, especially for the sedentary species. The only habitat available for these are the hard bottoms found mostly at shallow depths and these are patchily distributed and only available to species especially adapted to living in the sediment. The high diversity and regionality found in these forms is surprising: dozens of different species with similar morphology living closely sympatrically (e.g. 13 different species in 4 Van Veen grab samples at station 11 in the south, and 15 different species in 4 Van Veen grab samples at station 63 in the north, but only 2 shared species!). A likely explanation for the differences in species composition between these similar habitats in the north and the south seems to be the restricted exchange of species and larvae between these areas caused by the unfavourable muddy area of transects IV–VI. It is quite conceivable that this unfavourable habitat is a barrier for southern species attempting to extend their range northwards, because of lack of settlement substrate and adverse currents from the north. Only 10 species known exclusively from West Africa occur in the northern area of these only two species are known from previous records, viz. *Burtonanchora myxilloides* Lévi (1959) and *Adreus micraster* Burton (1956); the others are undescribed species and have consequently few distributional data). Of these 10 species 5 are shared between the northern and the southern transects, and thus presumably have (had) the ability to cross the mud area in northern direction. In contrast there are 23 species in the southern area which have a predominantly Mediterranean-Atlantic distribution, pointing to the mud area being a less efficient barrier for species extending their range southwards. In view of the prevailing Canary Current, which may transport larvae southwards, this conforms to the expectation. One might conclude that the mud area off the Banc d'Arguin functions as a semi-permeable barrier allowing a good percentage of Mediterranean-Atlantic species to extend their distribution, but largely preventing the reverse range extension of West African endemics.

Compared to the sponge-rich areas of the rocky coasts in the north (Mediterranean, Canary Islands and Madeira Archipelago) and to the south (Cape Verde Islands and Dakar area) the entire north West African continental platform is probably impoverished and constitutes a large scale barrier to all sponges other than those specialized for mobile or infaunal substrates. For the latter, the Banc d'Arguin run-off area is an important barrier

## References

Briggs, J. C., 1974. Marine zoogeography. McGraw-Hill, New York, 461 pp.

Boury-Esnault, N. & T. Lopes, 1985. Les Démosponges littorales de l'Archipel des Açores. Ann. Inst. océanogr. 61: 149–225.

Lévi, C., 1959. Campagne de la 'Calypso': Golfe de Guinée. Spongiaires. Ann. Inst. océanogr., 37: 115–141.

Topsent, E., 1900. Etude monographique des spongiaires de France. III. Monaxonida (Hadromerida). Arch. Zool. exp. gén., 8: 1–331.

Uriz, M.-J., 1989. Deep-water sponges from the continental shelf and slope off Namibia (Southwest Africa): Classes Hexactinellida and Demospongia. Monogr. Zool. Mar. 3: 9–157.

Van der Land, J., 1988. Shipboard report of the Tyro Mauritania-II Expedition 1988, 36 pp.

Van Soest, R. W. M., 1987. Biogeographic and taxonomic notes on some Eastern Atlantic sponges. In: W. C. Jones (ed.), European contributions to the taxonomy of sponges: 13–28 (Sherkin Island Marine Station, Ireland).

Van Soest, R. W. M., in press. Affinities of the marine Demosponge fauna of the Cape Verde Islands and tropical West Africa. In: J. C. den Hartog *et al.* (eds), Proc. 5th Symp. Biol. Cape Verde Islands. Courier Senckenberg Forschungs-Institut, 20 pp.

*Hydrobiologia* **258**: 107–117, 1993.
*W. J. Wolff, J. van der Land, P. H. Nienhuis & P. A. W. J. de Wilde (eds), Ecological Studies in the Coastal Waters of Mauritania.*




# The benthic infauna and benthic respiration off the Banc d'Arguin (Mauritania, Northwest Africa)

G. C. A. Duineveld, P. A. W. J. de Wilde, E. M. Berghuis & A. Kok
*Netherlands Institute for Sea Research, P.O. Box 59, 1790 AB Den Burg, Texel, The Netherlands*

*Key words:* macrofauna, meiofauna, benthic respiration, Mauritanian shelf

## Abstract

In May 1988, a study was made of the benthic community structure and benthic respiratory activity along two transects across the Mauritanian shelf off Banc d'Arguin. The main emphasis of the present paper is put on the macrobenthos, but results of a gross analysis of the meibenthos are discussed as well. Macrofaunal and meiofaunal density showed no consistent decrease with distance from the shore. The composition of the macrofauna taxa appeared to be closely correlated with sediment parameters and less so with depth. The highest macrofaunal biomass was found at the northern edge of the Banc d'Arguin, and the lowest biomass along the outer shelf. The biomass levels on the shelf match the ones reported for the northern Cap Blanc area where, in contrast to the seasonal upwelling in the investigated area, upwelling takes place all year round.

Benthic respiration rates on the offshore shelf were relatively high at the nearshore stations and low near the shelf break. Experimental evidence suggests that the low $O_2$ levels in the upwelling water covering part of shelf, inhibited benthic respiration. The high respiratory activity at some northern inshore stations coincided with the presence of oxygen-rich coastal water. In this area benthic respiration surpassed the level previously reported for the enriched Cap Blanc area. On the basis of our respiration data, an estimate is made of the total carbon demand of the benthic community on the shelf in May and this quantity is compared with the measurements of daily primary production.

## Introduction

The shelf along the coast of Mauritania (Northwest Africa; Fig. 1), can be divided into two parts. The most inshore part, bordering the Sahara, comprises a large area (> 12000 km$^2$) with shallows and tidal flats known as the Banc d'Arguin. The tidal flats of Banc d'Arguin owe their importance to the fact that they form the principal wintering habitat for boreal waders (Wolff & Smit, 1990). Approximately 85% of the tidal flats, as well as large parts of the subtidal areas on the Banc, are covered with seagrass. The shelf offshore of the Banc d'Arguin, with depths between 10 and 100 m, is broad in the north and becomes progressively narrower towards the south. Beyond the 100 m isobath, the seafloor abruptly drops to depths over 1000 m at the continental slope. The shelf edge, as shown by the irregular shape of the isobaths, is indented by steep submarine canyons. Because of the intense upwelling along the Mauritanian shelf, this area has been the subject of many studies. These have, for instance, shown that the duration of this phenomenon and the composition of the upwelling water differ with latitude (Minas *et al.*, 1982).

The Banc d'Arguin and the offshore shelf seem to constitute two different ecosystems. The ecosystem within the realm of the Banc is supposedly fueled by the detritus of seagrass, which apparently gives rise to a rather moderate macrofaunal biomass in comparison to boreal intertidal estuaries (Wolff & Smit, 1990). The offshore ecosystem, on the other hand, benefits from the enhanced primary production caused by the upwelling of nutrient rich water (Minas *et al.*, 1982) which sustains an intensive fishery along the shelf (Ansa-Emmin, 1982). The benthos is also positively affected by the upwelling, as shown by the high biomasses and activity rates near Cap Blanc where upwelling occurs throughout the year (Christensen & Packard, 1977; Nichols & Rowe, 1977).

The extend of a regular exchange of organic matter between the Banc and the upwelling region on the offshore shelf is unknown (Wolff & Smit, 1990). Several studies have established that Banc water is discharged onto the shelf and into the deep sea (e.g. Peters, 1976), but this event is subject to irregularities caused by tides and winds. Seagrass which is transported onto the shelf by the outflowing Banc water, could lead to a supplementary enrichment of the local benthos. It is conceivable that conversely the Banc could profit from the production in the upwelling region when this water reaches the shallower areas. Minas *et al.* (1982) argue that this could be effected by the cyclonic circulation pattern on the Banc as described by Mittelstaedt (1974, 1976).

During the Dutch 'Tyro' expedition in May 1988 on the offshore Mauritanian shelf, the composition and activity of the shelf benthos were studied along two transects perpendicular to the coast in order to detect gradients reflecting changes in food supply from the Banc to offshore waters. The analysis of the macrofaunal composition and biomass formed a part of this study. Because of their relatively long lifespan, macrofauna will more closely reflect the average food supply than short-living meiofauna and protozoans. In addition, benthic $O_2$-consumption was measured, representing the instantaneous carbon mineralization by the entire community. Since this activity is closely related to the presence of degradable organic material, it can be used to compare the actual food availability among stations as well as shed light on the origin and mechanisms of food supply in the Banc d'Arguin area.

## Material and methods

The position of the stations where infaunal samples were taken and respiration measurements were conducted, are shown in Fig. 1. The main abiotic characteristics of the bottom water and

*Table 1.* The main abiotic features of the bottom water and the sediment at the sampling stations on the Mauritanian shelf (see Fig. 1 for positions).

| | Southern transect | | | | Northern transect | | | |
|---|---|---|---|---|---|---|---|---|
| | Off ······> inshore | | | | Off ······> inshore | | | |
| Station no. | 2 | 1 | 4 | 3 | 5 | 7 | 8 | 6 |
| Depth in m | 103 | 62 | 42 | 22 | 113 | 50 | 33 | 19 |
| Bottom temp °C | 13.6 | 14.5 | 15.3 | 14.7 | 14.0 | 14.8 | 16.7 | 16.5 |
| Bottom % oxygen | 24 | 24 | 25 | 22 | 23 | 28 | 89 | 84 |
| Bottom salinity | 35.5 | 35.5 | 35.6 | 35.5 | 35.5 | 35.5 | 35.9 | 35.9 |
| Median grain $\mu$m | 298 | 40 | 40 | 35 | 220 | 160 | 125 | 1500 |
| % Silt <63 $\mu$m | 13 | 88 | 88 | 86 | 14 | 16 | 21 | <1 |
| % Carbon | 5.65 | 3.59 | - | 2.96 | 4.22 | 3.19 | 5.13 | 9.22 |
| % Nitrogen | 0.08 | 0.14 | - | 0.09 | 0.10 | 0.11 | 0.10 | 0.05 |
| C/N ratio | 76 | 25 | - | 33 | 45 | 39 | 50 | 203 |

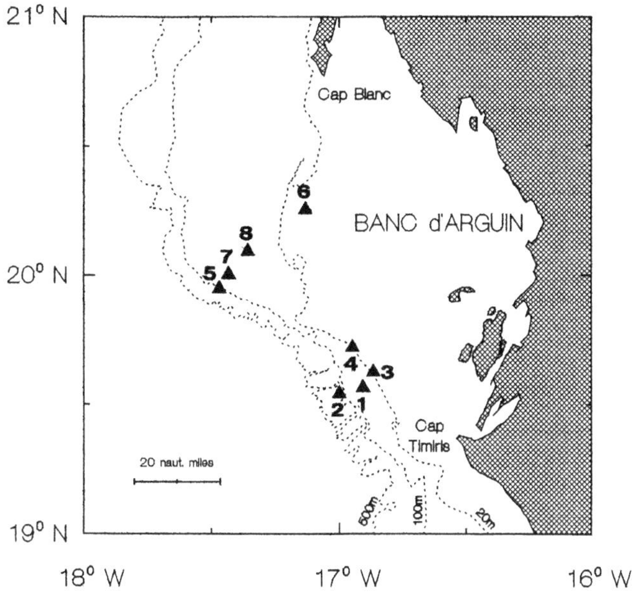


*Fig. 1.* Positions and codes of benthic stations sampled in May 1988.

the sediment at each of these stations have been summarized in Table 1. At each station in Fig. 1, at least two, but generally 5 samples were collected for macrofauna (≥ 1 mm) analysis with a Reineck's boxcorer (0.075 $m^2$). Samples for the analysis of sediment grainsize and meiofauna were extracted from additional boxcores. The macrofauna samples were washed on board over a 1 mm sieve and the residue was preserved in buffered formalin. In the laboratory, organisms were sorted under a stereo microscope and identified to the taxonomic family level. On the basis of the abundances of these taxa, stations were classified using TWINSPAN (Hill, 1979). The biomass of the macrofauna was determined by dehydration at 60 °C to a constant weight, and subsequent incineration at 550 °C for two hours. Two meiofauna samples were taken per station, with a corer 10 cm long and 2.5 cm in diameter. The samples were preserved in formalin for later elutration in the laboratory. A screen-size of 30 $\mu$m was used to retain the meiofauna. The animals in this residue were assigned to a gross taxonomical categories and counted.

Benthic respiration was measured in situ on the seabed and in incubated sediment cores on deck while using bottom water at in-situ temperatures. In-situ respiration was measured with stand-alone belljars. These belljars are basically Reineck's boxcorers where the coretube acts as measurement chamber after it has been inserted into the sediment and closed off by a valve equipped with a stirrer, micro-depth sounder and electrodes (YSI 5739). The coretube (0.075 $m^2$) is made of polyester in order to circumvent artefacts as described by Cramer (1989). Data from the electrodes were recorded every minute on an electronic memory which was attached to the frame of the corer. The frame also held the power supply and a sediment trap. At each station where in-situ measurements were conducted, two belljars were deployed, separated by a distance of less than one nautical mile. Each in-situ measurement lasted for at least six hours. Immediately after the retrieval of a belljar, the $O_2$-content of the overlying water was measured with the Winkler method. This information was used to calibrate the readings from the electrodes. Finally, the intact sediment sample from the corer was washed over a 1 mm sieve to retrieve the macrofauna for further analysis.

Two core sizes were used for the shipboard incubations, viz. 30 cm diameter polyester cores which could be mounted under the boxcorer and 10 cm diameter glass cores with which subsamples were taken from boxcore samples. One consequence of this, is that the methods differ in the degree to which the sediment is manipulated. The method using cores of 10 cm diameter has been described in detail by de Wilde *et al.* (1984). This particular method allowed us to determine the effect on benthic $O_2$-consumption of the undersaturated bottom water that covered a major part of the shelf (Table 1). For this purpose, cores which had first been incubated with in-situ bottom water, were re-incubated with fully saturated bottom water (air-bubbling). The 30 cm diameter cores, after being removed from the boxcorer, were sealed with a lid containing a stirrer and electrodes. Care was taken that the overlying water was not lost in this process. During the measurement, the cores were contained in a thermostatically controlled incubator on the ship's deck. Data from the electrodes were stored in an electronic memory. The macrofauna from all the incubation samples was analyzed.

## Results

*Macrofauna. Density, composition and biomass*

The density of the macrofauna (> 1 mm), after being separated into gross taxonomic categories, is shown in Table 2. Total density was on average higher on the northern transect, but no clear longitudinal gradient emerges in either two transects. Polychaetes invariably formed either a major or the dominating fraction of the macrofauna at all stations. Occasionally other phyla were found to be nearly as numerous or more abundant than polychaetes. At stations 1 and 4, which have a high silt content (> 80%), burrowing ophiuroids formed an important fraction of the macrofauna, whereas at stations 3 and 5, crustaceans appeared to be very numerous.

Table 3 shows the composition of the macrofauna in more detail, with the stations arranged into groups according to the TWINSPAN analysis (Hill, 1979). At the first division of TWINSPAN, the shallow station 6 with very coarse sediment (cf. Table 1), was split off. The composition of the macrofauna at station 6 was quite different from that found at the other stations. Characteristic were large numbers of interstitial polychaetes, oligochaetes and archiannelids in addition to a relatively large number of suspension-feeding epibenthic bivalves (mainly *Chama*). At the second division level, the southern stations having a high silt content, viz. 1, 3 and 4, were split from the remaining ones 2, 5, 7 and 8. The first group is characterized by a lower number of taxa. Most striking are the absence or low abundance of errant polychaetes belonging to the Eunicidae and Orbiniidae, of tubicolous polychaetes belonging to the Terebellidae and Ampharetidae, and of oligochaetes and epibenthic bivalves. Within this group of stations, stations 1 and 4 are more alike because of their high *Amphiura* density. Station 3, on the other hand, is characterized by large numbers of Tanaidacea (> 14000 ind. $m^{-2}$). The third station group (2, 5, 7 and 8) was neither homogeneous. Subdividing this group with TWINSPAN showed that the deepest, outer shelf stations (2, 5) could be separated from stations 7 and 8 on the basis of their higher density of tubicolous polychaetes belonging to the Chaetopteridae and the Oweniidae.

The total biomass and that of the principal categories are shown in Table 4. Similar to, but less explicit than with densities (Table 2), the total biomass of the northern transect is, on average, higher than the southern one. This is mainly caused by the high biomass at the shallow station 6, due to aggregates of the bivalve *Chama*.

*Table 2.* Number of individuals per 0.1 $m^2$ of the principal taxa at the shelf stations (see Fig. 1). Only specimens retained on a 1 mm sieve have been included.

| | Southern transect | | | | Northern transect | | | |
|---|---|---|---|---|---|---|---|---|
| | Off ·········> inshore | | | | Off ·········> inshore | | | |
| Station no. | 2 | 1 | 4 | 3 | 5 | 7 | 8 | 6 |
| Crustaceans | 125 | 88 | 4 | 1479 | 573 | 245 | 261 | 74 |
| Echinoderms | 1 | 110 | 150 | 0 | 2 | 50 | 3 | 2 |
| Molluscs | 17 | 10 | 72 | 26 | 60 | 25 | 48 | 96 |
| Polychaetes | 625 | 78 | 179 | 127 | 360 | 789 | 257 | 605 |
| Miscellaneous | 38 | 15 | 10 | 6 | 27 | 28 | 18 | 311 |
| Total | 806 | 301 | 415 | 1638 | 1022 | 1137 | 587 | 1088 |

*Table 3.* Abundances of the most common taxa among the macrofauna (> 1 mm) from the shelf. Polychaete families are printed in italics. The suffix 'epib.' stands for epibenthic species, 'inf.' for infaunal species, 'err.' for errant (free-living) species and 'sed.' for sedentary (tubicolous) species. Densities are represented by logarithmic abundance classes, viz. 1 = 1–9; 2 = 10–99; 3 = 100–999; 4 = 1000–9999; 5 = more than 10 000 ind $m^{-2}$. Stations within the two areas have been grouped according to the TWINSPAN classification. High densities characteristic for a station or a group of stations are in bold. Some abiotic parameters (Table 1) and the macrofaunal biomass (Table 4) have been summarized in the bottom section.

| TAXON | Station no. | 6 | 2 | 5 | 8 | 7 | 1 | 3 | 4 |
|---|---|---|---|---|---|---|---|---|---|
| Archiannelida | | **4** | – | – | – | – | – | – | – |
| *Pisionidae* | | **3** | – | – | – | – | – | – | – |
| *Serpulidae* | | **3** | – | – | – | – | – | – | – |
| *Amphinomidae* | | **3** | – | 1 | – | 1 | – | – | – |
| Bivalvia epib. | | **3** | 2 | 1 | 1 | 1 | – | – | – |
| Oligochaeta | | **4** | 3 | 3 | 1 | 3 | – | – | – |
| *Syllidae* | | **4** | 3 | 2 | 3 | 3 | – | 1 | 1 |
| *Eunicidae* err. | | **4** | 3 | 2 | 3 | 3 | 1 | 1 | 1 |
| *Glyceridae* | | 3 | 2 | 2 | 3 | 3 | 2 | 2 | 2 |
| Gastropoda epib. | | 3 | 1 | 3 | 3 | 2 | 1 | 3 | 3 |
| *Capitellidae* | | 2 | **3** | **2** | 3 | **3** | 1 | 2 | 1 |
| *Orbiniidae* | | 1 | **2** | **2** | **3** | **2** | – | – | – |
| *Ampharetidae* | | – | **3** | **1** | **3** | **3** | – | – | – |
| *Sabellidae* | | – | **2** | **1** | **3** | **3** | – | 1 | 1 |
| *Chaetopteridae* | | – | **4** | **3** | 1 | 2 | – | 1 | 2 |
| *Terebellidae* | | 2 | **3** | **2** | **2** | **3** | 1 | – | 1 |
| *Maldanidae* | | 2 | 3 | 3 | 2 | 4 | 3 | 3 | 3 |
| *Eunicidae* sed. | | 1 | 3 | 3 | 3 | 4 | 1 | 3 | 3 |
| Amphiura | | – | 1 | 1 | 1 | 1 | **4** | – | **4** |
| *Sigalionidae* | | – | – | – | – | 1 | **2** | **1** | **1** |
| *Flabelligeridae* | | – | 1 | 1 | 3 | 2 | – | 1 | 2 |
| *Oweniidae* | | – | 3 | 3 | 1 | 3 | – | 3 | 3 |
| Gammaridea | | 3 | 3 | 4 | 3 | 4 | 3 | 3 | 2 |
| Cumacea | | 1 | 2 | 3 | 2 | 3 | 3 | 2 | 1 |
| Isopoda | | 1 | 2 | 3 | 3 | 2 | 3 | 2 | 1 |
| Tanaidacea | | 2 | 2 | 2 | **4** | 2 | 2 | **5** | 2 |
| Bivalvia inf. | | 3 | 3 | 3 | 3 | 3 | 2 | 2 | 3 |
| *Paraonidae* | | 2 | 3 | 3 | 3 | 2 | – | 2 | 2 |
| *Spionidae* | | 2 | 3 | 2 | 3 | 3 | 2 | 3 | 2 |
| *Polynoidae* | | 3 | 2 | 1 | 1 | 1 | 3 | – | 2 |
| Median grainsize μm | | 1500 | 298 | 220 | 125 | 160 | 40 | 35 | 35 |
| Depth m | | 19 | 103 | 113 | 33 | 50 | 62 | 22 | 42 |
| Macrofauna biomass | | 42. | 9.4 | 6.2 | 11. | 21. | 7.5 | 7.9 | 29. |

The lowest biomass along the northern transect was found at the outer shelf station, where the biomass of polychaetes fell short as compared to the other stations. The biomass pattern on the southern is more erratic, with elevated levels at station 4 due to *Amphiura*. The deepest, outer shelf station along this transect did not yield the lowest biomass.

## *Meiofauna*

The density of the meiofaunal taxa are shown in Table 5. Along the northern transect no distinct gradient in total numbers is found. At the coarse sand station 6, nematodes are relatively scarce but copepods are more abundant, probably because of the interstitial space present in the sedi-

*Table 4*. Biomass of macrofauna (> 1 mm) at the shelf stations (see Fig. 1) expressed in g AFDW $m^{-2}$.

| | Southern transect | | | | Northern transect | | | |
|---|---|---|---|---|---|---|---|---|
| | Off ·····> inshore | | | | Off ·····> inshore | | | |
| Station no. | 2 | 1 | 4 | 3 | 5 | 7 | 8 | 6 |
| Crustaceans | 0.3 | 0.2 | 1.0 | 3.0 | 1.2 | 4.5 | 1.5 | 0.2 |
| Echinoderms | 0.1 | 5.9 | 16.8 | 0.0 | 0.1 | 1.8 | 0.4 | 0.0 |
| Moluscs | 0.5 | 0.1 | 3.7 | 0.7 | 0.6 | 0.6 | 1.0 | 31.5 |
| Polychaetes | 7.8 | 1.2 | 5.3 | 3.8 | 3.8 | 13.0 | 6.4 | 5.7 |
| Miscellaneous | 0.7 | 0.1 | 2.3 | 0.3 | 0.5 | 1.1 | 1.6 | 4.7 |
| Total | 9.4 | 7.5 | 29.1 | 7.9 | 6.2 | 21.0 | 10.9 | 42.1 |

ment (cf. Heip *et al.*, 1983). Among the remaining stations at the northern transect, which have sediments that are more or less alike (Table 1), no conspicuous differences in nematode and copepod numbers were found. Only at station 8 do foraminiferan numbers appear to be markedly lower. No explanation can be given for this apparent aberration.

Along the southern transect, nematodes being the dominant group and hence also total meiofauna, show depressed numbers at the shelf edge (station 2). This, however, does not seem to be part of a gradual decline in meiofauna density across the shelf as shown by the data from the other southern stations. The data in Table 5, furthermore, show that there is no consistent difference in meiofauna density between the northern and southern transect.

*Benthic community respiration*

The results of the benthic respiration measurements are listed in Table 6. The data from the paired in-situ belljars are fairly close. Along the northern transect they show a decreasing activity in an offshore direction with a slight increase again at the shelf edge. A similar offshore decrease was observed along the southern transect, in this case with the lowest activity at the shelf edge. The respiration estimates from the incubation of 30 cm diameter cores likewise display decreased uptake rates at the two outermost stations on the northern transect. On the southern transect this particular method was only applied at two of the four stations. It is, therefore, not possible to confirm the pattern found with the in-situ method for this transect. The incubation experiments with the

*Table 5*. Density of principal meiofaunal groups at the shelf stations expressed in millions of individuals per $m^2$.

| | Southern transect | | | | Northern transect | | | |
|---|---|---|---|---|---|---|---|---|
| | Off ·····> inshore | | | | Off ·····> inshore | | | |
| Station no. | 2 | 1 | 4 | 3 | 5 | 7 | 8 | 6 |
| Nematodes | 1.5 | 3.4 | 3.0 | 5.5 | 3.1 | 2.5 | 3.2 | 1.3 |
| Copepods | 0.2 | 0.2 | 0.7 | 0.2 | 0.3 | 0.5 | 0.2 | 1.0 |
| Foraminiferans | 1.1 | 1.7 | 2.5 | 1.1 | 4.6 | 6.3 | 0.3 | 4.0 |
| Total | 3.0 | 5.6 | 6.5 | 7.6 | 8.3 | 9.5 | 3.6 | 6.9 |

*Table 6.* Benthic community respiration ($\mu$mol $O_2$ $m^{-2}$ $h^{-1}$) measured respectively in-situ, and in incubated cores of 30 cm diameter. Values preceeded by ± are cases where, due to technical failures, the consumption is estimated by subtracting the final $O_2$ concentration in the vessel from the initaial one. The macrofaunal biomass (g AFDW $m^{-2}$) of the cores are shown in brackets.

| | Northern transect | | | |
|---|---|---|---|---|
| Station no. | 5 | 7 | 8 | 6 |
| In-situ | 600 (8.) | 390 (35.) | 1285 (17.) | 2225 (44.) |
| | ± 600 (4.) | 221 ( ) | ± 725 (10.) | 2350 (131.) |
| **mean** | **600** | **306** | **1005** | **2288** |
| Incubation 30 cm | 842 (10.) | 1064 (28.) | 1357 (8.) | 1284 (52.) |
| | 488 (6.) | 705 (19.) | 1159 (13.) | 821 (38.) |
| **mean** | **665** | **885** | **1258** | **1053** |
| Incubation 10 cm | 350 (10.) | 775 (22.) | 960 (4.) | 1410 (63.) |
| | 590 (30.) | 775 (32.) | 910 (11.) | – |
| | 850 (17.) | 325 (10.) | 1315 (12.) | – |
| | 290 (7.) | 425 (10.) | – | – |
| **mean** | **520** | **575** | **1062** | **1410** |
| | Southern transect | | | |
| Station no. | 2 | 1 | 4 | 3 |
| In-situ | 125 (9.) | 650 (8.) | – | 1210 (6.) |
| | 372 (–) | 518 (8.) | – | 1325 (11.) |
| **mean** | **249** | **584** | – | **1268** |
| Incubation 30 cm | – | – | 512 (32.) | – |
| | 288 (8.) | – | 843 (34.) | – |
| **mean** | **288** | – | **678** | – |
| Incubation 10 cm | 505 (8.) | 945 (49.) | 595 (0.3) | 410 (27.) |
| | 785 (10.) | 760 (18.) | – | 730 (6.) |
| | 575 (13.) | 745 (25.) | – | 650 (18.) |
| | 725 (10.) | 1350 (133.) | – | 835 (16.) |
| **mean** | **648** | **950** | **595** | **656** |

10 cm diameter cores again revealed a depressed activity at the outer shelf stations 7 and 5 on the northern transect. In contrast to the in-situ method, no inshore-offshore gradient in respiratory activity was found on the southern transect. This was mainly due to apparently increased rates at station 2 and depressed rates at station 3.

A second series of measurements using 10 cm cores was conducted to see what effect the under-saturated bottom water on the shelf had upon community respiration. Table 7 shows the respiration rates in cores that were first incubated with under-saturated water and a second time with fully saturated bottom water. In most cases, the $O_2$-consumption doubled in response to saturating the water in the circuit, except for station 2 where only a slight increase was found.

*Table 7.* Respiration ($\mu$mol $O_2$ $m^{-2}$ $h^{-1}$) measured in shipboard incubated 10 cm cores with in-situ and fully saturated bottom water.

| Station | $O_2$-saturation | Respiration | | | |
|---|---|---|---|---|---|
| 2 | 40% | 505 | 785 | 575 | 725 |
| | 100% | 700 | 1000 | 670 | 750 |
| 3 | 25% | 410 | 730 | 650 | 835 |
| | 100% | 830 | 1520 | 1270 | 1620 |
| 5 | 23% | 350 | 590 | 850 | 290 |
| | 100% | 870 | 1580 | 1115 | 990 |
| 7 | 28% | 775 | 775 | 325 | 425 |
| | 100% | 1525 | 1470 | 910 | 965 |

## Discussion

The preceeding sections indicate that no straightforward longitudinal gradient in density and biomass of macrofauna can be distinguished long the transects. On the basis of the macrofaunal composition, several station groups could be discriminated, but this classification seems to be more a reflection of the sediment type rather than of latitude, depth or macrofaunal biomass and density (Table 3). The levels of the biomass, however, become significant when compared to those in adjacent areas. For the shelf south of the present study area, Domain (1982) reports a biomass range of 1–4 g Dry Weight $m^{-2}$, with an extreme value of about 18 g on very coarse sand. These values hold for the summer; in winter the biomass diminishes by a factor of 2.5. Biomass data for the shelf (< 100 m depth) off Cap Blanc are derived from Nichols & Rowe (1977), who found spring values varying between 2 and 50 g Wet Weight $m^{-2}$ with one extreme of 94 g. The authors furthermore explicitly mention the absence of any decrease in biomass with increasing distance from the shore. Converting these figures into ash-free dry weight by taking 15% of the wet weight and 75% of the dry weight values (Rumohr *et al.*, 1987), shows that the biomass levels off the Banc d'Arguin are much higher than the ones reported for the southern area. This is likely due to the lower primary production in the southern region in relation to less frequent upwelling (Minas *et al.*, 1982). More surprising is that the Cap Blanc shelf, with its permanent upwelling, has a lower average biomass as well, viz. 4 g AFDW $m^{-2}$.

The fact that the meiofauna densities we found surpass the value that Thiel (1982) found on the Cap Blanc shelf is further support for an enhanced benthic standing stock of the Banc d'Arguin. Moreover, Thiel (1982) reports coinciding minima of meiofauna density and sediment phytopigment concentrations at the shelf edge near Cap Blanc. The author attributes these phenomena to adverse conditions for settlement of organic material on the steep upper slope and shelf edge, thus leading to an impoverished food supply for the fauna. On our southern transect, the lowest pigment concentration (Berghuis *et al.*, 1993) and lowest meiofauna density were likewise found at shelfedge station 2. This station lies on the narrow part of the shelf where on several earlier occasions (e.g. Peters, 1976) a downflow of 'heavy' Banc d'Arguin water has been observed. Such a cascade of Banc water did not occur during our period of observations, but there may well be a relation between earlier events and the presently low levels of meiofauna and phytopigments. In this respect it is important to note that at the outermost station (5) on the broadest part of the shelf, the meiofauna density and pigment content of the sediment were respectively of the same level and much higher than at the other stations along the northern transect.

The elevated biomass off the Banc d'Arguin points to an enrichment of the benthic system in comparison to the Cap Blanc area. Albeit single observations, the primary production data from the 'Tyro'-cruise support this assumption for the northern transect. Daily primary production along the northern transect (4–9 g C $m^{-2}$) clearly exceeds the spring levels near Cap Blanc (1–3 g C $m^{-2}$; Huntsman & Barber , 1977), which in turn lie in the range found on the southern transect (2–4 g C $m^{-2}$; Berghuis *et al.*, 1993). An enrichment of the Banc d'Arguin shelf relative to Cap Blanc, should also be reflected in increased benthic respiration rates. Unfortunately, the only comparable measurement available for Cap Blanc is the one by Smith *et al.* (1976) who found an in-situ $O_2$-uptake of about 800 $\mu$mol $O_2$ $m^{-2}$ $h^{-1}$ at a 30 m deep station in spring. Concurrent ETS measurements, showed that this particular station had a far higher sediment ETS activity than adjacent shallow (< 150 m) stations (Christensen & Packard, 1977). Taking into account the positive relation between benthic ETS-activity and $O_2$-uptake that Pfannkuche *et al.* (1983) found off Morocco, we can assume that $O_2$-uptake rates at the other Cap Blanc stations were lower than the 800 $\mu$mol $O_2$ $m^{-2}$ $h^{-1}$ found by Smith *et al.* (1976). Part of the respiration rates that we measured off the Banc d'Arguin may, on the other hand, be underestimates for the potential rates because of the undersaturated bottom water. In any case, our actual respiration data only substantiate the hypothesis of enrichment insofar as the shallowest stations (3, 6 and 8) on both transects are concerned.

Although all of our respiration measurements reveal a relative increase in activity at the aforementioned inshore stations, there are in many cases considerable variation among the estimates by the different methods. Unlike Cramer (1990), who found no significant differences between simultaneous in-situ measurements and shipboard incubations with 30 cm cores, we did find such discrepancies at station 6 and 7 (Table 6). We can explain this difference only for stations 6. The extremely coarse sand at this site prevented the formation of a good seal between the spade and the coretube, thus allowing partial drainage of the water overlying the core. This may have resulted in loss of the reactive layer on top of the sediment and, consequently, a lower $O_2$-consumption. Large deviations between results from incubating 10 cm cores and in-situ measurements are found at stations 2 and 3, for which we have no explanation. The differences with the in-situ data are in these cases not of the same sign and bear no relation with macrofaunal biomass. In general, there was a larger variation in biomass among the replicate 10 cm cores than among the 30 cm or the in-situ cores (Table 6) caused by the incidental inclusion or exclusion of larger organisms. This additional source of variation makes the method less suitable for estimating community respiration than the other two methods. Cramer (1990) also reported significant differences between respiration estimates obtained with 10 cm cores and the in-situ method.

In spite of a high macrobenthic biomass coinciding with an elevated respiratory rate at station 6, it is evident that there is no clear overall correlation between these two variables. This is true regardless of the techniques that we used (Table 6). It should, however, be pointed out that macrofauna is responsible for a variable, and usually moderate, share of the community metabolism (Banse *et al.*, 1971; Cramer, 1991). Moreover, individual respiratory activity is not linearly related to individual biomass (Banse, 1982; Peters, 1983) Therefore, an increase of biomass from one station to another, will not likely yield a proportional increase in community respiration, unless the animals are of the same size. In an attempt to assess the contribution of the shelf macrofauna to community respiration, we calculated the respiratory activity of the macrobenthos from the in-situ and 30 cm cores on the basis of the allometric relation given by Banse (1982). A correction for in-situ temperatures was made by means of a $Q_{10}$ of 2 (Schwinghamer *et al.*, 1986). Figure 2 shows that the relation between our macrofaunal respiration estimates and commu-

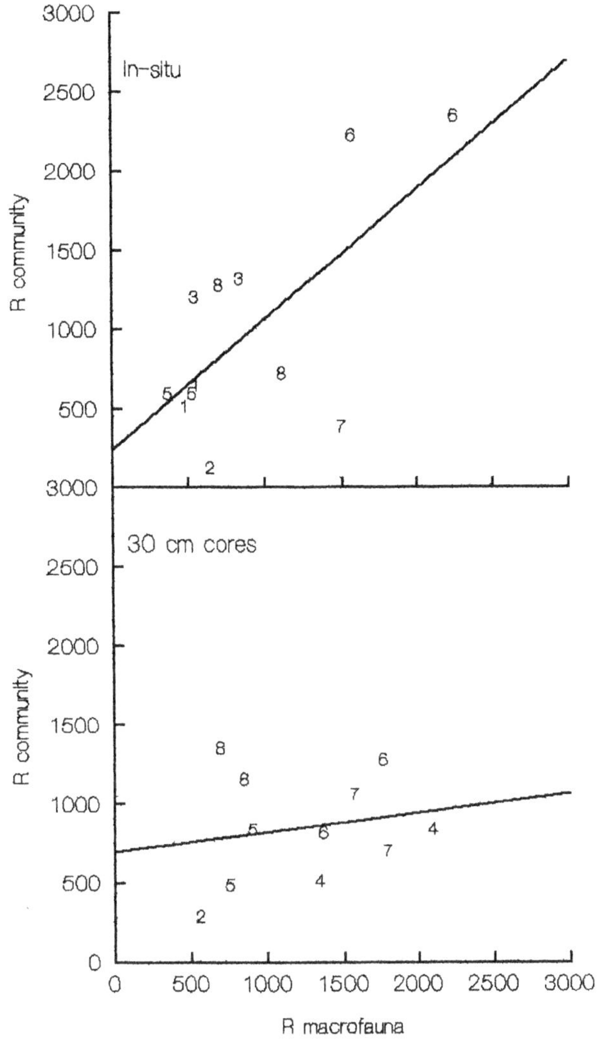


*Fig. 2.* Relation between community respiration and estimated macrofauna respiration (both in $\mu mol\ O_2\ m^{-2}\ h^{-1}$) for the data from the in-situ cores (top) and the 30 cm incubation cores (bottom). Macrofaunal respiration was calculated on basis of the allometric relation between individual biomass and respiration given by Banse (1982). Numbers in graphs refer to stations. Lines represent least-squares linear regressions. Only the line in the upper graph is significant ($P \leq 0.05$).

*Table 8.* Estimated respiratory activity of the shelf macrofauna expressed in $\mu mol\ O_2\ m^{-2}\ h^{-1}$. This quantity was calculated on the basis of the relation between individual weight and respiration given by Banse (1982).

| | Southern transect | | | | Northern transect | | | |
|---|---|---|---|---|---|---|---|---|
| | Off ··········> inshore | | | | Off ··········> inshore | | | |
| Station no. | 2 | 1 | 4 | 3 | 5 | 7 | 8 | 6 |
| | 977 | 709 | 1910 | 965 | 768 | 1666 | 993 | 1963 |

nity respiration is weak. Large deviations between in-situ measurements and macrofauna activity are found at station 2 and 7. The data from the 30 cm cores produced an even more scattered plot with most of the values for macrofauna respiration exceeding the actual respiration. The failure to establish any clear relation between community and macrofaunal respiration may be partly due to our admittedly rough calculations which did not allow a change in parameter values for the different taxonomic groups. Another complication is the apparent inhibition of respiration by the under-saturated bottom water that covered the majority of the stations (see Table 1 & 7). Although provisional, the estimates for macrofaunal respiration can be used as a comparative measure of the carbon that is required to sustain the macrofauna at the different stations. The average values for the stations in Table 8 show that the respiratory demand of the macrofauna on the southern transect is in the same range as that of the northern transect. This is in contrast with the pattern of primary production, which on the northern transect was twice as high as on the southern one (Berghuis *et al.*, 1993). This would suggest that the pattern we found in May does not hold for the whole year, and large fluctuations in primary production over the shelf probably occur.

The question as to whether the benthic community requires an import of allochthonous organic material or whether the shelf can export material, for instance to the Banc d'Arguin, can only be answered with supplementary knowledge of the pelagic carbon demand. However, the margin between primary production and benthic carbon demand does shed some light on the likelihood of export and import for the spring situation. Converting the in-situ respiration data into carbon equivalents yields a daily respiratory consumption by the benthic community of between 60 and 550 mg C $m^{-2}$, with an average of 210. The carbon required for production by the benthic community can only be approximated in a very rough way using published P/B ratios for boreal macrofaunal taxa (Brey, 1990), and a P/B ratio of 9 for the meiofauna (Gerlach, 1971). This leads to a daily production for the metazoan fauna of 35 to 85 mg C $m^{-2}$ when we do not take any seasonal variation of production into account. Assuming that the bacterial production is entirely recycled within the sediment community, this quantity will already be included in the total respiration. Van Duyl *et al.* (1991) have described the recycling of bacterial production in subtidal North Sea communities where daily production may rise to a value of 600 mg C $m^{-2}$. This quantity was measured in an enriched silty sediment with comparable abiotic features as the Mauritanian shelf, i.e. a daily primary production over 2 g C $m^{-2}$, a temperature of 16 °C and a macrofauna biomass of > 25 g AFDW $m^{-2}$. Support for a similar rapid recycling of bacterial production on the Mauritanian shelf is derived from Bak & Nieuwland (1993) who found that densities of heterotrophic flagellate densities on the Mauritanian shelf are in the same range as those in North Sea sediments. However, even if we assume that all bacterial production accumulates and we adopt a daily bacterial production of 600 mg C $m^{-2}$, total daily carbon demand would range from only 0.7 at station 2 to 1.2 g C $m^{-2}$ at station 6 (mean 0.9). Compared with a daily primary production between 2 and 9 g C $m^{-2}$ (mean 4.6), the benthic system on the Mauritanian shelf seems to consume only a moderate share of the primary production, leaving a substantial part for the pelagic system or for export.

**Acknowledgements**

We wish to thank M. S. S. Lavaleye, R. Witbaard, the technicians and the crew on board of the 'TYRO' for their assistance with the collection of the samples. We acknowledge the S.O.Z. for the financial support of the expedition.

**References**

Ansa-Emmim, M., 1982. Fisheries in the CINECA region. Rapp. P.-v. Réun. Cons. int. Explor. Mer 180: 405–422.

Banse, K., 1982. Mass-scaled rates of respiration and intrinsic growth in very small invertebrates. Mar. Ecol. Prog. Ser. 9: 281–297.

Banse, K., F. H. Nichols & D. R. May, 1971. Oxygen consumption at the sea bed. III – On the role of the macrofauna at three stations. Vie et Milieu 22: 31–52.

Bak, R. P. M. & G. Nieuwland, 1993. Patterns in pelagic and benthic nanoflagellate densities in the coastal upwelling system along the Banc d'Arguin, Mauritania. Hydrobiologia 258: 119–131.

Berghuis, E. M., G. C. A. Duineveld & J. Hegeman, 1993. Primary production and distribution of phytopigments in the water column and sediments on the upwelling shelf off the Mauritanian coast (northwest Africa). Hydrobiologia 258: 81–93.

Brey, T., 1990. Estimating productivity of macrobenthic invertebrates from biomass and mean individual weight. Meeresforschung 32: 329–343.

Christensen, J. P. & T. T. Packard, 1977. Sediment metabolism from the northwest African upwelling system. Deep Sea Res. 24: 331–343.

Cramer, A., 1989. A common artefact in estimates of benthic respiration by the use of stainless steel. Neth J. Sea Res 23: 1–6.

Cramer, A., 1990. Seasonal variation in benthic metabolic activity in a frontal system in the North Sea. In M. Barnes & R. N. Gibson (eds), Trophic Relationships in the marine environment. Proc. 24$^{th}$ Europ. Mar. Biol. Symp. Aberdeen University Press: 54–76.

Cramer, J., 1991. Benthic metabolic activity at frontal systems in the North Sea. Thesis Univ. Amsterdam: 93 p.

De Wilde, P. A. W. J., E. M. Berghuis & A. Kok, 1984. Structure and energy demand of the benthic community of the Oyster Ground, central North Sea. Neth. J. Sea Res. 18: 143–159.

Domain, F., 1982. Répartition de la biomasse globale du benthos sur le plateau continental ouest-africain de 17° à 12° N: densités comparées liées au différents types du fond. Rapp. P.-v. Réun. Cons. int. Explor. Mer 180: 335–336.

Gerlach, S. A., 1971. On the importance of marine meiofauna for benthos communities. Oecologia (Berlin) 6: 176–190.

Heip, C., R. Herman & M. Vincx, 1983. Subtidal meiofauna of the North Sea: a review. Biol. Jb Dodonaea 51: 116–170.

Hill, M. O., 1979. TWINSPAN – a Fortran program for arranging multivariate data in an ordered two-way table by classification of the individuals and their attributes. Ithaca, Cornell University, New York, 90 pp.

Huntsman, S. A., & R. T. Barber, 1977. Primary production off Northwest Africa: the relationship to wind and nutrient conditions. Deep Sea Res. 24: 25–33.

Minas, H. J., L. A. Codispoti & R. C. Dugdale, 1982. Nutrients and primary production in the upwelling region off Northwest Africa. Rapp. P.-v. Réun. Cons. int. Explor. Mer 180: 148–183.

Mittelstaedt, E., 1974. Some aspects of the circulation in the north-west African upwelling area off Cap Blanc. Téthys 6: 89–92.

Mittelstaedt, E., 1976. On the currents along the Northwest African Coast south of 22° North. Dt. hydrogr. Z. 29: 97–117.

Nichols, J. & G. T. Rowe, 1977. Infaunal macrobenthos off Cap Blanc, Spanish Sahara. J. mar. Res. 35: 525–536.

Peters, H., 1976. The spreading of the water masses of the Banc d'Arguin in the upwelling area off the northern Mauritanian coast. 'Meteor' Forsch.-Ergebn. A 18: 78–100.

Peters, R. H., 1983. The ecological implications of body size. Cambridge University Press, Cambridge, 330 pp.

Pfannkuche, O., R. Theeg & H. Thiel, 1983. Benthos activity, abundance and biomass under an area of low upwelling off Morocco, Northwest Africa. 'Meteor' Forsch.-Ergebnisse D 36: 85–96.

Rumohr, H., T. Brey & S. Ankar, 1987. A compilation of biometric conversion factors for benthic invertebrates of the Baltic Sea. The Baltic Marine Biologists Publication 9: 1–56.

Schwinghamer, P., B. Hargrave, D. Peer & C. M. Hawkins, 1986. Partitioning of production and respiration among size groups of organisms in an intertidal benthic community. Mar. Ecol. Prog. Ser. 31: 131–142.

Smith, K. L., Jr., C. H. Clifford, A. H. Eliason, G. Walden, G. T. Rowe & J. M. Teal., 1976. A free vehicle for measuring benthic community metabolism. Limnol. Oceanogr. 21: 164–170.

Thiel, H., 1982. Zoobenthos of the CINECA area and other upwelling regions. Rapp. P.-v. Réun. Cons. int. Explor. Mer 180: 323–334.

Van Duyl, F., R. P. M. Bak, A. J. Kop, G. Nieuwland, E. M. Berghuis & A. Kok, 1991. Mesocosm experiments: mimicking seasonal developments of microbial variables in North Sea sediments. In F. C. van Duyl (ed.), The applicability of mesocosms in North Sea eutrophication studies. NIOZ – Rapport 1991 – 5: 50–71.

Wolff, W. J. & C. J. Smit, 1990. The Banc d'Arguin, Mauritania, as an environment for coastal birds. Ardea 78: 17–38.

*Hydrobiologia* **258**: 119–131, 1993.
*W. J. Wolff, J. van der Land, P. H. Nienhuis & P. A. W. J. de Wilde (eds), Ecological Studies in the Coastal Waters of Mauritania.*




# Patterns in pelagic and benthic nanoflagellate densities in the coastal upwelling system along the Banc d'Arguin, Mauritania

R. P. M. Bak & G. Nieuwland
*Netherlands Institute for Sea Research (NIOZ), P.O. Box 59, 1790 AB Den Burg, Texel, The Netherlands*

*Key words:* protozoa, bacteria, nanobenthos, nanoplankton, upwelling, Mauritania

## Abstract

We studied spatial variation in abundance of marine benthic and pelagic heterotrophic nanoflagellates in relation to abundances of autotrophic flagellates, bacteria and cyanobacteria in an upwelling area off the Banc d'Arguin, Mauritania.

There was enormous spatial variation in densities. In the sediments these ranged from $8–219 \times 10^3$ cm$^{-3}$ for heterotrophic flagellates. Maximum values are in the range of those for temperate shallow marine bottoms. Low densities ($< 20 \times 10^3$) were confined to the deep stations (> 1000 m). Over the shelf (10–100 m depth) densities were high but related to grain size rather than to concurrent upwelling phenomena or to the abundance of benthic macrofauna.

Pelagic flagellate abundance appeared to be more indicative of contemporary hydrographic conditions, most obvious by an increase in the ratio heterotrophic/autotrophic nanoflagellates away from the area of most intense upwelling.

## Introduction

Concepts of the trophic structure of pelagic marine ecosystems changed substantially during the last decade. The discovery that there is an alternative to the classic algae-copepods-fish food chain, an alternative which is introducing an additional number of trophic levels, presents a considerable complication in our understanding of energy transfer through pelagic ecosystems (Azam *et al.*, 1983, Fenchel, 1987). In this alternative pathway for energy dissolved organic material, lost by algae though lysis and exudation, serves as a substrate for bacterial production. Small heterotrophic nanoflagellates (< 20 $\mu$m) are voracious predators on bacteria and form the next trophic level. At this level the flagellates may provide inorganic nutrients for increased bacterial activity at the basis of the microbial food web through mineralization (Bratbak, 1987), or they are consumed by larger protozoa such as ciliates (Rassoulzadegan & Sheldon, 1986; Sheldon *et al.*, 1986). The first process represents the microbial food web as an energy sink; in the second process the energy directed through the microbial loop becomes potentially available for consumption by metazoans such as copepods.

All available data concerning this 'microbial loop' are for pelagic organisms and systems, but we found recently that heterotrophic nanoflagellates can also be present in large numbers in marine sediments (Bak & Nieuwland, 1989). There are indications that these organisms are prominent as predators on bacteria in the sediment

(Hondeveld *et al.*, 1992), which suggests that relationships and processes similar to the small food web in pelagic systems also occur in the benthos.

The structure of the trophic relations, the biomass and numbers involved, depends strongly on the spatial distribution of environmental characteristics, e.g. the development of pelagic microbial food webs will differ in stratified, mixed or frontal waters (Van Duyl *et al.*, 1990). Areas in the ocean where environmental characteristics are likely to be extreme, and of particular interest, are regions of coastal upwelling; indeed such areas are often regarded as distinct systems. Optimal conditions of nutrient supply in the euphotic layer can be maintained here for long periods of time. The circulation patterns in such areas, bringing nutrient rich subsurface water to the surface, prevents the nutrient limitation of stratified waters as well as the light limitation of mixed waters (Barber & Smith, 1981) and they are stable enough in time and space to permit the forming of upwelling ecosystems.

The establishment of microbial communities in response to upwelling processes would, in view of the high growth rates of the organisms involved (Fenchel, 1986), be very early in the development

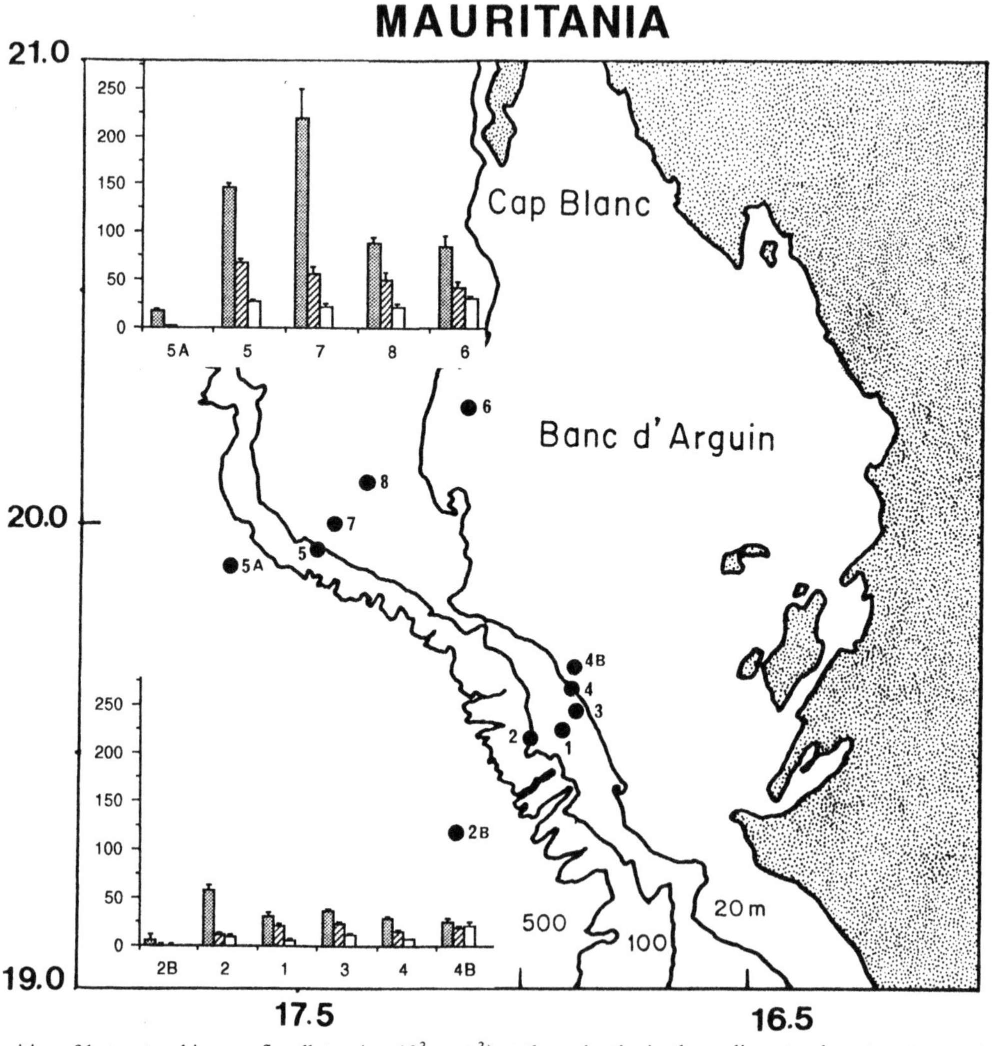


*Fig. 1.* Densities of heterotrophic nanoflagellates ($n \times 10^3$ cm$^{-3}$) at three depths in the sediments along two transects. Sediment depths of 0–3, 30–33 and 60–63 mm are represented by gray, hatched and white bars, respectively, s.e. indicated.

of a biotic response to the conditions of the abiotic environment. The spatial heterogeneity over larger distances along such coasts covering upwelling as well as other pelagic systems should show variation in components of the microbial food web. Such variation, reflecting characteristic processes, would be indicative of spatial variation in energy transfer and energy use in biological systems along the coast. If such variations and relations exist is unclear. It should be realized that at present there are very few data available on the components of the microbial food webs, such as numbers or size classes of the autotrophic and heterotrophic flagellates involved in such systems.

The region off the coast of Mauritania, between Cap Blanc and Cap Timiris, appears particularly suitable as an area for investigations regarding spatial variations of pelagic variables. It is known as a major coastal upwelling system (e.g. Barber & Smith, 1981, Minas *et al.*, 1982, Mittelstaedt, 1991) and part of this area is adjacent to the Banc d'Arguin, a large shallow inshore area of considerable productivity (Wolff & Smit, 1990). Here we report on the patterns in densities of benthic and pelagic nanoflagellates along transects off the

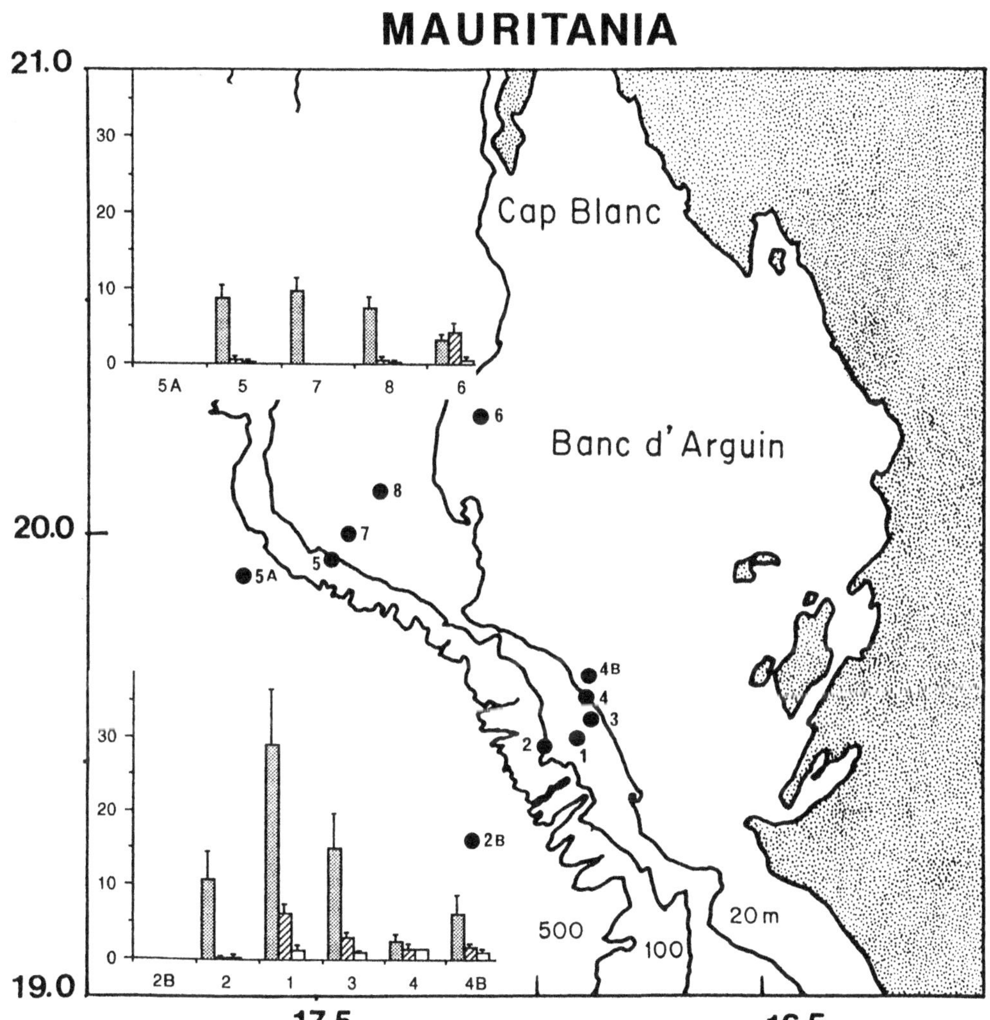


*Fig. 2.* Densities of autotrophic nanoflagellates ($n \times 10^3$ cm$^{-3}$) at three depths in the sediments along two transects. Sediment depths of 0–3, 30–33 and 60–63 mm are represented by gray, hatched and white bars, respectively, s.e. indicated.

coast in relation to biotic and abiotic environmental characteristics.

## Material and methods

Our data set was collected aboard the RV Tyro cruising the shelf off the Banc d'Arguin, Mauritania between May 12–28, 1988. It consists of two subsets, one collected in the water column sampling the pelagic system, and one collected in the sediments. The pelagic data were collected along six transects (coded II to VII) over the shelf from shallow ≈ 20 m to deep ≈ 100 m water. The sediment data were collected along two transects over the shelf from depths of ≈ 10 m to ≈ 1000 m. For locations of pelagic and benthic stations see Fig. 5 and Fig. 1, respectively, for exact geographical positions see Loktionov (1993).

We used epifluorescence microscopy to count and measure components of the microbial communities to obtain densities and biomass values. Our main interest was in heterotrophic nanoflagellates (<20 μm) but similar data were collected for autotrophic nanoflagellates, bacteria and cyanobacteria. Autotrophic nanoflagellates

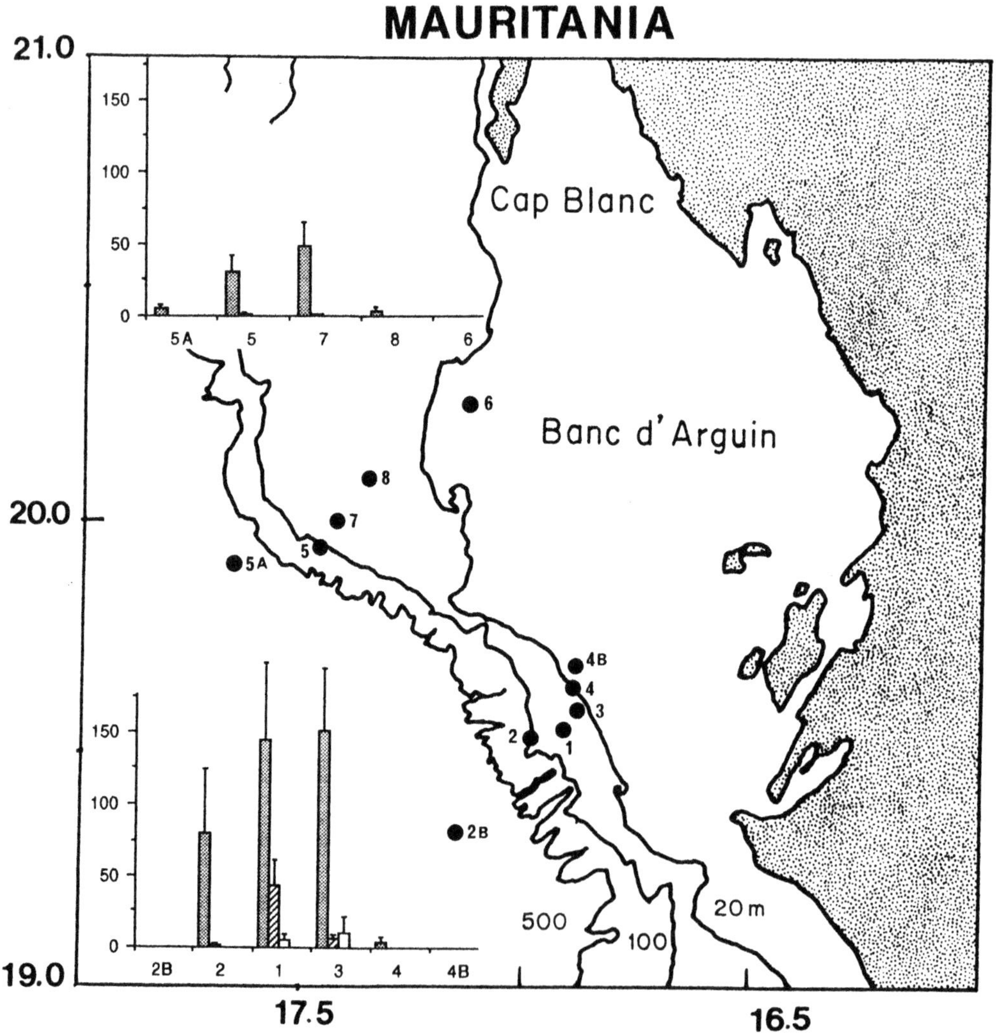


*Fig. 3.* Densities of cyanobacteria ($n \times 10^3$ cm$^{-3}$) at three depths in the sediments along two transects. Sediment depths of 0–3, 30–33 and 60–63 mm are represented by gray, hatched and white bars, respectively, s.e. indicated.

were distinguished through the presence of autofluorescence of photopigments and will possibly include mixotrophic organisms. Briefly, the samples were fixed immediately, stained, filtered and counted within one to three days. A model TXM Stabletop microscope table allowed counting aboard at all magnifications in all weather conditions. Exact procedures for pelagic samples are in Van Duyl *et al.* (1990), for sediment samples see Bak & Nieuwland (1989) and Bak *et al.* (1991). Cyanobacteria are processed along with the other groups and fluoresce brightly yellow/orange under our standard treatment (Stockner & Antia, 1986).

Pelagic samples at each station consisted of one sample of surface water (< 5 m) and one of bottom water (< 10 m). Sediment was collected with a 0.075 $m^2$ boxcorer at each benthic station which was subsampled to obtain three cores which were each subsampled at a sediment depth of 0–3 mm, 30–33 mm and 60–63 mm.

In statistical analyses we used Systat version 5.1 (Systat, Inc. Evanston, Il).

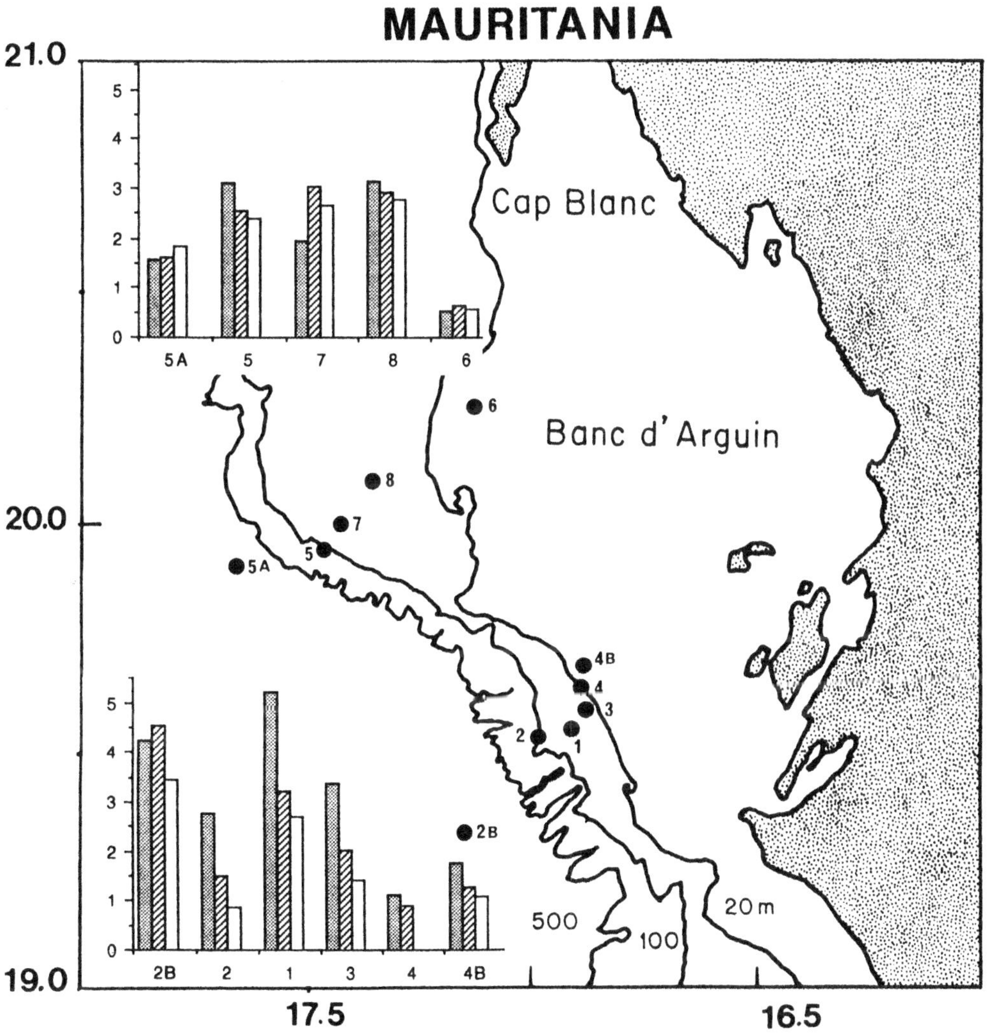


*Fig. 4.* Densities of bacteria ($n \times 10^9$ cm$^{-3}$) at three depths in the sediments along two transects. Sediment depths of 0–3, 30–33 and 60–63 mm are represented by gray, hatched and white bars, respectively.

## Results

There was enormous variation in densities of heterotrophic nanoflagellates between the benthic stations with a range of 8–219 × $10^3$ cells $cm^{-3}$ (Fig. 1). Along the northern and southern transects highest numbers were reached in the sediment surface around the 100 m isobath and lowest densities were found in the deep-sea sediments (depths 1080–1300 m). Another similarity between the two transects was that flagellate densities deeper in the sediment (30–33 mm) reflected surface numbers. They varied from 25–70% at the shelf stations but were much lower, ≈10%, in the deep-sea.

Much more remarkable than the similarity, however, was the dissimilarity between the two transects. There was a significantly higher density along the northern transect (Fig. 1) in the surface as well as in deeper sediment layers. There was an additional difference between the northern and southern transects in the size distribution of the heterotrophic flagellates (Table 1). Cells were smaller, significantly more being included in the size class <2 μm, along the northern transect (paired t-test, $p = 0.041$). Also the data show that

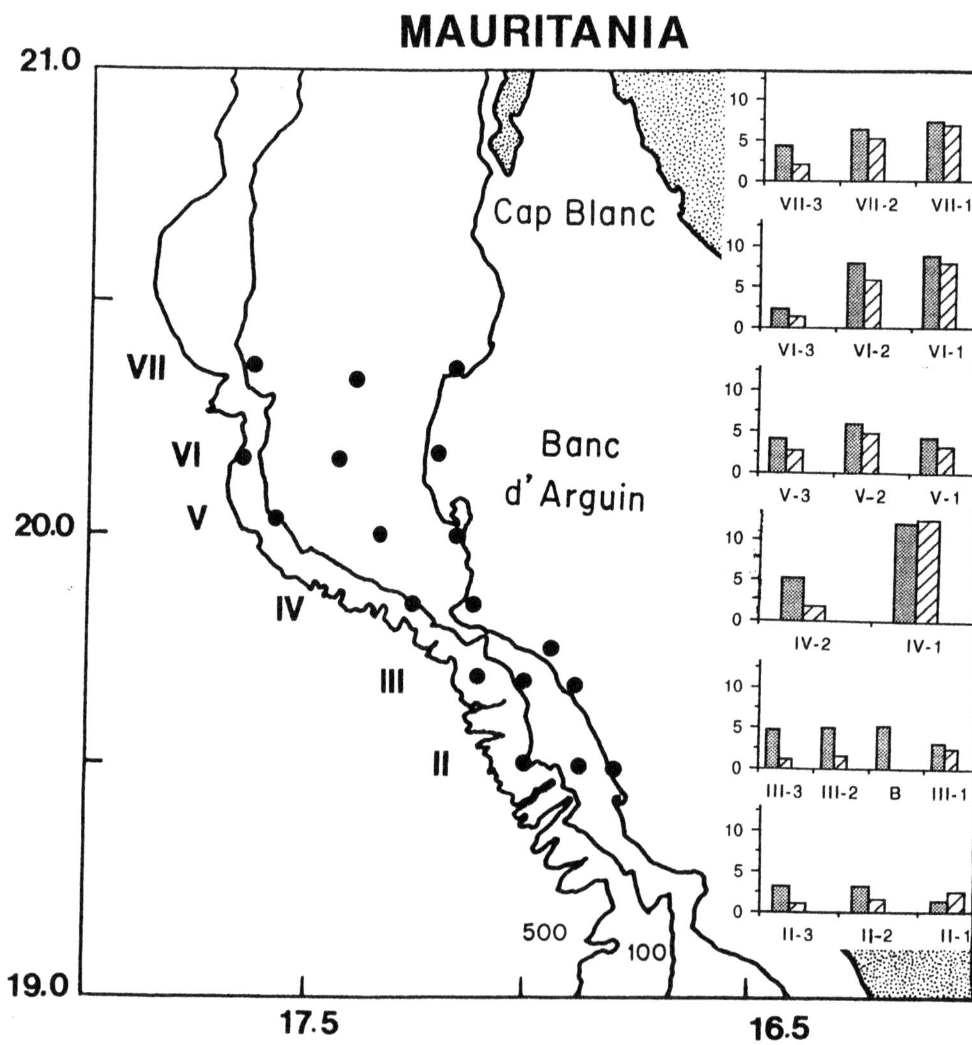


*Fig. 5.* Densities of heterotrophic nanoflagellates ($n \times 10^3$ $cm^{-3}$) in surface and bottom waters (gray and hatched bars respectively) along six transects.

*Table 1.* Size classes (% of total) of heterotrophic benthic flagellates in the sediment surface layer (0–3 mm), shelf of Banc d'Arguin, Mauritania).

| | | $<2\ \mu m$ | $2–5\ \mu m$ | $5–10\ \mu m$ | $10–20\ \mu m$ |
|---|---|---|---|---|---|
| Southern transect | | | | | |
| station | 2b | 0 | 82 | 18 | 0 |
| | 2 | 13 | 75 | 13 | 1 |
| | 1 | 6 | 73 | 19 | 2 |
| | 3 | 29 | 69 | 2 | 1 |
| | 4 | 29 | 69 | 4 | 0 |
| | 4b | 40 | 48 | 12 | 0 |
| Northern transect | | | | | |
| station | 5a | 35 | 49 | 16 | 0 |
| | 5 | 45 | 48 | 5 | 1 |
| | 7 | 36 | 53 | 10 | 1 |
| | 8 | 33 | 59 | 8 | 0 |
| | 6 | 51 | 40 | 7 | 2 |

benthic nanoflagellates are generally tiny creatures, more than 98% being smaller than 10 $\mu m$ in all samples.

Autotrophic nanoflagellates were absent at the deep-sea stations but present in relatively low numbers, $2–28 \times 10^3$ cells $cm^{-3}$, in the surface sediment over the shelf (Fig. 2). In contrast to the distribution of the heterotrophic flagellates it appears that highest densities were recorded along the southern transect. Other small photosynthesizing organisms, the cyanobacteria ($<2\ \mu m$), showed the same pattern (Fig. 3). Bacteria varied from $0.5–5.2 \times 10^9$ cells $cm^{-3}$ in the upper sediment layer (Fig. 4) and there was a high correlation in the occurrence of cocci and rods (ANOVA $p<0.001$). There was no clear pattern in the distribution of total bacterial numbers.

The distribution of pelagic heterotrophic flagellates is shown in Fig. 5. Densities range from $1.4–12.0 \times 10^3$ cells $cm^{-3}$ with highest numbers occurring in the northern part of the area. Values from bottom water generally reflect surface numbers, especially at high density stations. Autotrophic flagellates show a lower abundance, $0.6–6.2 \times 10^3$ cells $cm^{-3}$, and there is a more obvious effect of water depth (Fig. 6). At deeper stations autotrophic cells were hardly present in the bottom water. That autotrophic flagellates were relatively abundant in the southern part of the study area is shown by the low ratio of hetero- to autotrophic flagallates in the south (Fig. 7). Also the phototrophic cyanobacteria are practically confined to the southern area (Fig. 8).

Densities of pelagic bacteria displayed little variation, ranging from $1.6–3.9 \times 10^3\ cm^{-3}$ in surface waters and from $0.7–2.9 \times 10^3\ cm^{-3}$ close to the bottom (Fig. 9). There was an obvious difference in the ratio of cocci and rods in surface and bottom water. At the surface rods and cocci occurred in densities of the same order of magnitude but in bottom waters densities of rods were relatively low ($y = 0.059 + 0.339x$, $p<0.001$).

## Discussion

Densities of heterotrophic nanoflagellates show substantial differences in different water types

*Table 2.* Densities of bacteria and heterotrophic nanoflagellates in pelagic marine systems.

| Locality | Bacteria ($\times 10^6\ cm^{-3}$) | Flagellates ($\times 10^3\ cm^{-3}$) | Author |
|---|---|---|---|
| Oligotrophic tropical | 0.05–0.5 | 0.5–3.0 | Sorokin, 1981 |
| Atlantic (off Georgia) | 2–10 | 0.3–3.3 | Sherr *et al.*, 1984 |
| Bering Sea | 0.6–2.4 | 0.01–0.1 | Fenchel, 1986 |
| Coral reef | 0.8–1.5 | – | Sorokin, 1981 |
| North Sea | 1.2–2.7 | 0.9–6.6 | Nielsen & Richardson, 1989 |
| North Sea | 0.2–4.5 | 0.7–10.8 | van Duyl *et al.*, 1990 |
| Eutrophic Japan Sea | 1–5 | 0.7–9.5 | Sorokin, 1977, 1981 |
| Peru upwelling | 1–5 | 0.7–13.0 | Sorokin, 1981 |
| Mauritania | 1.6–3.9 | 1.4–12.0 | This paper |

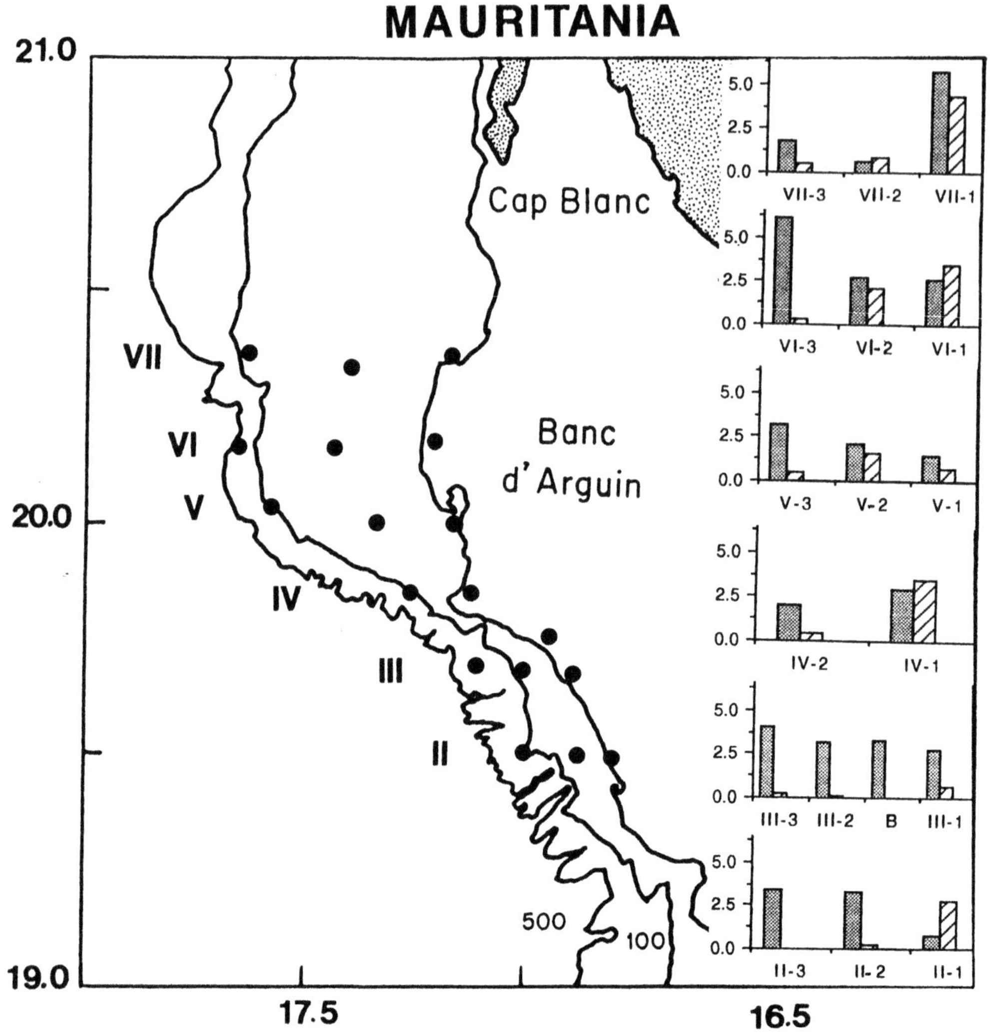


*Fig. 6.* Densities of autotrophic nanoflagellates ($n \times 10^3$ cm$^{-3}$) in surface and bottom waters (gray and hatched bars respectively) along six transects.

(Table 2). In oligotrophic tropical and oceanic conditions densities range from 0.01 to 3.0 cells $\times 10^3$ cm$^{-3}$. In mesotrophic temperate waters densities range from 0.3 to 7.0, but in eutrophic and estuarine conditions up to $11 \times 10^3$ cm$^{-3}$ have been counted. Bacterial numbers show a similar trend with low numbers in oceanic and oligotrophic waters (0.05–2.4 cells $\times 10^6$ cm$^{-3}$) and high densities, up to $10 \times 10^6$ cm$^{-3}$, in more eutrophic and estuarine areas (Table 2). There are hardly any such data from upwelling systems and part of the data available, including information from the shelf off northwest Africa, is incompatible because an altogether different methodology has been used (Rico, 1977; Vallespinos & Tejero, 1977).

Sorokin (1979) studied densities of heterotrophic nanoflagellates in eutrophic waters of coastal upwelling along the Peruvian coast. The protist communities consisted of small, 3–6 $\mu$m, *Bodo* species. He found variable but occasionally high densities of $0.2–13 \times 10^3$ cm$^{-3}$. The bulk of these occurred in the surface layer of water column. Sorokin (op. cit.; 1981) contends that the lower and higher densities represented different stages in the development of the upwelling water

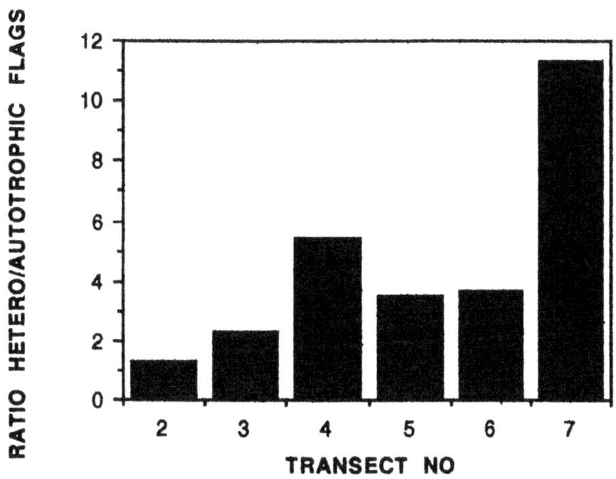


*Fig. 7.* Relative importance of heterotrophic and autotrophic nanoflagellates in six pelagic transects off the Banc d'Arguin. For position of transects see Fig. 5.

mass. At sites of intensive upwelling, numbers of heterotrophs were low, increasing at neighbouring stations. Similarly, bacterial abundance was generally high in the Peruvian upwelling area, $1\text{–}5 \times 10^6\ cm^{-3}$ (Sorokin, 1981), but low densities, $0.3\text{–}0.7 \times 10^6\ cm^{-3}$ were reported for freshly upwelled water in the center of upwelling (Sorokin & Kogelschatz, 1979). There was a succession in the upwelling water, from an autotrophic to a heterotrophic phase, in time and, when this development is related to the local currents, in space. During the autotrophic phase a surplus of organic material was produced which was subsequently used during the following heterotrophic phase. Sorokin (*op. cit.*) suggested that surplus organic production is advected from the upwelled water to the frontal area at the boundary of the upwelling where it is used by bacteria and protozoa.

The range in pelagic heterotrophic flagellate densities in the shelf area off the Banc d'Arguin is similar to that off the Peruvian coast, 1.4–12.0 and $0.2\text{–}13.0 \times 10^3\ cm^{-3}$, respectively. The interpretation of these differences in relation to hydrographic conditions is, however, more complicated because coastal waters flowing off the Banc protrude into the shelf upwelling area. Nevertheless it is clear that during our cruise upwelling was most intense along the two southern transects (II and III), but also occurred at the outer stations of the northern transects (IV-2, V-3, VI-3, see Berghuis *et al.* (1993) and Kuipers *et al.* (1993)). In agreement with Sorokin's (1981) theory densities of heterotrophic flagellates are relatively low at all these stations (Fig. 5). The first phase of the succession he proposed to occur in centres of upwelling, the autotrophic phase of succession, is not so easy to detect because the non-upwelling waters originating from the outer edge of the shallow Banc d'Arguin showed relatively high rates of

*Table 3.* Densities of bacteria and heterotrophic nanoflagellates in the surface of marine sediments.

| Locality | Bacteria ($\times 10^9\ cm^{-3}$) | Flagellates ($\times 10^3\ cm^{-3}$) | Author |
|---|---|---|---|
| Wadden Sea (intertidal) | 0.5–4.0 | 36–280 | Bak & Nieuwland, 1989 |
| Frisian Front (North Sea) | 2.0–3.5 | 10–50 | Bak *et al.*, 1991 |
| Broad Fourteens (North Sea) | 0.5–1.0 | 50–350 | Bak *et al.*, 1991 |
| Organically rich sediment & detritus | 3–10 | – | Sorokin, 1981 |
| Weddell Sea (200–800 m) | 1.0–2.3 | 5–12 | Bak & Nieuwland, unpubl. |
| Weddell Sea (1000–2000 m) | 1.3–1.5 | 1–4 | Bak & Nieuwland, unpubl. |
| Mauritania | | | |
| < 1000 m | 0.5–5.2 | 26–219 | This paper |
| > 1000 m | 1.6–4.2 | 8–18 | This paper |

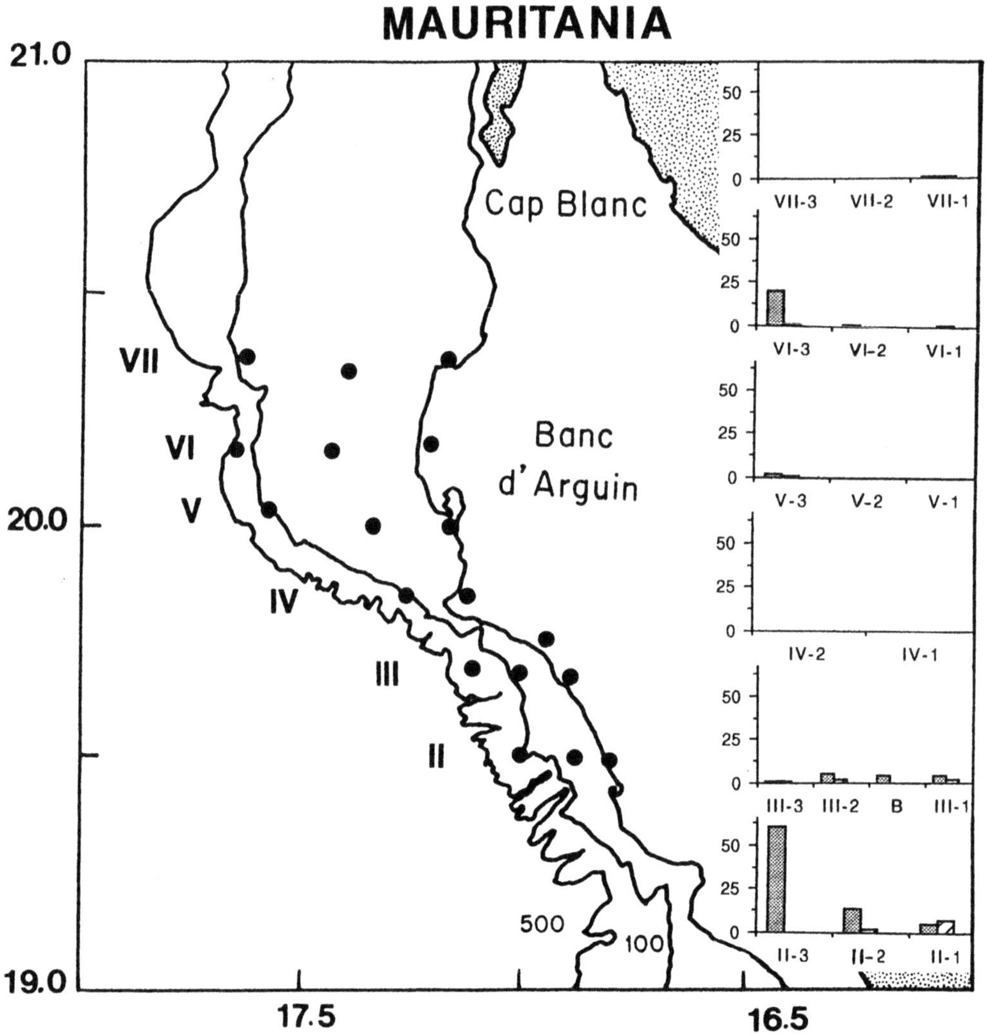


*Fig. 8.* Densities of cyanobacteria ($n \times 10^3$ cm$^{-3}$) in surface and bottom waters (gray and hatched bars respectively) along six transects.

primary production and high chlorophyll *a* values (Berghuis *et al.*, *op. cit.*). The autotrophic organisms which would be most rapid in showing a response to increased nutrient levels, autotrophic nanoflagellates and, even smaller, the cyanobacteria, are however prominent in the upwelling area of the two southern transects (Figs 7, 8). Small autotrophic flagellates were also numerous in coastal Banc water (shallow stations IV-1, VI-1, VII-1), but cyanobacteria were restricted to shelf waters.

Comparable data on densities of components of microbial communities in the benthos are scarce and the few available data are difficult to compare because of great differences in the methodologies used (Alongi, 1991). Dominant groups among the benthic heterotrophic flagellates, belonging to the Euglenozoa, are kinetoplastids such as *Bodo* species (Patterson *et al.*, 1989; Larsen & Patterson, 1990). High densities of such nanoflagellates, 8.5 to $300 \times 10^3$ cells cm$^{-1}$, have been recorded for the surface layers of intertidal and subtidal marine sediments (Bak & Nieuwland, 1989; Alongi, 1990). Similar data from deeper marine bottoms, North Sea 28–38 m depth, ranged from 10 to $350 \times 10^3$ cells cm$^{-3}$

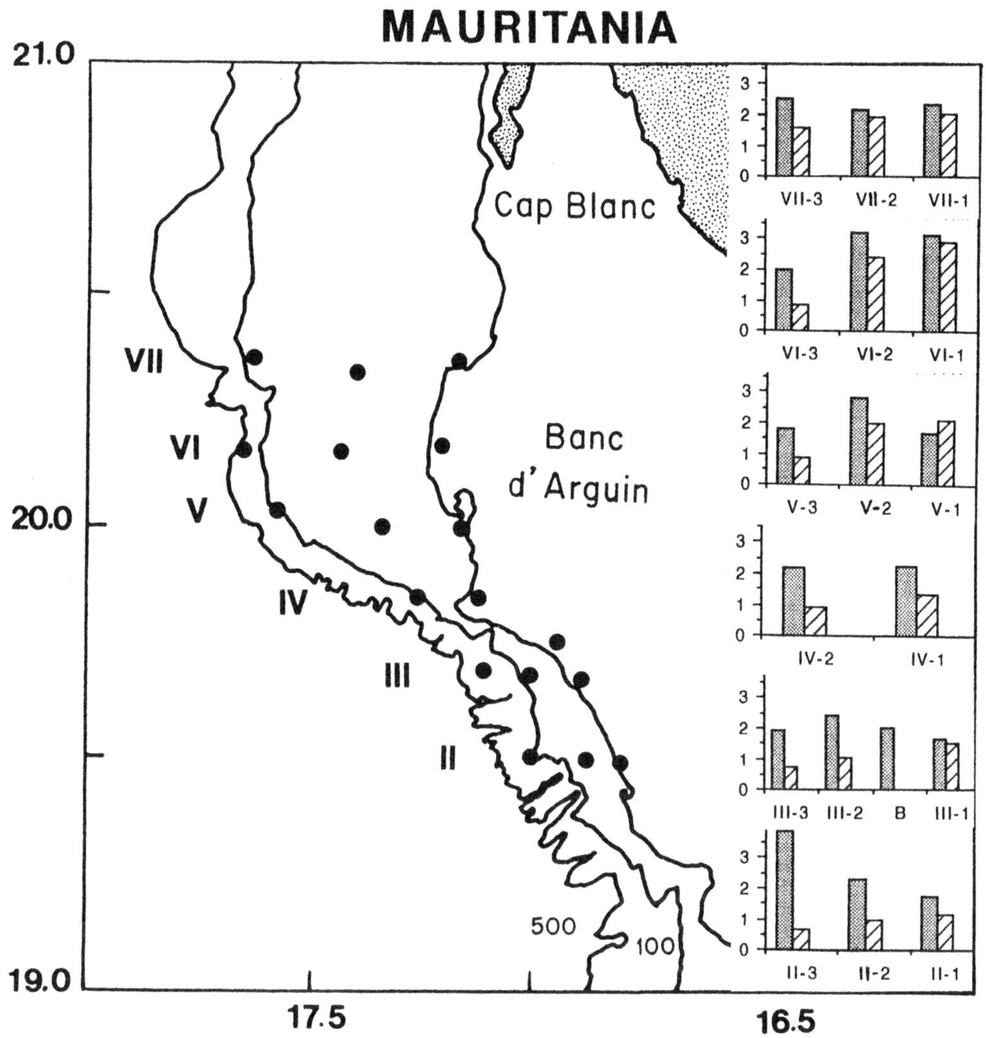


*Fig. 9.* Densities of bacteria ($n \times 10^6$ cm$^{-3}$) in surface and bottom waters (gray and hatched bars respectively) along six transects.

(Bak *et al.*, 1990). Bacterial abundance varied from $0.5–4 \times 10^9$ cm$^{-3}$ in temperate shallow seas and estuaries (Van Duyl & Kop, 1990; Van Duyl *et al.*, in press) to reach $3–10 \times 10^9$ cells cm$^{-3}$ in organically rich sediments and detritus (Sorokin, 1981). Table 3 shows that bacterial and flagellate densities off the Banc d'Arguin are entirely within the range found for shallow temperate seas.

There was an enormous variation in spatial distribution of the heterotrophic flagellates (Fig. 1). At the deepest stations (2B, 5A) low densities occurred comparable to those recorded for the deep Weddell Sea (Table 3) and the mid-Atlantic (Bak *et al.*, unpubl.). But there was also a great difference between the southern and the northern transect. At first glance this difference, a relatively low density of heterotrophic flagellates along the southern transect, reflects the characteristics of the pelagic community. We suggest that the explanation is not the same, related to upwelling, as that for the pelagic. In contrast with the phenomena occurring in the pelagic, where concurrent hydrographic conditions can be related to processes in the microbial communities, the benthos has an additional controlling factor which is sediment characteristics. Flagellate as well as bacterial densities are strongly influenced by sediment grain size. Bacterial densities are higher in silty sediments (Meyer-Reil, 1984) and densities of heterotrophic nanoflagellates are lower (Bak & Nieuwland, 1989; Bak *et al.*, 1991). The southern transect crossed a section of the shelf with very silty sediments compared with the northern transect (Domain, 1985) and this accounts for the relatively low flagellate and high bacterial densities in the sediment. The autotrophic organisms recorded (flagellates and cyanobacteria) originate from the water column and their presence probably reflects the densities of the pelagic system.

Biomass of benthic heterotrophic nanoflagellates varied considerably along the transects from $1.5 \times 10^{-3}$ g C $m^{-2}$ in deep-sea sediments (station 2B) to $33.6 \times 10^{-3}$ g C $m^{-2}$ at 50 m (station 7). The higher values are similar to such carbon data for temperate seas (Bak & Nieuwland, 1989). Biomass of flagellates is negligible compared to a macrobenthos biomass of several g C $m^{-2}$ in all shelf stations (Duineveld *et al.*, this volume). There was no relation between numbers or biomass of benthic macrofauna and heterotrophic flagellates along the transects (ANOVA: $p = 0.438$, $p = 0.921$). This is to be expected when time scales of biological processes are so dissimilar between populations.

We conclude that densities of nanoflagellates show large variation along the Banc d'Arguin but can reach comparatively high values in the water column as well as in the benthos. Spatial variation in the sediments relates to a difference beween deep sea and shelf bottoms and to sediment grain size. Spatial variation in the water column appears to be related to upwelling phenomena.


## Acknowledgements

We thank the crew of RV Tyro and all cruise members for their cooperation. Financial support was provided by the Netherlands Marine Research Foundation (SOZ).


## References


Alongi, D. M., 1990. Abundances of benthic microfauna in relation to outwelling of mangrove detritus in a tropical coastal region. Mar. Ecol. Prog. Ser. 63: 53–63.

Alongi, D. M., 1991. Flagellates of benthic communities: Characteristics and methods of study. In D. J. Patterson & J. Larsen (eds), The biology of free-living heterotrophic flagellates, The Systematics Association, Oxford, 45: 57–76.

Azam, F., T. Fenchel, J. G. Field, J. S. Gray, L. A. Meyer-Reil & F. Thingstad, 1983. The ecological role of water-column microbes in the sea. Mar. Ecol. Prog. Ser. 10: 257–263.

Bak, R. P. M. & G. Nieuwland, 1989. Seasonal fluctuations in benthic protozoan populations at different depths in marine sediments. Neth. J. Sea Res. 24: 37–44.

Bak, R. P. M., F. C. van Duyl, G. Nieuwland & A. J. Kop, 1991. Benthic heterotrophic nanoflagellates in North Sea field/mesocosm bottoms and their response to algal sedimentation. Ophelia 33: 187–196.

Barber, R. T. & R. L. Smith, 1981. Coastal upwelling ecosystems. In A. R. Longhurst (ed.), Analysis of marine ecosystems. Academic Press, London: 31–68.

Berghuis, E. M., G. C. A. Duineveld & J. Hegeman, 1993. Primary production and distribution of phytopigments in the water column and sediments on the upwelling shelf off the Mauritanian coast (Northwest Africa). Hydrobiologia 258: 81–93.

Bratbak, G., 1987. Carbon flow in an experimental microbial ecosystem. Mar. Ecol. Prog. Ser. 36: 267–276.

Domaill, F., 1985. Carte sedimentologique de plateau continental Mauritanien. Edition de l'Orstom, Collection Notice Explicative 105.

Duineveld, G. C. A., P. A. W. J de Wilde, E. M. Berghuis & A. Kok, 1993. The benthic infauna and benthic respiration off the Banc d'Arguin (Mauritania, Northwest Africa). Hydrobiologia 258: 107–117.

Fenchel, T., 1986. The ecology of heterotrophic flagellates. Adv. Microb. Ecol. 9: 57–97.

Fenchel, T., 1987. The ecology of protozoa. Science Tech Publishers, Madison (Wisconsin), 187 pp.

Hondeveld, B. J. M., R. P. M. Bak & F. C. cvan Duyl, 1992. Bacterivory by heterotrophic nanoflagellates in marine sediments measured by uptake of fluorescently labelled bacteria. Mar. Ecol. Prog. Ser. 89: 63–71.

Kuipers, B. R., H. J. Witte & S. R. Gonzalez, 1993. Zooplankton distribution in the coastal upwelling system along the Banc d'Arguin, Mauritania. Hydrobiologia 258: 133–149.

Larsen, J. & D. J. Patterson, 1990. Some flagellates (Protista) from tropical marine sediments. J. Nat. Hist. 24: 801–937.

Loktionov, Y., 1993. Hydrographical observations west of the Banc d'Arguin, Mauritania, in May 1988. Hydrobiologia 258: 21–32.

Meyer-Reil, L. A., 1984. Bacterial biomass and heterotrophic activity in sediments and overlying waters. In J. E. Hobbie & P. J. leB. Williams (eds), Heterotrophic activity in the sea. Plenum Press, New York: 523–546.

Minas, H. J., L. A. Codispoti & R. C. Dugdale, 1982. Nutrients and primary production in the upwelling region off Northwest Africa. Rapp. P.-v. Réun. Cons. int. Explor. Mer 18: 148–183.

Mittelstaedt, E., 1991. The ocean boundary along the northwest African coast: Circulation and oceanographic properties at the sea surface. Prog. Oceanog. 26: 307–355.

Nielsen, T. G. & K. Richardson, 1989. Food chain structure of the North Sea plankton communities: Seasonal variations of the role of the microbial loop. Mar. Ecol. Progr. Ser. 56: 75–87.

Patterson, D. J., J. Larsen & J. O. Corliss, 1989. The ecology of heterotrophic flagellates and ciliates living in marine sediments. Progr. Protistol. 3: 185–277.

Rassoulzadegan, F. & R. W. Sheldon, 1986. Predator-prey interactions of nanozooplankton and bacteria in an oligotrophic marine environment. Limnol. Oceanogr. 31: 1010–1021.

Rico, G., 1977. Contribution al estudio de los protozoarios marinos de los alrededores del cabo Blanco. Res. Exp. Cient.B/O Cornide 6: 245–290.

Sheldon, R. W., P. Nival & F. Rassoulzadegan, 1986. An experimental investigation of a flagellate-ciliate-copepod food chain with some observations relevant to the linear biomass hypothesis. Limnol. Oceanogr. 31: 184–188.

Sherr, B. F., E. B. Sherr & S. Y. Newell, 1984. Abundance and productivity of heterotrophic nanoplankton. J. Plankton Res. 6: 195–203.

Sorokin, Yu. I., 1977. The heterotrophic phase of plankton succession in the Japan Sea. Mar. Biol. 41: 107–117.

Sorokin, Yu. I., 1979. Zooflagellates as a component of the community of eutrophic and oligotrophic waters in the Pacific Ocean. Oceanol. 19: 316–319.

Sorokin, Yu. I., 1981. Microheterotrophic organisms in marine ecosystems. In A. R. Longhurst (ed.), Analysis of marine ecosystems. Academic Press, London: 293–342.

Sorokin, Yu. I. & J. E. Kogelschatz, 1979. Analysis of heterotrophic microplankton in an upwelling area. Hydrobiol. 66: 195–209.

Stockner, J. G. & N. J. Antia, 1986. Algal picoplankton from marine and freshwater ecosystems: a multidisciplinary perspective. Can. J. Fish. aquat. Sci. 43: 2472–2503.

Vallespinos F. & A. Tejero, 1977. Distribution de bacterias heterotrofas en la zona de afloramento del NW Africa. Res. Exp. Cient. B/O Cornide 6: 151–164.

Van Duyl, F. C. & A. J. Kop, 1990. Seasonal patterns of bacterial production and biomass in intertidal sediments of the western Dutch Wadden sea. Mar. Ecol. Prog. Ser. 59: 249–261.

Van Duyl, F. C., R. P. M. Bak, A. J. Kop & G. Nieuwland, 1990. Bacteria, auto- and heterotrophic nanoflagellates, and their relations in mixed, frontal and stratified waters of the North Sea. Neth. J. Sea Res. 26: 97–109.

Van Duyl, F. C., R. P. M. Bak, A. J. Kop & G. Nieuwland. Mesocosm experiments: mimicking seasonal developments of microbial variables in North Sea sediments. Hydrobiologia, in press.

Wolff, W. J. & C. J. Smit, 1990. The Banc d'Arguin, Mauritania, as an environment for coastal birds. Ardea 78: 17–37.

*Hydrobiologia* **258**: 133–149, 1993.
*W. J. Wolff, J. van der Land, P. H. Nienhuis & P. A. W. J. de Wilde (eds),*
*Ecological Studies in the Coastal Waters of Mauritania.*




# Zooplankton distribution in the coastal upwelling system along the Banc d'Arguin, Mauritania

B. R. Kuipers, H. J. Witte & S. R. Gonzalez
*Netherlands Institute for Sea Research (NIOZ), P.O. Box 59, 1790 AB, Den Burg, Texel, The Netherlands*

*Key words:* zooplankton, copepods, upwelling, Mauritania

## Abstract

In the framework of the Dutch Mauritania Expedition 1988, zooplankton was sampled in the beginning of the spring upwelling season at 44 stations along the Banc d'Arguin, Mauritania. Distribution patterns of oceanic and neritic copepod species and especially of the upwelling indicator *Calanoides carinatus* (Kröyer) were analysed in search for direct proof of circulation of ocean upwelling water over the inshore banks. The vertical temperature distribution at the stations pointed to upwelling in two different parts of the shelf area. According to the copepod data the two upwellings were of a different nature. A restricted area at the shelf edge near Cap Timiris appeared to represent a relatively stable coastal upwelling situation. Upwelled water present near Cap Blanc was found to have traveled quite recently over a considerable part of the shelf. In the central area between the two upwellings, 'Banc-water' extended from the shelf into the ocean. The possible impact of the observed circulation of ocean water on productivity of the shelf benthic system is discussed.

## Introduction

The North-West African upwelling area is one of four main regions on the earth where surfacewater is driven away from the land by easterly tradewinds and cold, nutrient-rich water ascends from the deep to fertilize the euphotic layer. Locally enhanced primary production and high secondary- and fish production are the direct consequences of the nutrient supply.

Upwelling areas have, during the last 20 years in particular, been the subject of many oceanographic expeditions. In the seventies alone, some 80 scientific cruises were aimed at the Canary Current system (Smed, 1982). The ocean region off Cap Blanc (Mauritania) has been identified as a convergence zone between two main deep watermasses: the North-Atlantic Central Water (NACW) and the South Atlantic Central Water (SACW), which follow the continental slope until they meet in the area and bend in westerly directions. Over these deep currents, and separated from them by a 5 to 100 and sometimes even 1000 m deep thermocline, the relatively cold Canary-current flows from the north along the African continent until during the upwelling seasons it is, in the Cap Blanc region driven from the coast by the North-Easterly trade winds (Reyssac, 1977). Deep water, rich in nutrients, rises to replace the surface water (Pavlov, 1968). It depends on meteorological conditions far outside the upwelling region, at what actual position the convergence between northern and southern central waters will take place, and what quantity and type of watermixture will rise to replace a deficit at the surface. The origin, the composition and the biota of the ocean surface water in the Cap Blanc region are, therefore, complicated and variable in time and space. Governed by the annual ocean surface temperature cycle and the trade winds, North-West African upwelling is generally strongest in the periods February to June and October to December. In an irregular band *ca* 40–50 miles offshore such upwelling, indicated mainly by low temperatures and high chlorophyll concentrations in the ocean surface, is most pronounced.

The connection between the high productivity of the shallow inshore Banc d'Arguin to the South of Cap Blanc and the ocean upwelling phenomena further offshore is not well understood yet. Since there is no river input into the system it seems evident that the high benthos production at the actual bank is related some way to upwelling against the continental slope at the outer side of the shelf sea some 50 to 100 km away. Moreover, a flow of extremely saline 'Banc-water' with a load of organic material tends to leave the Banc d'Arguin along the seabed of the continental slope, currently exporting nutrients with the debris (Margalef, 1975). To explain the richness of the banks and to counterbalance the export of the system, nutrients from the oceanic upwelling region off Cap Blanc must quite substantially enrich the shallow coastal sea between Cap Blanc and Cap Timiris.

The aim of the present paper is, to check whether the zooplankton data collected during the Dutch Mauritania expedition in 1988 in the ocean region off Banc d'Arguin and inshore to the minimum depth of 7 metres, together with measurements of temperature, salinity, $O_2$-saturation, chlorophyll-*a*, POC and phaeopigment (which were carried out on board the R.V. TYRO by the participants together for general use), point to enrichment of the inshore area by circulating oceanwater from the upwelling zone indeed.

## Methods

### *Hydrography*

On 10, 11 and 12 May 1988, samples were taken at 40 CTD stations further referred to as the 'Hydrographic Survey', in order to get preliminary information on the distribution of different watermasses in the research area. In the framework of the present paper, this data set will be used only to a limited degree, the complete material being published in Loktionov (this volume).

From 13 to 27 May, after the 'Hydrographic Survey', sampling was concentrated on a smaller grid of stations during the 'Pelagic Survey', which consisted basically of six transects crossing the continental slope in right angles at various latitudes (Fig. 1b). At each of the transects there were three stations, with depths of 10–20, 20–50 and 100–300 metres respectively.

At all stations vertical profiles were made with the CTD, mounted with an oxygen-probe, of temperature, salinity and $O_2$-saturation (%), whereas chlorophyll-*a* concentration (in $\mu$g Chl*a* $l^{-1}$), phaeopigment-concentration ($\mu g\, l^{-1}$) and P.O.C. content of the water ($\mu$g carbon $l^{-1}$) were determined in watersamples taken by a Rosette sampler in the watersurface (0–1 m), in the middle of the watercolumn, and as close to the bottom as possible. Details on the methods are given in Cadée (1986).

### *Zooplankton*

In order to obtain a quick overview over the composition and density distribution of mesozooplankton before the start of the 'Pelagic Survey', vertical net hauls were made at 26 of the 40 stations of the 'Hydrographic Survey' grid (Fig. 1a). For this purpose a 300 $\mu$m vertical plankton net was used (Baars *et al.*, 1990), to make hauls from 200 m depth to the surface (at shallower stations from 1 m above the seabed to the surface).

After that, the 18 stations of the 'Pelagic Survey' grid (Fig. 1b) were sampled in much more detail. At each of the stations 4 different nets were used to sample the whole zooplankton size spectrum adequately from 80 $\mu$m to body sizes of several centimeters:

1. a 50 $\mu$m vertical net (Daan, 1987; Fransz, 1988) for the smallest mesozooplankton stages;

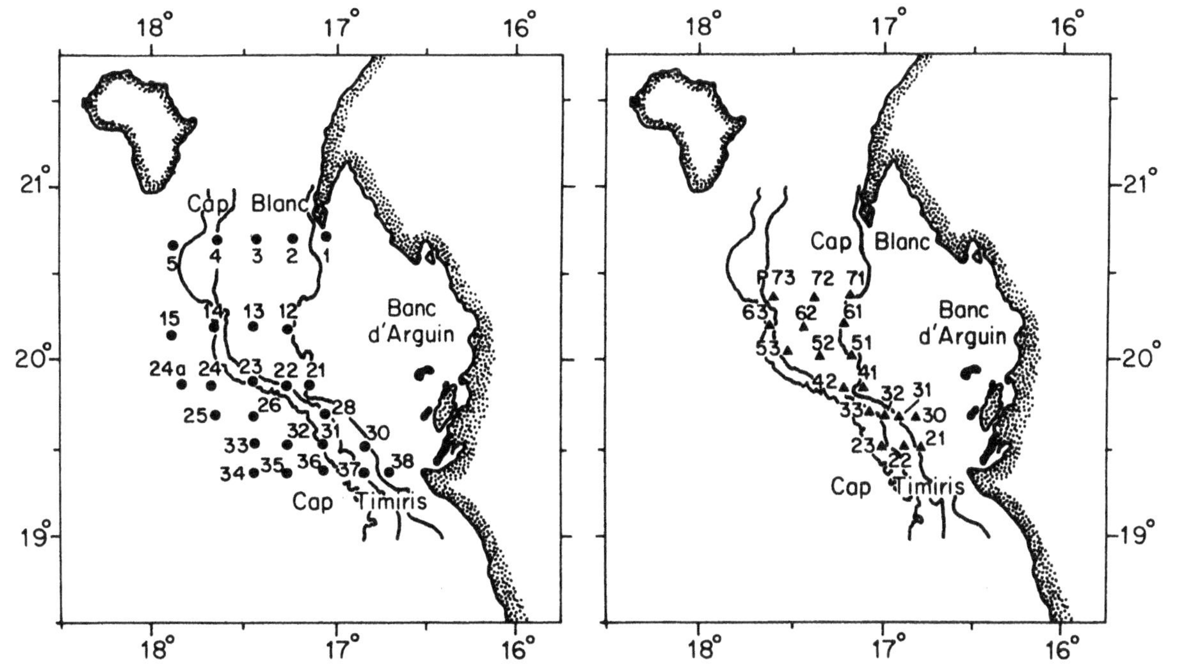


*Fig. 1.* Banc d'Arguin region with the position of sampling stations between Cap Blanc and Cap Timiris, including the 20 m, 100 m and 500 m isobaths. a. stations of the 'Hydrographic Survey' (10–20 May 1988); b. stations of the 'Pelagic Survey' (13–27 May 1988).

2. a 300 $\mu$m vertical net (also used in the Hydrographic Survey) for adult mesozooplankton;
3. a modified Gulf-III high speed sampler for large and fast mesozooplankton and small macrozooplankton (Fransz, 1976);
4. an Isaacs-Kidd midwater trawl (IKMT) (opening 7.3 $m^2$; meshes 1.4 mm) for the macroplankton (Isaacs & Kidd, 1953).

The volumes filtered by the vertical nets are derived from the line-length and net opening; plankton torpedo and IKMT (which made double oblique hauls) were fitted with electronic sensors for depth, temperature and waterflow with reading on deck. The oblique hauls reached down to just above the seabed on the shelf, and to 200 m depth in the ocean.

All samples were preserved in 4% formaldehyd solution in seawater for later analysis. Subsampling, sorting, counting and measuring was done according to Fransz (1976, 1988), Fransz & van Arkel (1980) and Fransz *et al.* (1984). Biomasses were estimated by means of length- (or diameter-) weight relationships given in Nassogne (1972), Fransz & van Arkel (1980), Klein-Breteler *et al.* (1982) and different unpublished measurements by the Netherlands Institute for Sea Research (NIOZ) of zooplankton ADW and carbon weight.

The different nets sampled with some overlap different areas of the zooplankton size continuum. After determining the separation point where of a given size-category one net began to catch less than the following, a combined density estimate per station was calculated for each of the species or life stages. Because the 300 $\mu$m vertical net and 300 $\mu$m plankton torpedo did not yield significantly different numbers per $m^3$, their catches were averaged.

## Results

### *Hydrography*

Figure 2 presents the vertical distribution of temperature measured with the CTD along the six transects of the 'Pelagic Survey' (see 2.1). At the southernmost transects II and III, the ascent

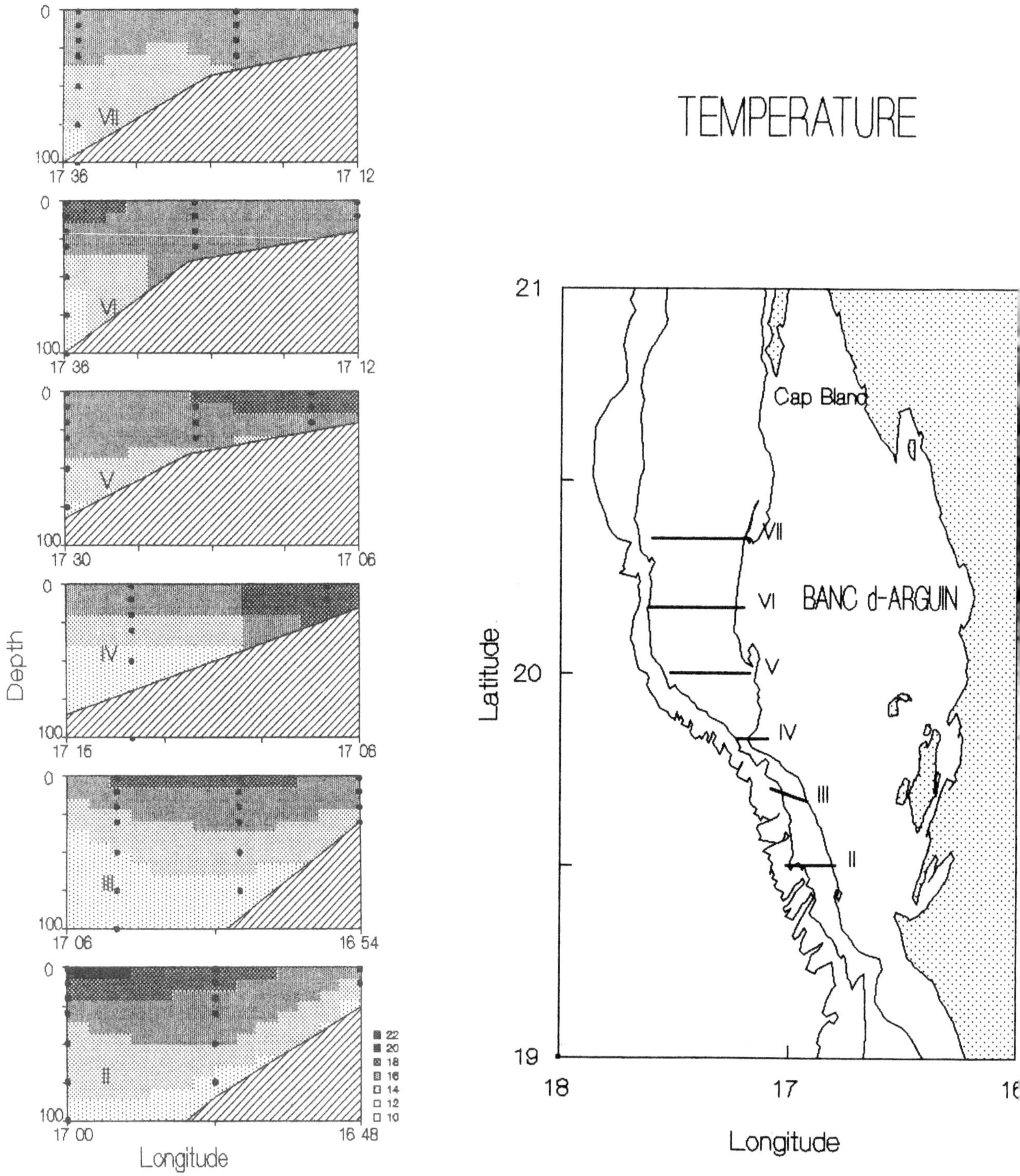


*Fig. 2.* Vertical distribution of temperature (°C) along the 6 transects of the 'Pelagic Survey'.

along the seabed of deep and cold oceanwater (13–15 °C) to ca. 20 metres under the sea surface can be recognized. Low surface-water temperatures locally on the shelf complete the picture of this typical example of coastal upwelling – further referred to as the 'southern upwelling area'. Whereas the surface water of the ocean has a temperature of 18–20 °C, surface temperatures at the upper slope are only 15–17 °C as a result of local 'erosion' of temperature stratification from underneath.

The central transects IV and V (Fig. 2) show

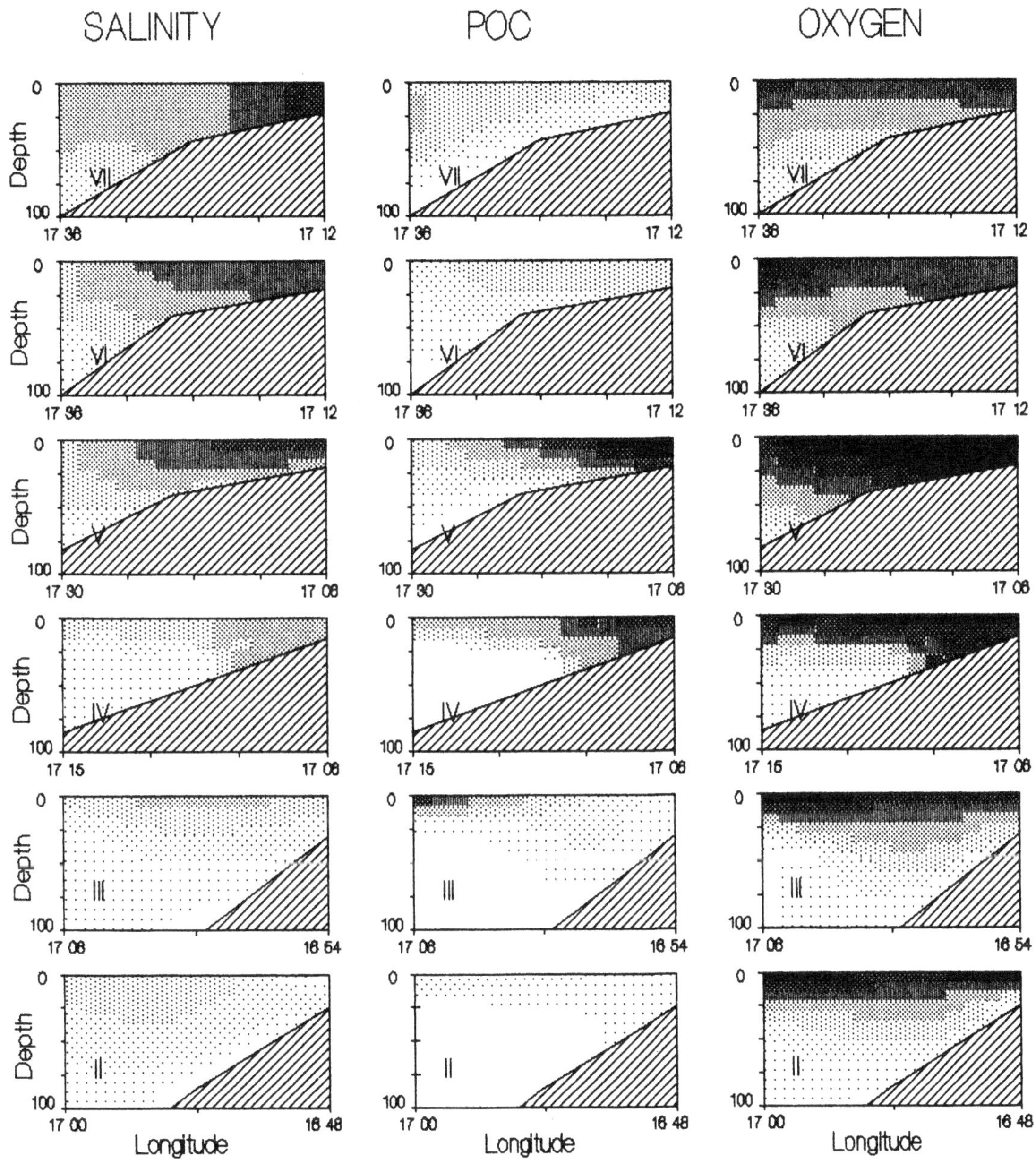


*Fig. 3.* Vertical distribution (in five equal linear steps) of salinity (‰), POC ($\mu g\ C\ l^{-1}$) and oxygen saturation (%) along the transects of Fig. 2. The minimum (white) and maximum (black) values for salinity are 35.2 and 36.2, for POC 0 and 1 and for oxygen saturation 0 and 100% respectively.

more or less the opposite picture of the southern transects II and III. Here (and especially at V) warm water (18–20 °C) extends oceanwards over the slope whereas upwelling of cold water stays further offshore, as is illustrated by relatively low surface water temperatures (16–17 °C) at the ocean sides of transects IV and V.

Vertical temperature distributions at the northern transects VI and VII (Fig. 2) differed considerably from the rest at the time of sampling. This part of the shelf was obviously covered with completely vertically mixed cold water.

Figure 3 gives the vertical distributions of salinity, total POC and oxygen saturation at the transects of Fig. 2. The salinity data at transects II, III, IV and V seem to agree well with the temperature distributions (Fig. 2), assuming that higher salinities are basically due to increased evaporation by warming up in the shallow Banc area.

Distribution patterns of oxygen saturation (Fig. 3) seem to give the best indication about where upwelled water entered the system and, moreover, how long ago. Deep ocean water is poor in oxygen, as is illustrated by the deeper measurements at the ocean stations of transects II and III. Its climbing onto the continental slope in the 'southern upwelling area' leads to locally (weakly) lower oxygen saturations in the surface water at the southern transects.

At transects IV and V, in the centre of the area, where warm Bancwater was flowing towards the ocean (Fig. 2), the surface water over the slope shows high oxygen saturations (Fig. 3). This

*Table 1.* Density estimates (n $m^{-3}$) of adult copepods at the sampling stations of the 'Hydrographic Survey', given in Fig. 1a.

| Copepod adults | Station nr. | | | | | | | | | | | | |
|---|---|---|---|---|---|---|---|---|---|---|---|---|---|
| | 1 | 2 | 3 | 4 | 5 | 12 | 13 | 14 | 15 | 21 | 22 | 23 | 24–1 |
| *Acartia clausi* | 47.7 | 549.0 | 228.0 | 43.4 | 31.6 | 43.4 | 154.0 | 21.3 | 10.9 | 19.1 | 19.1 | | |
| *Calanoides carinatus* | 0.4 | 12.0 | 51.0 | | 1.6 | | 0.1 | 13.6 | 2.1 | 0.3 | 9.7 | 2.2 | 7.3 |
| *Candacia armata* | | | | 2.3 | 1.6 | | | 1.0 | 1.4 | 0.2 | | | |
| *Centropages chierchiae* | 4.5 | | | 75.8 | 34.5 | 0.4 | 2.5 | 53.8 | 60.0 | 41.6 | 28.8 | 53.8 | 1.5 |
| *Eucalanus spec.* | | | | 2.3 | 0.8 | | | 2.0 | 0.7 | | 0.3 | 0.4 | 1.2 |
| *Metridia lucens* | | | 2.4 | | | | | 111.0 | 23.2 | 0.3 | | 7.0 | 11.7 |
| *Paracalanus parvus* | | 142.0 | | 43.4 | | 17.4 | 163.0 | 383.0 | 119.0 | 153.0 | 86.8 | 81.0 | 40.4 |
| *Rhincalanus nasutus* | | | | | 3.3 | 0.7 | | 10.6 | | | | | 1.8 |
| *Temora turbinata* | | | 40.5 | 330.0 | 107.0 | | 34.2 | 63.8 | 5.4 | 124.0 | 26.0 | 2.9 | |
| *Oncaea spec.* | 10.2 | 57.1 | 17.4 | 18.9 | 8.7 | | 63.8 | 54.3 | 9.6 | | 23.2 | 34.7 | 14.3 |
| Other species | 4.3 | | 4.9 | 29.0 | | 9.1 | | 11.2 | 17.3 | 0.2 | 11.7 | 11.8 | 16.7 |

| | Station nr. | | | | | | | | | | | | |
|---|---|---|---|---|---|---|---|---|---|---|---|---|---|
| | 24–2 | 25 | 26 | 28 | 30 | 31 | 32 | 33 | 34 | 35 | 36 | 37 | 38 |
| *Acartia clausi* | | 7.0 | | | | | | | | | | | |
| *Calanoides carinatus* | 5.0 | 194.0 | 0.6 | 29.4 | 187.0 | 46.8 | 10.0 | 10.3 | 4.3 | 4.3 | 270.0 | 10.0 | 38.3 |
| *Candacia armata* | | 4.4 | 1.9 | 2.2 | | | 1.1 | 2.3 | | | | 0.8 | 0.8 |
| *Centropages chierchiae* | 43.7 | 83.2 | 21.5 | 25.1 | 111.0 | 120.0 | 41.4 | 49.9 | 13.3 | 9.1 | 142.0 | 33.7 | 155.0 |
| *Eucalanus spec.* | 3.3 | | 1.1 | | | | | | 1.0 | | | | 1.4 |
| *Metridia lucens* | 3.3 | 4.4 | 0.8 | 0.7 | 5.9 | 5.9 | | 9.1 | 27.0 | 3.4 | 3.4 | 8.2 | 8.2 |
| *Paracalanus parvus* | 52.1 | 52.1 | 69.4 | 39.6 | 103.0 | 122.0 | 65.1 | 36.6 | 29.8 | 16.9 | 273.0 | 137.0 | 1737.0 |
| *Rhincalanus nasutus* | | 13.3 | 0.8 | 3.6 | 5.9 | 11.1 | 0.3 | 2.3 | 15.3 | 0.4 | | | |
| *Temora turbinata* | | 34.7 | | | 11;4 | | | 9.1 | 37.2 | 37.2 | | | 43.4 |
| *Oncaea spec.* | | | 17.4 | 5.7 | 11.4 | 26.0 | 8.7 | 9.1 | 27.9 | 16.9 | 156.0 | 19.5 | 21.7 |
| Other species | 44.4 | 105.5 | 19.7 | 2;2 | 5.9 | 74.1 | 0.6 | 19.2 | 8.5 | 17.2 | 72.6 | 29.1 | 27.3 |

*Table 2.* Density estimates (n $m^{-3}$) of copedpod eggs, nauplii, copepodites and adult copepods at the sampling stations of the 'Pelagic Survey', given in Fig. 1b.

| Species and stage | Station | | | | | | | | | | | | | | | | | |
|---|---|---|---|---|---|---|---|---|---|---|---|---|---|---|---|---|---|---|
| | P21 | P22 | P23 | P30 | P31 | P32 | P33 | P41 | P42 | P51 | P52 | P53 | P61 | P62 | P63 | P71 | P72 | P73 |
| Copepod eggs | 37377 | 9998 | 10748 | 20284 | 32981 | 19997 | 13582 | 10869 | 3746 | 18333 | 3238 | 14574 | 15484 | 9621 | 25381 | 12056 | 42125 | 21614 |
| Copepod nauplii | | | | | | | | | | | | | | | | | | |
| *Acartia clausi* | | 188 | 1131 | | 564 | | | | | 5000 | 4048 | 809 | 1327 | 566 | 3435 | 2127 | 2362 | 2793 |
| *Centropages chierchiae* | 5424 | 1698 | 19519 | 255 | 1129 | 3396 | 3679 | | 212 | | 1619 | 6477 | | 5849 | 190 | | | |
| *Paracalanus parvus* | 4245 | 4056 | 23762 | | | | | | | | | | | | | | | |
| *Temora turbinata* | 2358 | | 1131 | 3816 | 16384 | 7169 | 3773 | | | | 809 | | | 377 | 190 | | 6299 | |
| *Olthona similis* | 9669 | 6603 | 16407 | 19662 | 28248 | 27169 | 12075 | 9782 | 5586 | 13333 | 43724 | 27530 | 15486 | 9811 | 30343 | 21276 | 17716 | 27001 |
| *Oncaea spec.* | 9433 | 2169 | 3111 | 5142 | 1129 | 4905 | 3396 | 3260 | | 833 | 5668 | 5668 | 2654 | 2075 | 11068 | 7092 | 6692 | 9310 |
| All other species | 2357 | 1319 | 9334 | 1081 | 5646 | 886 | 3772 | | 352 | | 1619 | 3238 | | 377 | 381 | | 787 | 931 |
| Copepodites | | | | | | | | | | | | | | | | | | |
| *Calanoides carinatus* | 6.9 | 26 | 21.4 | | 4.5 | 53.5 | 37.5 | 19.6 | 37.4 | 1.1 | 3.8 | 22.5 | | 0.4 | 26.2 | 1.1 | 0.3 | 24.3 |
| *Temora turbinata* | 235 | | 585 | 1846 | 1129 | | | | | | 1619 | 1619 | | | | | | 931 |
| *Euterpina acutifrons* | 4952 | 94 | | 11968 | 12994 | 7169 | 5094 | 4347 | 777 | 2500 | 10526 | 809 | 3982 | 566 | 4389 | 4964 | 4330 | |
| *Paracalanus parvus* | 7547 | 2924 | 8203 | 5979 | 12994 | 754 | 6603 | 3260 | 636 | | 809 | 5668 | 442 | 2452 | 381 | | | 1862 |
| All other species | 1747 | 489 | 815 | 1653 | 6550 | 1057 | 1218 | 373 | 337 | 156 | 218 | 234 | 21 | 140 | 238 | 63 | 83 | 235 |
| Copepod adults | | | | | | | | | | | | | | | | | | |
| *Acartia clausi* | 0 | 8.7 | | 71.4 | | | | 25 | | 3.8 | 5.6 | 6.6 | 64;4 | 35.1 | 9.2 | 129 | 49.5 | 53 |
| *Calanoides carinatus* | 65.8 | 142 | 150 | 13.7 | 13.6 | 79.1 | 20.5 | 3.8 | 34.8 | 0.5 | 1.4 | 20.7 | 0.2 | 0.4 | 3.3 | 0.1 | 0.1 | 57.8 |
| *Candacia armata* | | 0.5 | 1.6 | | 0.3 | 1.5 | 0.9 | 0.1 | 0.7 | | | 0.3 | | | 1.2 | | | 0.4 |
| *Centropages chierchiae* | 325 | 95.2 | 55.1 | 80.6 | 164 | 140 | 214 | 28.7 | 54.8 | 2.3 | 3.2 | 70 | 0.1 | | 58.2 | 1.1 | 0.3 | 21.2 |
| *Eucalanus spec.* | 0.2 | 0.4 | 0.6 | | 0.1 | 0.5 | 0.8 | | 0.1 | | 0.1 | 0.5 | | 0.3 | 0.5 | | | |
| *Metridia luscens* | | 0.1 | 0.1 | | | | 0.6 | | | | | | | | 0.2 | | | |
| *Paracalanus parvus* | 2122 | 943 | 564 | 2281 | 6778 | 377 | 1697 | 611 | 424 | 809 | | 1251 | 819 | 566 | 190 | | 393 | |
| *Rhincalanus nasutus* | | 0.7 | 2.1 | 0.1 | 0.2 | 0.8 | 0.8 | 0.2 | 0.8 | 0.2 | | | | | 0.1 | | | |
| *Temora turbinata* | 142 | 13.9 | 501 | 39.1 | 3.2 | | 169 | | 45.5 | 5.6 | | 4.7 | 35.9 | | | | 22.4 | 88.5 |
| *Oithona similis* | 6367 | 3490 | 15257 | 7142 | 16949 | 6792 | 7169 | 5434 | 1131 | 833 | 17004 | 5668 | 7079 | 4339 | 7251 | 2127 | 8661 | 14897 |
| *Oncaea spec.* | 1650 | 2830 | 5374 | 4162 | 5649 | 2641 | 3962 | 108 | 1485 | | 2429 | 1619 | 5309 | 1698 | 5152 | 3546 | 4724 | 25139 |
| All other species | 12 | 108 | 881 | | 33788 | 1141 | 568 | | 156 | 860 | | 830 | 884 | | 1074 | 709 | 1973 | 1090 |

stands in contrast to the situation in the North, where low temperatures point to recent inflow. The rather low oxygen saturations in the surface water at transects VI and VII (Fig. 3) indicate this area to be covered with fresh upwelling water indeed.

*Copepod distribution*

Density estimates (n. $m^{-3}$) for the adults of the main copepod species encountered in the 'Hydrographic Survey' zooplankton samples are given in Table 1. In total 35 copepod species were found, 6 species of which occurred in very varying abundances almost everywhere whereas 16 were found at very few stations only. The adult copepod densities at the 18 stations of the restricted but more detailed 'Pelagic Survey' can be found in Table 2, together with densities of copepod eggs, nauplii and copepodites.

The distributions of adult copepods of the dominating species, based on all samples, are given in Figs 4 and 5. They are listed in order from truly oceanic (Oc) species to mainly neritic (Ne) ones, according to the specifications given in Vives (1982). Figure 6 presents the distributions of nauplii and copepodites of some of the species.

A number of copepod species (*Rhincalanus nasutus* (Giesbrecht), *Metridia lucens* (Boeck), *Eucalanus* spec. and *Candacia armata* (Boeck), listed by Vives (1982) as purely oceanic, appear to be confined to the farest and deepest stations on the research grids indeed; they were only accidently found inside the 200 m isobath (Fig. 4).

*Centropages chierchiae* (Giesbrecht), *Paracalanus parvus* (Claus) and *Temora turbinata* (Dana) (Fig. 5), are known as oceanic/neritic (Vives, 1982), and appeared to occur predominantly at the continental slope, with the emphasis on the area inside the 20 m isobath off Cap Timiris.

Further inshore, the only copepods generally met were species of the genus *Oncaea*, and *Acartia clausi* (Giesbrecht). *Oncaea* spec. (Fig. 5) are real neritic surface dwellers; *A. clausi* has been described by Vives (1982) as both oceanic and neritic. We observed *A. clausi* exclusively at the most shallow shelf stations. It is therefore regarded as a neritic copepod, in accordance with Schulz (1982), who mentions that *A. clausi* found in the ocean can be driven with surface water far from the coast by the trade winds.

Regarding the abundance of adult copepods, the central part of the research area (transects IV and V), where saline and warm Bancwater dominates the continental slope area (see 3.1), draws attention because of its striking minimum in the distributions of almost all species presented in Figs 4 and 5. While some of the oceanic species (Fig. 4) were encountered in limited numbers on the slope, they never occurred in the central area. The neritic/oceanic species with concentrations on the slope (Fig. 5) were found in the South and North, but never in the central area. The neritic genus *Oncaea* and the species *A. clausi*, which inhabit the continental shelf (Fig. 5), were scarce in the central area as well.

The distribution of the oceanic species *Calanoides carinatus* (Kröyer) is of special interest, because its adults and particularly its copepodites stage V are upwelling indicators (Bainbridge, 1960). According to Fig. 4 the adults occurred, apart from some deeper ocean stations – where they live normally at 300–450 metres depth (Schulz, 1982) – in two shallower areas: they were found at the steeper part of the continental slope in the South of the survey-area, and on the actual shelf close to Cape Blanc. The distribution of *C. carinatus* copepodites (Fig. 6) shows basically three similar concentrations, with the emphasis on the northern shelf area close to Cap Blanc.

*Other plankton*

Table 3 gives the biomass of the different categories of zooplankton on the 18 stations of the Pelagic Survey. The distribution of dominating systematic groups is summarized in Table 4. The areal distribution of the non-copepod zooplankton was, similar to that of the copepods, rather clear-cut in the southern upwelling area. The first six groups of Table 4 were not found in the south-

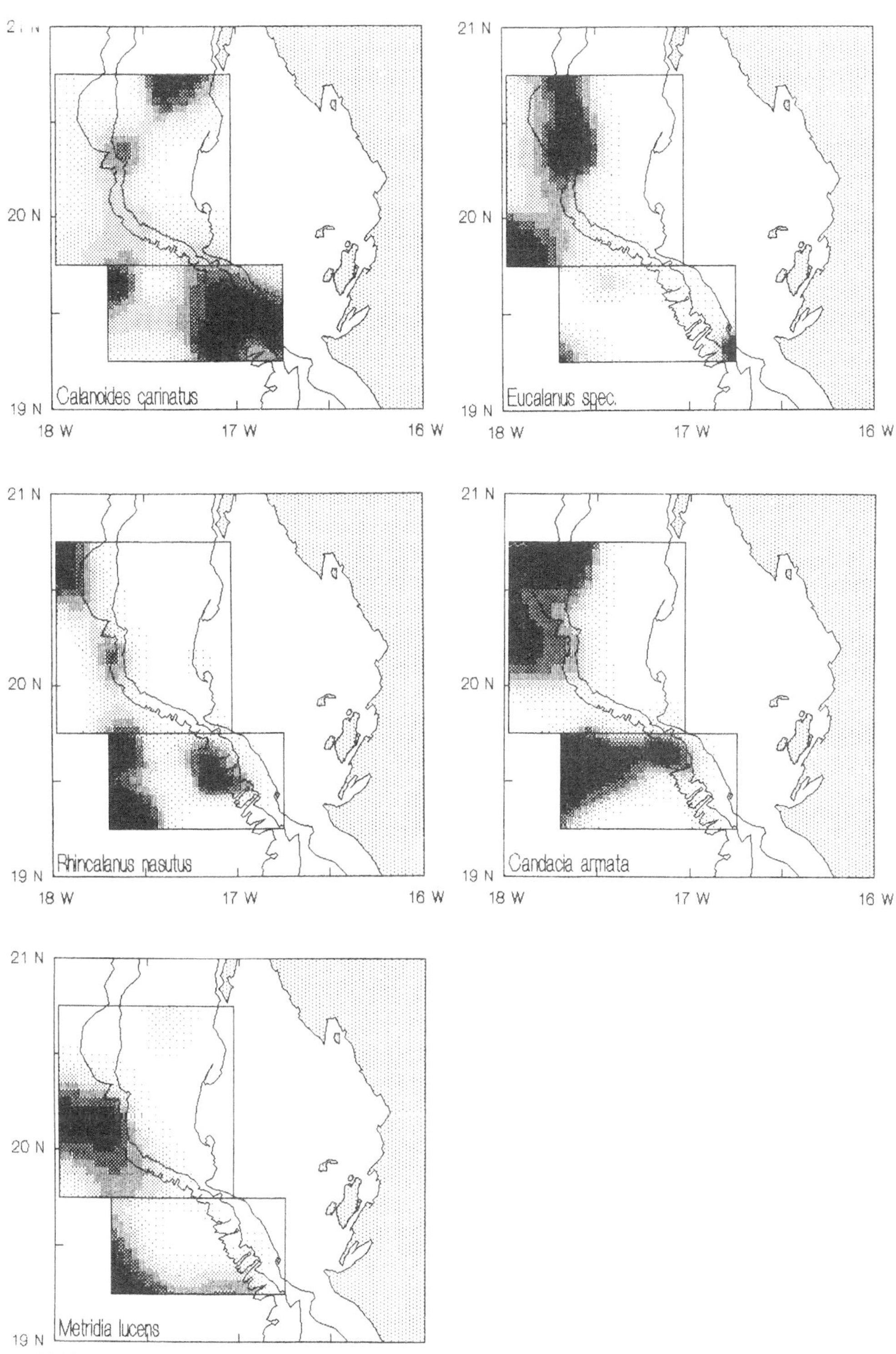


*Fig. 4.* Density distribution of adult oceanic copepods over the sampling grids (all stations). Scales are logarithmic with 10 equal steps between the maximum (black) and zero (white). The species are: *Calanoides carinatus* Kröyer (Oceanic and Neritic, normally occurring from 0–2000 m depth); *Rhincalanus nasutus* Giesbrecht (Oc, 0–1000 m); *Metridia lucens* Boeck (Oc, 0–1000 m); *Eucalanus* spec. (Oc, 0–1000 m) and *Candacia armata* Boeck (Oc, 0–1000 m).

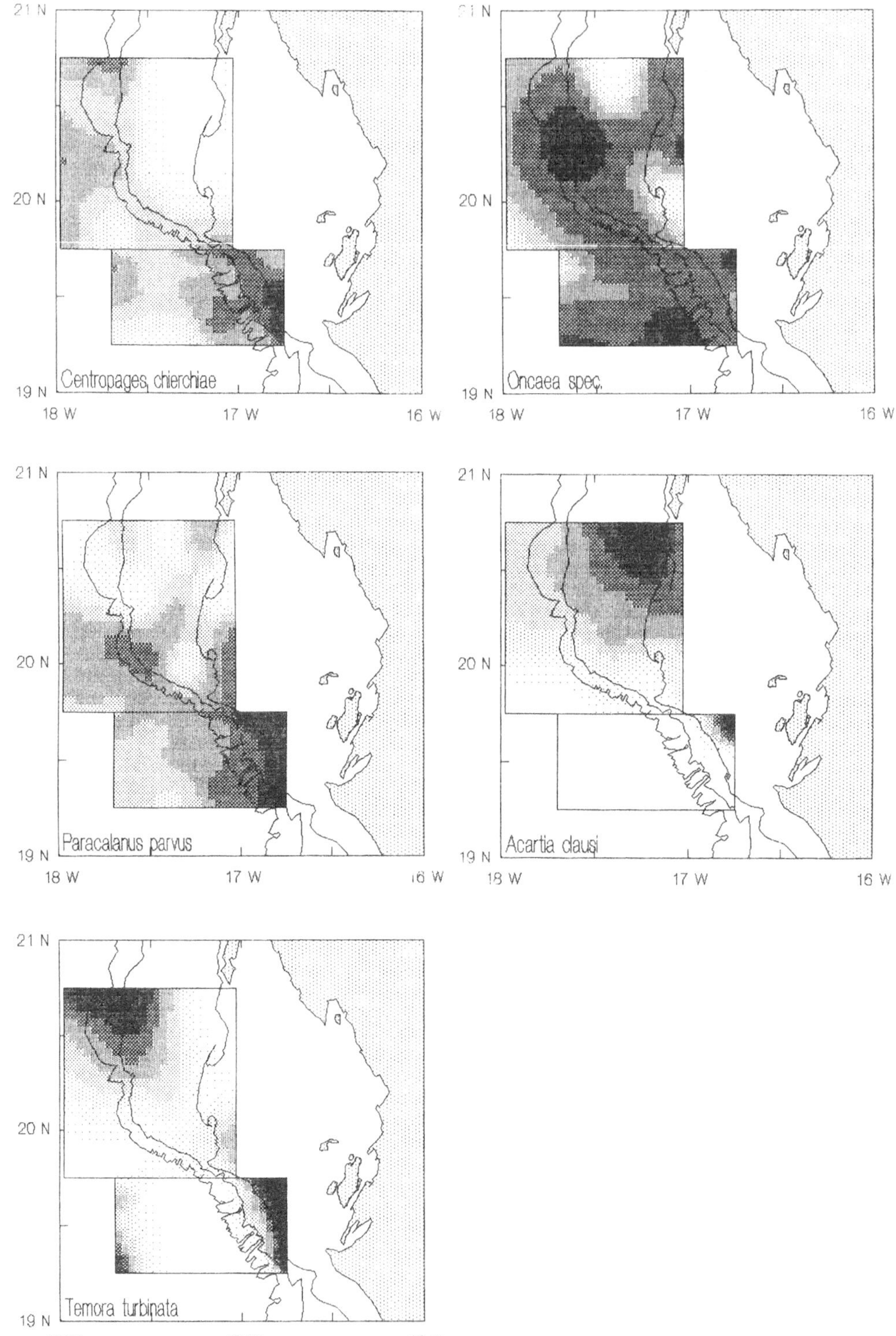


*Fig. 5.* Density distribution of adults of oceanic/neritic and neritic copepods over the sampling grids (all stations). Scales as in Fig. 4. Shown are: *Centropages chierchiae* Giesbrecht (Oc and Ne, 0–1000 m); *Paracalanus parvus* Claus (Ne and Oc, = 0–500 m); *Temora turbinata* Dana (Ne and Oc, 0–200 m); *Oncaea* spec. (Ne and Oc, 0–50 m) and *Acartia clausi* Giesbrecht (Ne and Oc, 0–500 m).

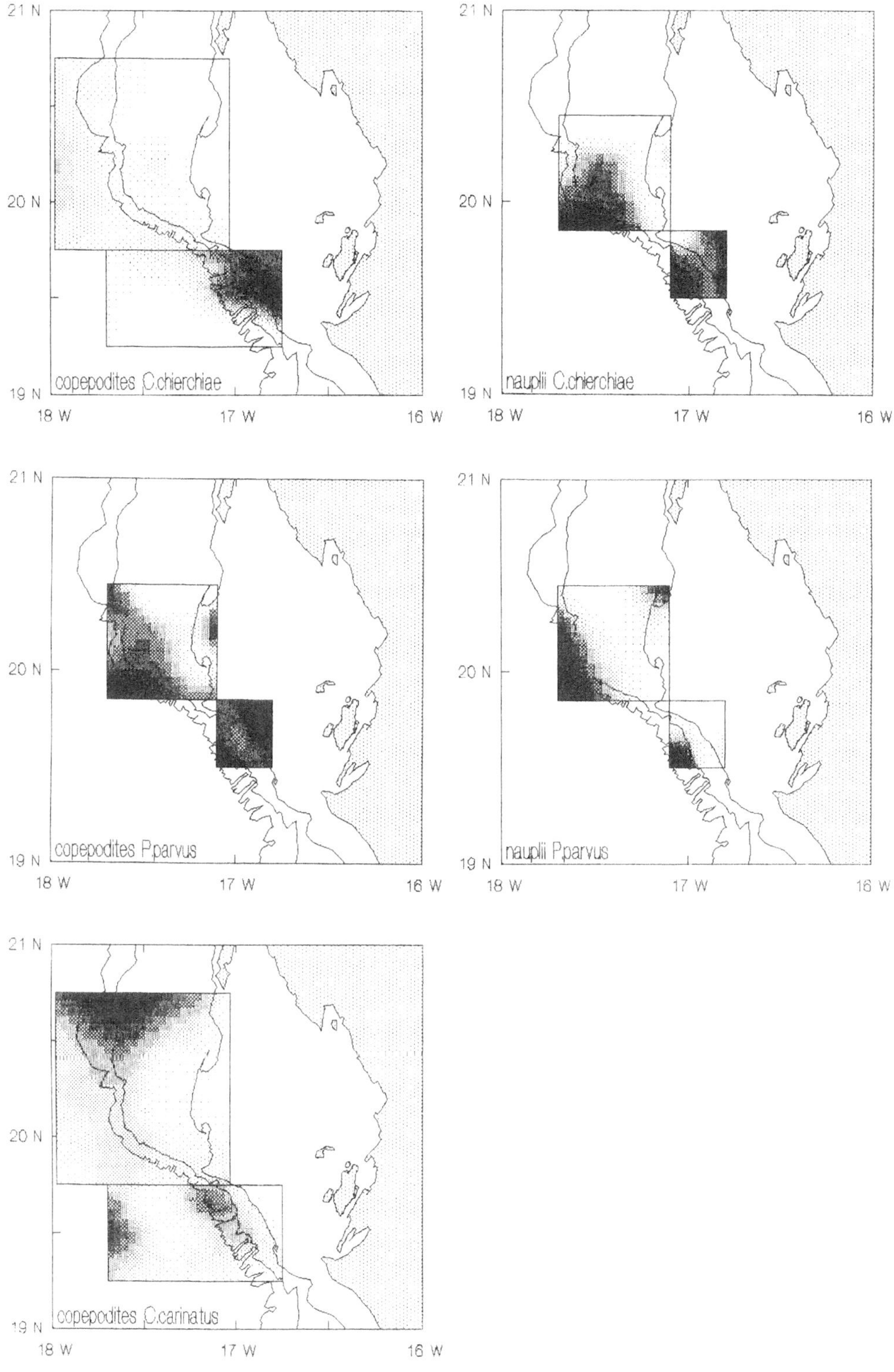


*Fig. 6.* Density distribution of copepodites and nauplii of *C. chierchiae* and *P. parvus*, and copepodites of *C. carinatus*. For scales see Fig. 4.

*Table 3.* Biomass estimates in ashfree dryweight of zooplankton at the stations of the 'Pelagic Survey' given in Fig. 1b, for different developmental stages of copepods and species of 'other mesozooplankton' and macrozooplankton separately.

| Species or group | Station | | | | | | | | | | | | | | | | | |
|---|---|---|---|---|---|---|---|---|---|---|---|---|---|---|---|---|---|---|
| | P21 | P22 | P23 | P30 | P31 | P32 | P33 | P41 | P42 | P51 | P52 | P53 | P61 | P62 | P63 | P71 | P72 | P73 |
| Copepods (mg per $m^{-3}$) | | | | | | | | | | | | | | | | | | |
| Eggs | 3.9 | 1.8 | 1.6 | 8.2 | 10.1 | 4.3 | 2.6 | 1.1 | 1.3 | 8.0 | 0.3 | 1.9 | 5.8 | 1.1 | 7.3 | 1.2 | 14.8 | 2.7 |
| Nauplii | 12.8 | 6.7 | 48.3 | 9.5 | 22.8 | 21.7 | 23.7 | 3.0 | 2.0 | 5.2 | 21.5 | 17.3 | 8.9 | 8.5 | 19.1 | 11.9 | 12.4 | 13.6 |
| Copepodites | 54.2 | 17.2 | 46.3 | 47.7 | 108.9 | 19.9 | 42.2 | 16.7 | 5.6 | 3.2 | 21.7 | 27.8 | 9.3 | 11.6 | 9.1 | 5.3 | 4.7 | 23.4 |
| Adults | 59.4 | 45.8 | 74.0 | 68.5 | 118.3 | 43.2 | 54.5 | 22.0 | 17.2 | 11.4 | 39.6 | 46.0 | 41.8 | 21.1 | 39.6 | 19.2 | 42.8 | 132.4 |
| Other mesozooplankton (mg per $m^{-3}$) | | | | | | | | | | | | | | | | | | |
| *Noctiluca* sp. | 4.5 | 2.7 | 0.6 | 58.9 | 47.5 | 9.8 | 1.3 | 9.8 | 7.3 | 10.0 | 7.3 | 15.4 | 2.2 | 4.0 | 13.8 | 1.4 | 4.3 | 0.9 |
| *Polychaetes* | 4.7 | | | 71.6 | 124 | | 3.8 | 21.7 | 4.2 | 33.3 | 16.2 | | 44.2 | 19.1 | | 42.6 | 23.6 | 18.6 |
| *Gastropod larvae* | | | | 3.3 | 4.5 | 0.8 | 0.4 | | 0.7 | | 4.9 | 1.6 | 1.8 | 3.8 | 0.8 | 2.8 | 0.8 | 1.9 |
| *Lamellibranch larvae* | 0.9 | | 0.6 | 0.3 | | | | 2.2 | 1.3 | 6.3 | 17.8 | | 15.9 | 13.7 | | 5.7 | 1.6 | 27.9 |
| *Cladocerans* | | | | | 0.1 | 0.1 | | 0.5 | 0.1 | | 0.1 | | | | | 0.1 | | |
| *Balanidea* | | | | | 28.2 | | | | | | | | | | | | | |
| *Amphiodus lanceolatus* | | | | 0.0 | 0.0 | | | 0.0 | 0.0 | | | | 0.0 | | | | 0.1 | 3.5 |
| *Appendicullarlans* | 4.7 | 3.3 | 28.3 | 0.7 | 11.3 | 7.5 | 6.6 | 54.3 | 5.0 | 217 | 101 | 85.0 | 22.1 | 25.8 | 21.7 | 7.1 | 9.8 | |
| Macrozooplankton (mg per 1000 $m^{-3}$) | | | | | | | | | | | | | | | | | | |
| *Hydrozoa* | 275 | 0.1 | | 3157 | 195 | | 76.5 | 306 | 37.5 | 241 | 10.5 | 116 | 756 | 121 | 81.0 | 1.0 | | 238 |
| *Siphonophores* | 1.1 | 33.0 | 16.5 | | 68.2 | 2.5 | 91.8 | 13.5 | 125 | 444 | 12.5 | 4.8 | 202 | 12.7 | 14.0 | 2482 | 895 | 152 |
| *Cephalopods* | | 0.3 | | | | | | | | | 0.1 | 8.0 | 0.1 | | | | | |
| *Amphipods* | 2.0 | 0.1 | | | 0.1 | 0.2 | 2.0 | 52.5 | | | | 3.0 | 0.8 | | 9.0 | 1.0 | | 16.0 |
| *Mysids* | | | | | | | | 16.0 | 271 | | | | 1.0 | 1.0 | | 0.3 | 9.0 | 57.0 |
| *Euphausids* | 0.4 | 11.0 | 30.0 | | | 64.0 | 72.0 | 5.0 | 3.0 | | 0.5 | 43.0 | 1.0 | 0.2 | 380 | 0.8 | 0.5 | 82.0 |
| *Meroplankton* | 119 | 266 | 234 | | 103 | 2379 | 89.0 | 2786 | 1288 | 214 | 16.0 | 191 | 97.1 | 68.5 | 370 | 384 | 41.6 | 1627 |
| *Decapods* (adults) *pelagic* | | 0.3 | | | | | | | | | | | | | 342 | | | |
| *Chaetognaths* | 757 | 1029 | 800 | 23458 | 5885 | 3502 | 945 | 3209 | 1232 | 4433 | 183 | 225 | 209 | 251 | 451 | 44.3 | 259 | 2307 |
| *Salps* | | 0.2 | | | | | 17.0 | | 0.2 | | | | | | | | | |
| *Myctophidae* | | | | | | | | | | | | | | | 18.0 | | | |
| *Gobiidae* | 0.3 | | | | | | | | | | | | | | | | | |
| | | | | | | | | | | | | | | | | | | 0.1 |
| *Demersal fish* (larvae) | 1.0 | 0.0 | 0.2 | 0.3 | 5.0 | 7.0 | 0.2 | 4.0 | 8.0 | 3.0 | 3.0 | | | 0.3 | 0.1 | | 0.0 | 1.0 |
| *Anguilla anguilla* (elver) | | | | | | | | | | | | 1.0 | | | 1.0 | | | |
| *Pelagic fish* (larvae) | 139 | 15.0 | 11.8 | 1.0 | 34.0 | 46.6 | 9.4 | 178 | 98.0 | 127 | 155 | 16.0 | 12.4 | 26.4 | 8.1 | 33.7 | 2.6 | 10.4 |

*Table 4.* Occurrence of zooplankton other than copepods at different parts of the Mauritanian coastal upwelling system.

| Plankton group | Southern grid | | | Northern grid | | |
|---|---|---|---|---|---|---|
| | Shelf | Slope | Ocean | Shelf | Slope | Ocean |
| *Noctiluca* | ++ | – | – | + | – | + |
| *Polychaetes* | ++ | – | + | + | +– | – |
| *Gastropod larvae* | ++ | – | +– | + | + | – |
| *Appendicularians* | + | – | + | ++ | + | + |
| *Hydrozoa* | ++ | – | + | + | +– | + |
| *Siphonophora* | ++ | – | + | ++ | +– | – |
| *Bivalve larvae* | – | – | – | + | + | – |
| *Euphausids* | – | ++ | – | – | – | ++ |
| *Crustacean larvae* | – | ++ | – | + | + | ++ |
| *Chaetognaths* | ++ | + | + | + | + | + |
| *Demersal fish larvae* | – | ++ | – | – | ++ | – |
| *Pelagic fish larvae* | – | ++ | – | + | ++ | – |

ern upwelling area on the slope, whereas they were relatively abundant on the shelf. In the North the distribution patterns are less distinct. In contrast to copepods, most species of other zooplankton groups were present on the slope as well as on the shelf.

The *C. carinatus* patch in the South (Fig. 4) appeared to be also the domain of euphausids, fish-larvae of crustaceans. In the North on the shelf, however, euphausids were not found together with *C. carinatus.*

## *Zooplankton biomass and grazing*

With some assumptions the grazing of presumably mainly herbivorous part of the meso- and macrozooplankton can be estimated on basis of biomass. For this purpose predators like chaetognaths were left out of the macroplankton and errant polychaetes out of the mesozooplankton. Biomass estimates, based on measured densities, size distributions, and length/weight-relationships (see 2.2) are given in Fig. 7 for five groups of herbivores, for all stations in ascending order of bottom-depth.

A rough estimate of food-consumption in mg per $m^3$ under ad libitum feeding conditions could be obtained by assuming that the zooplankton (with a C to ADW ratio of *ca* 0.4) produces *ca* 10% of its body- weight per day (an estimate based on an average generation time of about one month, with a food conversion efficiency of roughly 30% (Fransz *et al.*, 1991)), leading to an ingestion roughly equal to one third of the grazer biomass per day. These maximal grazing estimates are summarized in Table 5. Although total biomass, as well as the share of copepods, appears to be rather fluctuating, four situations can be recognized:

a. shallow stations on the shelf to 21 m depth, poor in copepods, but with abundant other

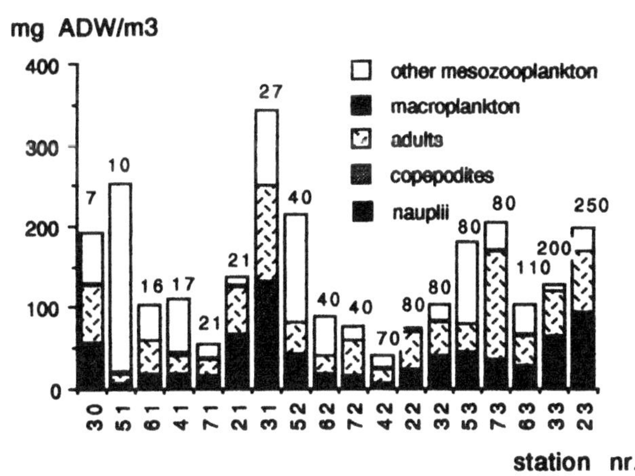


*Fig. 7.* Biomass (mg AFDW $m^{-3}$) of different zooplankton components at the stations of the 'Pelagic Survey' (Fig. 1b). Station depth (m) indicated above the bars.

*Table 5.* Maximum estimates of grazing by all herbivores in mg C m$^{-3}$ d$^{-1}$ at the 'Pelagic Survey' sampling stations given in Figure 1b. Stations listed in order of depth.

| | | | | | | | | | |
|---|---|---|---|---|---|---|---|---|---|
| Station: | 3–0 | 5–1 | 6–1 | 4–1 | 7–1 | 2–1 | 3–1 | 5–2 | 6–2 |
| Depth in m | 7 | 10 | 16 | 17 | 21 | 21 | 27 | 40 | 40 |
| Max. grazing: | 25.7 | 33.8 | 13.7 | 14.9 | 7.5 | 18.2 | 45.6 | 28.5 | 11.8 |
| Station: | 7–2 | 4–2 | 2–2 | 3–2 | 5–3 | 7–3 | 6–3 | 3–3 | 2–3 |
| Depth in m | 40 | 70 | 80 | 80 | 80 | 80 | 110 | 200 | 250 |
| Max. grazing: | 10.3 | 5.4 | 10.0 | 13.7 | 24.2 | 27.3 | 13.8 | 18.2 | 26.4 |

mesozooplankton (i.e. according to Table 3: Appendicularia, larvae of different benthic invertebrates, and *Noctiluca spec.*), where total maximal grazing on average is approximately 19 mg C m$^{-3}$ d$^{-1}$. This estimate seems to represent 'Bancwater'.

b. stations near the shelf edge (21–40 m depth) with high copepod biomasses, where grazing estimates are 18, 45 and 28 mg C m$^{-3}$ d$^{-1}$ with an average of 30 mg, representing upwelling enriched shelfwater.
c. stations on the shelf edge (40–80 m depth) with low copepod- as well as other zooplankton biomasses, with an average grazing estimate of 10 mg C m$^{-3}$ d$^{-1}$, representing the freshly upwelled water itself,
d. stations, where conditions are increasingly oceanic (80–250 m depth), where total zooplankton biomass, dominated by copepods, increases gradually with depth, and average grazing comes to ca. 22 mg C m$^{-3}$ d$^{-1}$.

When grazing in vertically mixed shallow waters occurs over the whole watercolumn, maximum estimated grazing in 'Bancwater' (a) mentioned above is roughly 280 mg C m$^{-2}$ d$^{-1}$; on the upwelling enriched shelf edge (b) 900; in relatively new upwelling water (c) 600 mg C; and that in oceanic stations (d) – when grazing is equally distributed over the upper 50 metres of the watercolumn – *ca* 1100 mg C m$^{-2}$ d$^{-1}$.

## Discussion

In contrast to productivity and biomass of the plankton, which in systems with strong seasonal fluctuations depend rather directly on boundary conditions, the benthic biomass can be regarded as an average reflection of fluctuating import and sedimentation of organic matter. In the most extreme case, one heavy pulse per year could be enough to raise the biomass of the macrobenthos to a steady high level. In the case of the Banc d'Arguin, where a relatively high benthos biomass is one of the key-factors in the attractiveness of the area as a habitat for migrating and wintering waders (Wolff & Smit, 1990), energy-pulses to the benthos must quite logically be expected to occur during the upwelling seasons. It is, however, not clear how during the periods of the strongest offshore tradewinds water could be transported from the ocean over the shelf to the inshore banks. Although for the crucial observations the present transects should have covered the entire shelf, the present restricted 1988 datasets on hydrography and zooplankton distribution together point to one possible way along which offshore upwelling can enhance inshore benthic productivity.

Temperature profiles at four of the six transects (Fig. 2) show that the water on the shelf (as far as visited) was entirely vertically mixed, which might be expected in a sea area less than 25 m deep during strong wind. This excludes the possibility of a vertical circulation of ocean upwelling water, flowing along the bottom into the shelf sea to compensate the deficit of surfacewater driven away from the shore – at least at the time of observation. As the temperature profiles suggest, it seems more likely that a large scale horizontal exchange had taken place shortly before the time of sampling. Bancwater driven away from inshore areas under the prevailing Nort-Easterly wind, seems to haven been replaced by ocean water risen to the surface along the continental talud West or North-West of Cap Blanc. Although the low temperatures of the surface water at the two Northern transects of Fig. 2 point to upwelling, they do not prove the deep oceanic origin of this water. However, low oxygen saturations (Fig. 3), make it very unlikely that we are dealing with a mass of cold Northern surface water from the Canary Current. Moreover, the presence of adults and copepodites of the species *Calanoides carinatus* (Fig. 4 and 6) can only be explained, when the cold watermass covering the shelf South of Cap Blanc to a considerable degree originates from a recent upwelling. Apart from the presence of the upwelling indicator (Bainbridge, 1960; Vives, 1974; Weikert, 1982), the cold watermass is characterized by very low POC concentrations (Fig. 3), which stresses its recent ascent from below the euphotic zone. Otherwise, according to the generally accepted view on planktonic developments after upwelling (see Boje & Tomczak, 1978), algal blooms would have started to increase the organic content.

The 'Banc-water', replaced by the cold watermass entering the Northern part of the shelf area, can be recognized most distinctly by its extremely high POC content and oxygen saturation to dominate the two central transects of Fig. 3, where it protrudes considerably over the shelf-edge into the ocean. This water is substantially warmer than the ocean itself, and therefore forms a distinct surface layer at stations deeper than 25 metres. Zooplankton sampling showed the distribution of this watermass to coincide with a minimum in the density of all copepod species (see 3.2). Although it might be quite acceptable to assume that the warm surfacewater has been driven by the strong offshore wind from the shelf into the ocean, the bay between Cap Blanc and Cap Timiris seems to small to assume such wind-effect only at the central transects in order to explain the inflow of cold water in the North as a compensatory current. Therefore, we prefer the hypothetical explanation that during the upwelling season the waterlevel in the bay as a whole is affected negatively by strong easterly trade winds, but that there are irregular intervals with less wind as well during which water from the ocean enters the shelf area. During the our expedition, the wind came predominantly from the North-East and varied in strength between 4 and 7 on the scale of Beaufort indeed. Since the Canary Current flows outside the shelf in a Southern direction, the deficit in the Bay will be compensated at times of less wind-stress to a certain degree with Canary Current water, especially in its Northern part. Moreover, it seems likely as well that, shortly after periods with heavy offshore wind, fresh upwelling of deeper water has reached a maximum outside Cap Blanc. Hence, compensatory movements of Canary Current water into the Banc d'Arguin region might well be expected to bring freshly ascended deep water into the bay. Although there is a lack of direct evidence, it would according to the above logic not be the trade wind itself, but rather its variability which offers an explaination for the flow of cold nutrient rich ocean water in the direction of the Banc d'Arguin.

Once considerable bodies of deep oceanwater have penetrated the shelf sea (the case described above refers to a mass of several km$^3$ as indicated by the size of the *C. carinatus* copepodite patch shown in Fig. 6) the most crucial step in the enriching process seems to be taken. During further circulation and mixing by wind and tides the new nutrients will be incorporated into the actual 'Banc-water' system. Such mixing is suggested by the gradual increase of salinity along transects VI and VII in Fig. 3. New phytoplankton production will follow in the shallow water and contribute to the high oxygen-saturation and POC-content observed at the central transects (Fig. 2).

Of the grazing-estimates given in 3.4, those for 'Banc-water' are considerably lower than the other estimates, which is according to Figs. 4, 5 and 7 especially due to low copepod density and biomass. Most of the grazing in 'Banc-water' must be attributed to the category 'other mesozooplankton' in which appendicularians, small polychaetes and *Noctiluca* spec. have the main share (see Table 3). Together, the grazers were estimated to remove only 280 mg C m$^{-2}$ d$^{-1}$ from water where a primary production was measured

– although only at one station – of approximately 8 gram C $m^{-2}$ $d^{-1}$ (Hegeman, pers. comm.).

To our opinion the shallow depth, and especially the high temperature and salinity attained by 'Banc-water' during its stay in the inshore area are the factors responsible for the low pelagic grazing pressure. Such surrounding will be completely unfit for all copepodes of the deeper and colder water of the Canary Current (Weikert, 1977). Further, the domination of the grazers by the small group of mesozooplankton mentioned above suggests that it may be doubted whether the phytoplankton species blooming in a rare water type like that covering the Banc d'Arguin are suitable as copepod food.

The marked concentration of copepod species at the shallow stations of the Southern transects II and III (Fig. 4 and 5) can be explained (following the generally accepted paradigm) by upwelling of nutrient rich water against the slope to depth of less than 25 metres, leading to enhanced production of the favourable diatoms. The presence of copepod naupli and copepodites (Fig. 6), proof of successful local reproduction and hence, of good feeding conditions, supports this view. 'Banc-water' was not encountered here.

When during the upwelling seasons the shallow sea between Cap Blanc and Cap Timiris receives substantial nutrient inputs from the ocean indeed, and the resulting new production is hardly consumed in the watercolumn itself due to unfavourable conditions for grazers, it might well be the benthos which profits greatly from the fertilizing effect of entering upwelling water. Among the macrobenthos, it will in the first place be the suspension feeding molluscs which could benefit from the high food-concentration. Deposit feeders could, however, only profit from the enhanced phytoplankton production when a substantial sedimentation took place. This seems not very likely during turbulent periods of strong wind. Parts of the inshore area with a dence vegetation of seagrass might, however, act as sediment traps, and play a key role in the incorporation of energy subsidies from the ocean by the Banc d'Arguin's benthic system, which as a whole could – concluding – be characterized as a 'benthic upwelling system'.

## Acknowledgements

The present study was carried out under financial support of the Netherlands Marine Research Foundation, SOZ.

## References

Baars, M. A., A. B. Sutomo, S. S. Oosterhuis & O. H. Arinardi, 1990. Zooplankton abundance in the eastern Banda Sea and northern Arafura Sea during and after the upwelling season, August 1984 and February 1985. Neth. J. Sea Res. 25: 527–543.

Bainbridge, V., 1960. Occurrence of *Calanoides carinatus* Kröyer in the plankton of the Gulf of Guinea. Nature, 188 (4754): 932–933.

Boje, R. & M. Tomzcak, 1978. Ecosystem Analysis and the Definition of Boundaries in Upwelling Regions. In R. Boje & M. Tomzcak (eds), Upwelling Ecosystems, Springer-Verlag, Berlin Heidelberg New York, 1978: 3–11.

Cadée, G. C., 1986. Organic carbon in the water column and its sedimentation, Fladen Ground (North Sea), May 1983. Neth. J. Sea Res. 20: 347–358.

Daan, R., 1987. Impact of the egg predation by *Noctiluca miliaris* on the summer development of copepod population in the Southern North Sea. Mar. Ecol. Prog. Ser. 37: 9–17.

Fransz, H. G., 1976. The spring development of calanoid copepod populations in the Dutch coastal waters as related to primary production. In G. Persoone & E. Jaspers (eds), Proc. 10th E.M.B.S., Oostende. Universal Press, Wetteren; Belgium: 247–269.

Fransz, H. G., 1988. Vernal abundance, structure and development of epipelagic copepod population of the Eastern Weddell Sea (Antartica). Polar Biol. 9: 107–114.

Fransz, H. G. & W. G. van Arkel, 1980. Zooplankton activity during and after the phytoplankton spring bloom at the central station in the FLEX box, Northern North Sea, with special reference to the calanoid copepod *Calanus finmarchicus* (Gun). 'Meteor' Forsch. Ergebn. A 22: 113–121.

Fransz, H. G., J. M. Colebrook, J. C. Gamble & M. Krause, 1991. The Zooplankton of the North Sea. Neth. J. Sea Res. 28 (1/2): 1–52.

Fransz, H. G., J. C. Miquel & S. R. Gonzalez, 1984. Mesozooplankton composition, biomass and vertical distribution, and copepod production in there stratified central North Sea. Neth. J. Sea Res. 18: 82–96.

Isaacs, J. D. & Kidd, L. W., 1953. Isaacs-Kidd midwater trawl. Final report. Scripps Institution of Oceanography Ref. 53.-3, Oceanographic equipment report 1.

Klein Breteler, W. C. M., H. G. Fransz & S. R. Gonzalez, 1982. Growth and development of calanoid copepods under experimental and natural conditions. Neth. J. Sea Res. 16: 195–207.

Margalef, R., 1975. Composicion y distribucion del fitoplancton marino en la region de afloramiento del NW de Africa, en marzo de 1973 (Campana 'Atlor II' del 'Cornide de Saavedra'). Res. Exp. Cient. B/O Comide. 4: 131–144.

Nassogne, A., 1972. Etudes préliminaires sur le rôle du zooplankton dans la constitution et le transfert de la matière organique au sein de la chaîne alimentaire marine en mer Ligure. Thesis Univ. Amsterdam: 1–237.

Pavlov, V. Y., 1968. Plankton distribution in the Cape Blanc region. Oceanology (Moscow), 8: 381–387.

Reyssac, J., 1977. Hydrologie, phytoplancton et production primaire de la baie du Lévrier et du Banc d'Arguin. Bull. Inst. Franç. Afrique Noire 39 A (3): 488–554.

Schulz, K., 1982. The vertical distribution of calanoid copepods north of Cape Blanc. Rapp. P.-v. Réun. Cons. int. Explor. Mer, 180: 280–288.

Smed, J., 1982. The oceanographic data base for the CINECA region. Rapp. P.-v. Réun. Cons. int. Explor. Mer, 180: 11–28.

Vives, F., 1974. Le zooplankton et les masses d'eau des environs du Cap Blanc. Thetys 6: 313–318.

Vives, F., 1982. Sur les copepodes de la région CINECA (Parties nord et centrale). Rapp. P.-v. Réun. Cons. int. Explor. Mer, 180: 289–296.

Weikert, H., 1977. Copepod Carcasses in the Upwelling Region South of Cap Blanc, N.W. Africa. Mar. Biol. 42: 351–355.

Weikert, H., 1982. Some features of zooplankton distribution in the upper 200 m in the upwelling region off Northwest Africa. Rapp. P.-v. Réun. Cons. int. Explor. Mer, 180: 280–288.

Wolff, W. J. & C. J. Smit, 1980. The Banc d'Arguin, Mauritiana, as an environment for coastal birds. Ardea 78: 17–38.

*Hydrobiologia* **258**: 151–163, 1993.
*W. J. Wolff, J. van der Land, P. H. Nienhuis & P. A. W. J. de Wilde (eds), Ecological Studies in the Coastal Waters of Mauritania.*




# Biomass of macrobenthic tidal flat fauna of the Banc d'Arguin, Mauritania

W. J. Wolff[1,2], A. G. Duiven[2,3], P. Duiven[4,5], P. Esselink[4], Abou Gueye[6], A. Meijboom[3], G. Moerland[2,3] & J. Zegers[3]
[1]*DLO – Institute for Forestry and Nature Research, P.O. Box 46, 3956 ZR Leersum, The Netherlands;* [2]*Dept. of Nature Conservation, Agricultural University, P.O. Box 8080, 6700 DD Wageningen, The Netherlands;* [3]*DLO – Institute for Forestry and Nature Research, P.O. Box 167, 1790 AD Den Burg, Texel, The Netherlands;* [4]*Stichting Internationaal Wad- en Watervogelonderzoek c/o Driebergseweg 16c, 3708 JB Zeist, The Netherlands;* [5]*Netherlands Institute for Sea Research, P.O. Box 59, 1790 AB Den Burg, The Netherlands;* [6]*Parc National du Banc d'Arguin, P.O. Box 124, Nouadhibou, Mauritania*

*Key words:* Banc d'Arguin, Mauritania, tidal flats, zoobenthos, biomass

## Abstract

The biomass of the benthic macrofauna of the tidal flats of the Banc d'Arguin, Mauritania, retained by a sieve of 0.6 mm, has been estimated. The total biomass amounts to about 17.0 g ash-free dry weight (AFDW) $m^{-2}$ in the area sampled (about 300 $km^2$). The large bivalve *Anadara senilis* is responsible for 8.1 g AFDW $m^{-2}$, over 100 other species for another 8.9 g $m^{-2}$. The largest biomasses are found near the continental shore; at the offshore tidal flats benthic filter-feeders are lacking. Seagrass beds, which occupy the larger part of the tidal flats, have an under average biomass. Sebkhas, extremely saline flats above mean high-water mark, contain no live animals. To sustain the consumption of benthic prey by the very abundant wintering shorebirds relatively low P/B ratios are sufficient.

## Introduction

The Banc d'Arguin in Mauritania, West-Africa, is an area of tidal flats and shallow inshore waters bordering the Sahara desert (Fig. 1).

Recent ornithological studies have shown that the Banc d'Arguin is the most important wintering site for shorebirds migrating along the eastern seaboard of the Atlantic (Smit & Piersma, 1989). Altenburg *et al.* (1982) arrived at a total of about 2250000 waders in the winter of 1979–80.

Altenburg *et al.* (1982) collected also some data on the benthic macrofauna of the tidal flats, being the food of the wintering waders. They found, contrary to expectation, that the benthic biomass at the tidal flats of the Banc d'Arguin in February 1980 amounted to 7.6 g ash-free dry weight (AFDW) $m^{-2}$ only. In this figure 4.7 g is included for a large bivalve, the West African bloody cockle *Anadara senilis*, which can only be fed on by Oystercatchers *Haematopus ostralegus*, a relatively rare wader species on the Banc d'Arguin (Swennen, 1990). Hence, over two million waders seemed to be dependent on a benthic biomass less than 3 g AFDW $m^{-2}$ occurring in a relatively small tidal flat area of just over 500 $km^2$. Compared to other areas visited by the same wader populations, for instance the Wadden Sea area in Denmark, Germany and The Netherlands with 4000 $km^2$ of tidal flats, a benthic biomass of about

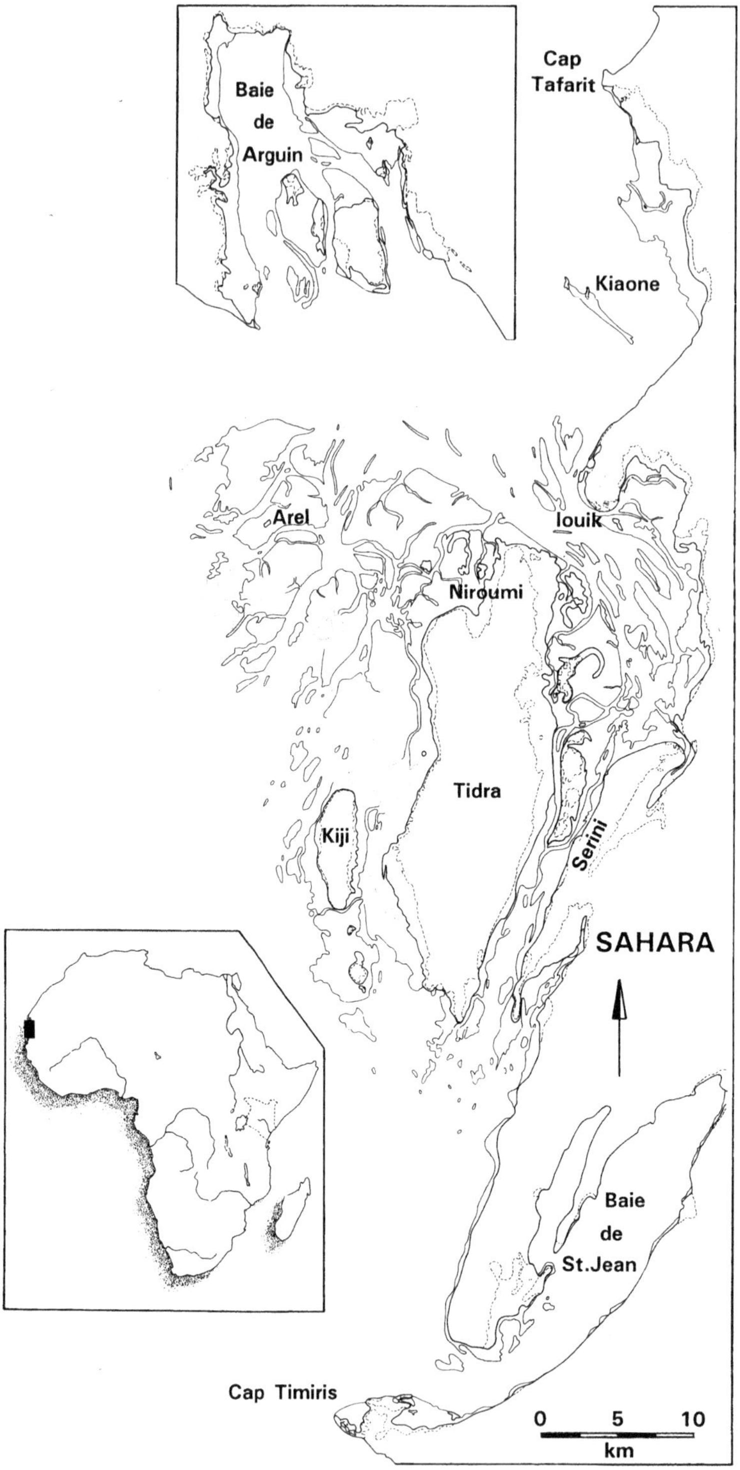


*Fig. 1.* The tidal flat area of the Banc d'Arguin, Mauritania.

25 g AFDW $m^{-2}$, and a maximum number of waders of about 2 million (Smit & Wolff, 1981; Wolff, 1991), the Banc d'Arguin seems to be very poor in food relative to its large numbers of birds and small area (see also Zwarts, 1988).

This paper reports on the biomass of the benthic macrofauna of the tidal flats of the Banc d'Arguin and discusses some aspects of the productivity of this fauna, based on observations made mainly in late winter and early spring of 1986.

## Material and methods

### *Description of the area*

A general description of the tidal flats of the Banc d'Arguin has been given by Altenburg *et al.* (1982) and Wolff & Smit (1990). A detailed report published by Ens *et al.* (1989) on the ornithological research carried out in 1985–86 also contains many data on the abiotic conditions of the Banc d'Arguin. The information relevant for this study is summarized below.

Average monthly air temperatures at the Banc d'Arguin range from about 18 °C in winter to about 22 °C in summer. The temperature of the waters of the Banc d'Arguin tidal flat area fluctuates between about 20 and 30 °C. Salinity is higher than in the adjacent open ocean, with values of 38–42 in the larger part of the tidal flat area, but much higher values close to the shore and in land-locked bays. The tidal range is about 1 m at neap tides and 2 m at spring tides; maximum current speeds are about 1 m $sec^{-1}$. Precipitation is rare at the Banc d'Arguin (< 50 mm $yr^{-1}$); there is also no freshwater land drainage. Winds are normally from northerly directions and are most of the time 4–6 on the Beaufort scale (Smit *et al.*, 1989a, 1989b; Wolff & Smit, 1990; Ould Dedah, 1993; Sevrin-Reyssac, 1993).

The larger part of the Banc d'Arguin area consists of soft sediments. Only locally rocky outcrops occur, predominantly at and in the vicinity of capes and headlands. Scattered over the tidal flats concentrations of sandstones occur, but the area covered by these is very small.

At the tidal flats sandy and muddy areas alternate. In general the areas covered with seagrasses are muddy, whereas unvegetated areas are sandy. Sandy areas occur especially at places where the tidal currents enter and leave large tidal flat areas. Also at the highest tidal levels, where due to a larger fetch of the wind at high tide, larger waves reach the shore, sandy areas are found. The sebkhas, situated above high tide level, are in general sandy, as are the adjacent parts of the desert. Table 1 shows some characteristics of the sediment in different habitats.

Based on a LANDSAT – 4 MSS image an estimate was made of the area of tidal flats and sebkhas between Cap St. Anne and Cap Timiris (Table 2). North of Cap St. Anne the area of tidal flats is negligible, but the sebkhas cover extensive areas.

Extensive parts of the tidal flats are covered with seagrasses. We estimate that at least 193 $km^2$ of the tidal flats south of Cap St. Anne are covered by dense (cover 100%) seagrass beds. Less dense seagrass vegetations occur on the larger part of the 219 $km^2$ of muddy flats. Three sea-

*Table 1.* Sediment characteristics (ranges) of the tidal flats of the Banc d'Arguin, based on samples collected in February-April 1986. After Wolff & Smit (1990).

| Type of tidal flat | Median grain-size ($\mu$m) | Perc. < 16 u % | Perc. org. mat. % |
|---|---|---|---|
| Sebkha | 98–144 | 3.5–11.7 | 0.8–4.0 |
| Beach | 101–265 | 0.7–8.3 | 0.2–3.4 |
| High sandy flat | 99–256 | 2.7–8.2 | 0.6–3.7 |
| Low sandy flat | 111–212 | 2.5–6.8 | 0.4–3.4 |
| Seagrass bed | 28–230 | 2.8–31.5 | 0.7–9.5 |

*Table 2.* Areas (in $km^2$) covered by various types of tidal flats based on LANDSAT data per subregion (see Fig. 1) of the Banc d'Arguin. Corrections have been made for tidal inundation (subregions Arel and SW Tidra) and terrestrial vegetation (Baie de St. Jean). The corrected values are given in brackets. After Wolff & Smit (1990).

| | Seagrass beds | Muddy flats | Sandy flats | Sebkhas |
|---|---|---|---|---|
| A Baie d'Arguin | 12.9 | 32.8 | 20.1 | 16.0 |
| B Cap Tafarit – Kiaone | 18.4 | 8.4 | 5.2 | 14.8 |
| C Arel – Kiji | 81.7 (109.0) | 39.4 (52.5) | 15.3 | 10.9 |
| D Iouik – Serini | 31.8 | 49.2 | 25.3 | 36.8 |
| E S of Tidra | 10.5 (21.0) | 31.1 (62.2) | 9.5 | 14.5 |
| F Baie de St. Jean | 2.8 (0.5) | 14.0 | 4.6 | 15.8 |
| Totals | 158.1 (193.4) | 174.9 (219.1) | 80.0 | 108.8 |

grass species were observed in 1985 and 1986, *viz. Zostera noltii*, *Halodule wrightii*, and *Cymodocea nodosa*. Of these *Z. noltii* is by far the most common species at the tidal flats. Especially at the more seaward flats it forms a dense monospecific cover. *H. wrightii* occurs more sporadically, but at some places it is the only species present. *C. nodosa* is largely restricted to places covered by water at all times. Most tidal flats covered with seagrass beds are interspersed with 0.2–1.0 m deep semi-circular pools with a diameter of 5–25 m. At high-level flats vegetations of *Vaucheria* species occur, often mixed with sparse *Z. noltii*. Altenburg *et al.* (1982) estimate the area covered by *Vaucheria* at about 3 $km^2$. At sandy flats a brown coloration, most likely due to benthic diatoms, was observed. At some places, especially near the islands of Nair and Niroumi, the cord grass *Spartina maritima* occurs in extensive monospecific vegetations or mixed with *Zostera noltii*. The *S. maritima* beds cover probably less than 25 $km^2$ at the Banc d'Arguin. Open vegetations of stunted mangroves (*Avicennia africana*) occur near Cap Timiris and at the northern tip of the island of Tidra. Altogether the mangroves cover less than 5 $km^2$ at the Banc d'Arguin.

## *Methods*

The sampling stations have been chosen at random. On a 1:170000 map of the tidal flat area based on satellite imagery arbitrary horizontal and vertical coordinates were drawn at distances of 1 mm. Sampling stations were determined by choosing coordinates by means of a table of random numbers. Over 200 stations were thus selected, but in practice only 82 stations were sampled (Fig. 2). The remainder either appeared not to be situated in the intertidal zone, or was difficult to visit due to restrictions on transportation by the authorities of the Parc National du Banc d'Arguin. The stations sampled cover a contiguous area; no stations in this area have been omitted.

All stations have been sampled during low tide. Some could be reached along the shore but most stations were visited by using small boats. Navigating on landmarks and the topography of the tidal flats the stations were approached as close as possible. Before the boat was left the number of paces to the station was estimated and this distance was walked over the flats. At the sampling station the very large bivalve *Anadara senilis* was sampled by hand using ranked sets sampling by means of 0.25 $m^2$ quadrats (Wolff *et al.*, 1987). All other species were sampled with a corer of 10 cm internal diameter which was pushed into the sediment to a depth of about 40–45 cm. At each station 5 core samples were randomly taken. The cores were sieved through a sieve with a mesh width of 0.6 mm (except for the samples from 2 stations with coarse sediment: 1 mm sieve) and put into polythene bags. All samples were sorted without visual aids in our 'laboratory' tent while the animals were still alive. Occasional rel-

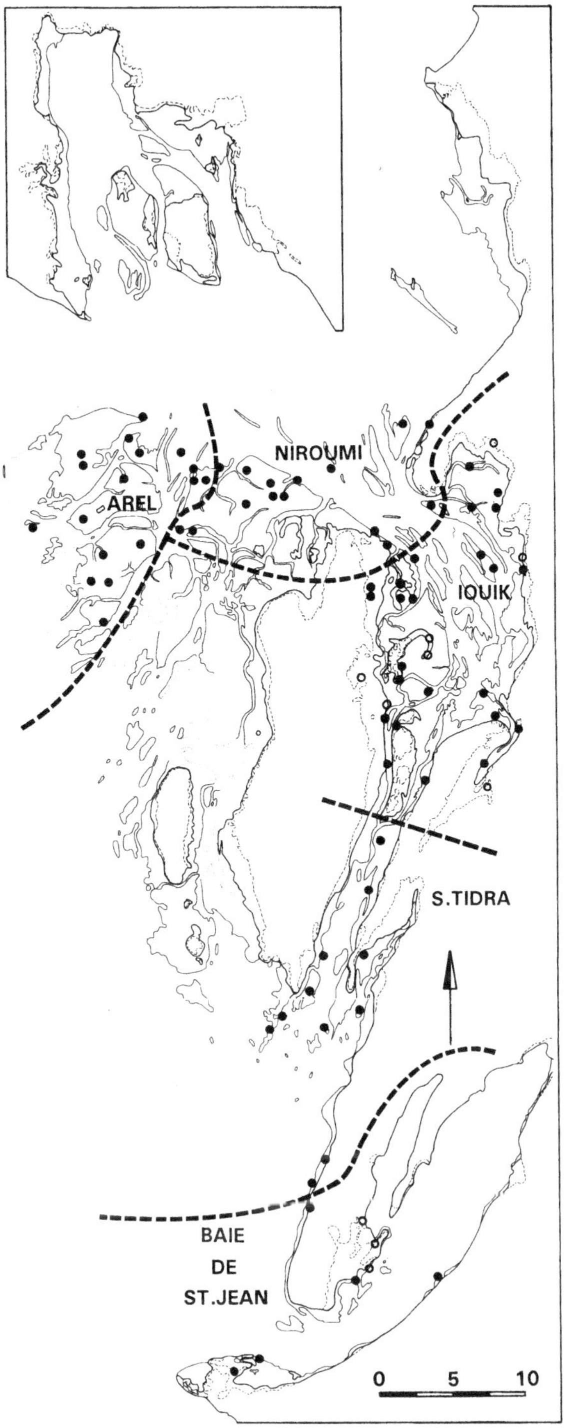


*Fig. 2.* Sampling stations at the tidal flats of the Banc d'Arguin. The dots denote stations in the intertidal zone; the open circles denote stations in the sebkhas without living animals. The dotted lines separate the subareas distinguished in Table 5.

atively large specimens of meiofauna groups, such as nematodes and ostracodes, and a few typical hard-bottom species, such as ascidians, were not considered for further analysis. All species were preserved in seawater with about 5% buffered formalin. Identification, measuring, and determining wet, dry, and ash-free dry weight (AFDW) have been carried out at the laboratory of the Institute for Forestry and Nature Research at Texel, The Netherlands, for all taxonomic groups except for crustaceans. For species from the latter group only wet weight was determined at the Institute of Taxonomic Zoology of the University of Amsterdam because of the large number of rare and undescribed species. Wet weight of these species was transformed into AFDW by multiplication with 0.15 (Lappalainen & Kangas, 1975; Ivell, 1983; Leuven *et al.*, 1985). AFDW of all other species was obtained by drying the specimens at 60 °C for at least 48 hours and afterwards incinerating them at 560 °C for 2 hours.

Some conspicuous large and/or very mobile species did not turn up in our core samples, *viz.* the decapod crustaceans *Uca tangeri* and *Callinectes marginatus*, and the gastropod molluscs *Cymbium cymbium*, *Conus pulcher papilionaceus*, and *Semifusus morio*. The bivalve *Anadara senilis* did occur in our core samples but was neglected, since the data required were obtained by a different way of sampling (see above).

At each sampling station the following data were noted: geographical position, tidal level, nature of the sediment, occurrence of macrophytes, occurrence of all macrofauna species recognizable in the field, and any pecularities. Also a sediment sample was taken and analysed in The Netherlands for grain-size distribution, contents of organic matter, contents of calcium carbonate, and pH. Salinity and water temperatures were only determined in the main tidal channels, not on the sampling stations.

For biomass calculations the 5 core samples taken at each station were pooled so that one sample was obtained from each station. These pooled samples were used for all further calculations.

## Results

### *Biomass per species, per taxonomic group and total biomass*

Of the 82 stations sampled 10 may be characterized as sebkhas, *i.e.* barren sandy flats situated above mean high water mark and containing large quantities of sodium chloride. We never encountered any live benthic animal in samples from this habitat, although terrestrial arthropods do occur at the surface. Hence, we have excluded these stations from our calculations.

Table 3 enumerates all species observed with their average biomass over the remaining 72 stations.

*Table 3.* List of species observed at the tidal flats of the Banc d'Arguin with their average biomass (g AFDW $m^{-2}$) and average number (per $m^2$) calculated over 72 stations at the Banc d'Arguin in February-April 1986. Frequency denotes the percentage of subsamples in which the species has been found (max. 72 * 5 = 360). At the end of the list species are enumerated which were not observed in the biomass samples, but encountered elsewhere. These include very large species occurring in low densities (L), very mobile species occurring in fairly high densities (M) and species encountered in other sampling programs (O).

| Species | Tax. group | Weight | Number | Frequency |
|---|---|---|---|---|
| *Anadara senilis* (Linné) | (Bivalvia) | 8.1 | (Wolff e.a. 1987) | |
| *Loripes lacteus* (Linné) | (Bivalvia) | 2.6002 | 90.5 | 29.7 |
| *Marphysa sanguinea* (Montagu) | (Polychaeta) | 1.2907 | 21.9 | 14.1 |
| *Petaloproctus terricola* Quatrefages | (Polychaeta) | 0.8874 | 35.0 | 11.4 |
| *Turritella torulosa* Kiener | (Gastropoda) | 0.8161 | 25.1 | 5.6 |
| *Euclymene oerstedii* (Claparède) | (Polychaeta) | 0.4518 | 105.7 | 32.2 |
| *Naineris laevigata* (Grube) | (Polychaeta) | 0.3865 | 97.6 | 22.2 |
| *Columbella rustica* (Linné) | (Gastropoda) | 0.2339 | 8.5 | 4.7 |
| *Clavatula bimarginata* Lamarck | (Gastropoda) | 0.2237 | 9.9 | 6.4 |
| *Diopatra neapolitana* (delle Chiaje) | (Polychaeta) | 0.1353 | 24.8 | 10.0 |
| *Lumbrineris* spec. | (Polychaeta) | 0.1338 | 4.2 | 3.3 |
| *Crenella dollfusi* Dautzenberg | (Bivalvia) | 0.1295 | 15.2 | 2.2 |
| *Arabella iricolor* (Montagu) | (Polychaeta) | 0.1148 | 3.9 | 3.1 |
| *Pontodrilus litoralis* (Grube) | (Oligochaeta) | 0.0995 | 11.7 | 1.4 |
| *Conus genuanus* Hwass | (Gastropoda) | 0.0963 | 3.2 | 2.2 |
| *Amyclina pfeifferi* (Philippi) | (Gastropoda) | 0.0923 | 4.2 | 1.4 |
| *Hydrobia ulvae* (Pennant) | (Gastropoda) | 0.0824 | 67.9 | 8.9 |
| *Abra tenuis* (Montagu) | (Bivalvia) | 0.0775 | 56.9 | 13.3 |
| *Terebella lapidaria* Linné | (Polychaeta) | 0.0741 | 3.5 | 2.5 |
| *Dosinia lupinus* (Poli) | (Bivalvia) | 0.0669 | 0.7 | 0.3 |
| *Bittium reticulatum* (da Costa) | (Gastropoda) | 0.0604 | 60.1 | 3.6 |
| *Capitella capitata* (Fabricius) | (Polychaeta) | 0.0525 | 282.2 | 17.2 |
| *Persicula chudeaui* Bavay | (Gastropoda) | 0.0517 | 1.8 | 1.1 |
| *Heteromastus filiformis* (Claparède) | (Polychaeta) | 0.0495 | 31.1 | 12.8 |
| *Cirriformia tentaculata* (Montagu) | (Polychaeta) | 0.0435 | 21.2 | 8.6 |
| *Prunum amygdala* (Kiener) | (Gastropoda) | 0.0425 | 3.9 | 3.1 |
| *Nassarius miga* Bruguière | (Gastropoda) | 0.0393 | 6.7 | 3.9 |
| *Sipunculida* (indet.) | (Sipunculida) | 0.0342 | 11.7 | 4.7 |
| *Anthuridae* (indet.) | (Isopoda) | 0.0327 | 62.2 | 19.2 |
| *Scoloplos capensis* (Day) | (Polychaeta) | 0.0273 | 25.5 | 10.8 |
| *Perinereis cultrifera* Grube | (Polychaeta) | 0.0265 | 19.8 | 7.8 |
| *Gibbula umbilicalis* (da Costa) | (Gastropoda) | 0.0255 | 1.8 | 1.1 |
| *Asterina gibbosa* (Pennant) | (Echinodermata) | 0.0252 | 1.1 | 0.8 |
| *Crepidula goreensis* (Gmelin) | (Gastropoda) | 0.0250 | 10.6 | 4.2 |
| *Tharyx* cf. *marioni* (St-Joseph) | (Polychaeta) | 0.0217 | 30.8 | 4.4 |
| *Scoloplos chevalieri* Fauvel | (Polychaeta) | 0.0202 | 14.5 | 6.9 |
| *Haminea orbignyana* (Férussac) | (Gastropoda) | 0.0181 | 3.5 | 1.9 |

*Table 3.* (Continued)

| Species | Tax. group | Weight | Number | Frequency |
|---|---|---|---|---|
| *Nephtys* spec. | (Polychaeta) | 0.0175 | 0.7 | 0.6 |
| *Bullaria adansoni* (Philippi) | (Gastropoda) | 0.0168 | 1.1 | 0.8 |
| *Ampithoidae* (indet.) | (Amphipoda) | 0.0145 | 4.2 | 1.9 |
| *Pisania d'orbignyi* Payraudeau | (Gastropoda) | 0.0143 | 2.5 | 1.7 |
| *Anaitides mucosa* (Oersted) | (Polychaeta) | 0.0142 | 0.4 | 0.3 |
| *Glycera rouxii* Audouin & Milne-Edwards | (Polychaeta) | 0.0133 | 2.5 | 1.4 |
| *Nemertinea* (indet.) | (Nemertinea) | 0.0132 | | 1.4 |
| *Nereis caudata* | (Polychaeta) | 0.0131 | 7.1 | 3.9 |
| *Nereis* spec. | (Polychaeta) | 0.0122 | 9.9 | 4.7 |
| *Idotea chelipes* (Pallas) | (Isopoda) | 0.0112 | 12.7 | 6.1 |
| *Polycirrus aurantiacus* Grube | (Polychaeta) | 0.0111 | 7.8 | 4.2 |
| *Amphipholis squamata* (delle Chiaje) | (Echinodermata) | 0.0108 | 20.2 | 6.4 |
| *Pista* cf. *cristata* (O. F. Müller) | (Polychaeta) | 0.0099 | 2.5 | 1.7 |
| *Lysianassidae* (indet.) | (Amphipoda) | 0.0096 | 14.8 | 5.3 |
| *Haminea elegans* (Gray) | (Gastropoda) | 0.0090 | 0.4 | 0.3 |
| *Driloneris filum* (Claparède) | (Polychaeta) | 0.0088 | 0.4 | 0.3 |
| *Typosyllis armillaris* (O. F. Müller) | (Polychaeta) | 0.0087 | 8.1 | 5.3 |
| *Aricidea fauveli* Hartman | (Polychaeta) | 0.0072 | 4.2 | 2.8 |
| *Glycera alba* (O. F. Müller) | (Polychaeta) | 0.0072 | 2.8 | 1.9 |
| *Oligochaeta* (indet.) | (Oligochaeta) | 0.0049 | 32.9 | 6.4 |
| *Polynoinae* (indet.) | (Polychaeta) | 0.0049 | 0.4 | 0.3 |
| *Cerastoderma edule* (Linné) | (Bivalvia) | 0.0044 | 1.8 | 1.4 |
| *Apseudes* spec. | (Amphipoda) | 0.0037 | 11.7 | 3.6 |
| *Sphaeromatidae* (indet.) | (Isopoda) | 0.0036 | 2.1 | 1.7 |
| *Calliostoma striatum* (Linné) | (Gastropoda) | 0.0036 | 2.5 | 0.8 |
| *Ampithoe rubricata* (Montagu) | (Amphipoda) | 0.0032 | 2.8 | 0.8 |
| *Polydora antennata* Claparède | (Polychaeta) | 0.0030 | 4.6 | 2.5 |
| *Nereis falsa* Quatrefages | (Polychaeta) | 0.0029 | 2.1 | 1.1 |
| *Scolelepis squamata* (O. F. Müller) | (Polychaeta) | 0.0027 | 2.8 | 0.6 |
| *Larvae Insecta* (indet.) | (Insecta) | 0.0023 | 1.4 | 1.1 |
| *Onuphys* cf. *eremita* Aud. & Milne-Edwards | (Polychaeta) | 0.0021 | 1.8 | 0.6 |
| *Bhawania goodei* Day | (Polychaeta) | 0.0020 | 0.4 | 0.3 |
| *Maricola* (indet.) | (Planaria) | 0.0019 | 1.1 | 0.8 |
| *Aricidea longobrachiata* Day | (Polychaeta) | 0.0018 | 1.1 | 0.8 |
| *Armandia intermedia* Fauvel | (Polchaeta) | 0.0017 | 2.1 | 1.7 |
| *Halcampa* spec. | (Coelenterata) | 0.0015 | 1.1 | 0.8 |
| *Melita palmata* (Montagu) | (Amphipoda) | 0.0012 | 2.5 | 1.4 |
| *Arcopsis afra* (Gmelin) | (Bivalvia) | 0.0012 | 0.7 | 0.6 |
| *Ampeliscidae* (indet.) | (Amphipoda) | 0.0010 | 1.4 | 0.6 |
| *Aloides laticosta* Ed. Lamy | (Bivalvia) | 0.0010 | 0.7 | 0.6 |
| *Eumida sanguinea* (Oersted) | (Polychaeta) | 0.0008 | 1.8 | 1.1 |
| *Holothurioidea* (indet.) | (Echinodermata) | 0.0007 | 0.4 | 0.3 |
| *Tricolia pulla* (Linné) | (Gastropoda) | 0.0007 | 1.1 | 0.8 |
| *Ampithoe vaillanti* Lucas | (Amphipoda) | 0.0005 | 0.7 | 0.6 |
| *Victoriopisa atlantica* Stock & Platvoet | (Amphipoda) | 0.0005 | 1.4 | 1.1 |
| *Edwardsiidae* (indet.) | (Coelenterata) | 0.0004 | 0.4 | 0.3 |
| *Calyptraea chinensis* (Linné) | (Gastropoda) | 0.0004 | 0.4 | 0.3 |
| *Dynamene* spec. | (Isopoda) | 0.0002 | 0.7 | 0.6 |
| *Microdeutopus chelifer* (Bate) | (Amphipoda) | 0.0002 | 2.5 | 1.1 |
| *Tellina* spec. | (Bivalvia) | 0.0002 | 1.1 | 0.6 |
| *Nucula nitida* Sowerby | (Bivalvia) | 0.0002 | 0.4 | 0.3 |

*Table 3.* (Continued)

| Species | Tax. group | Weight | Number | Frequency |
|---|---|---|---|---|
| *Isopoda* (indet.) | (Isopoda) | 0.0002 | 1.1 | 0.8 |
| *Prionospio malmgreni* Claparède | (Polychaeta) | 0.0001 | 1.1 | 0.8 |
| *Fabricia sabella* (Ehrenberg) | (Polychaeta) | 0.0001 | 0.7 | 0.6 |
| *Jasmineira elegans* St-Joseph | (Polychaeta) | 0.0001 | 0.7 | 0.6 |
| *Nematonereis unicornis* (Grube) | (Polychaeta) | 0.0001 | 0.7 | 0.6 |
| *Persicula cingulata* (Dillwyn) | (Gastropoda) | 0.0001 | 0.7 | 0.6 |
| *Cuna gambiensis* Nicklès | (Bivalvia) | 0.0001 | 1.1 | 0.8 |
| *Iraeidae* (indet.) | (Amphipoda) | 0.0001 | 0.4 | 0.3 |
| *Leucothoidae* (indet.) | (Amphipoda) | 0.0001 | 0.4 | 0.3 |
| *Microdeutopus* spec. | (Amphipoda) | 0.0001 | 2.1 | 1.7 |
| *Phoxocephalus* spec. | (Amphipoda) | 0.0001 | 1.8 | 0.6 |
| *Tanais* spec. | (Tanaidacea) | 0.0001 | 3.2 | 1.7 |
| *Isolda whydhaensis* Day | (Polychaeta) | 0.0001 | 0.4 | 0.3 |
| *Prionospio* spec. | (Polychaeta) | 0.0001 | 1.1 | 0.3 |
| *Dosinia* spec. | (Bivalvia) | 0.0000 | 0.4 | 0.4 |
| *Aoridae* (indet.) | (Amphipoda) | 0.0000 | 0.4 | 0.3 |
| *Microdeutopus damnoniensis* (Bate) | (Amphipoda) | 0.0000 | 0.4 | 0.3 |
| *Pythina mactroides* Hanley | (Bivalvia) | 0.0000 | 0.4 | 0.3 |
| *Exogone verugera* (Claparède) | (Polychaeta) | 0.0000 | 0.4 | 0.3 |
| *Gastrana matadoa* (Gmelin) | (Bivalvia) | 0.0000 | 0.4 | 0.3 |
| *Venus verrucosa* (Linné) | (Bivalvia) | ? | 0.4 | 0.3 |
| Fragments of Polychaeta | (Polychaeta) | 0.0446 | | |
| Fragments of Echinodermata | (Echinodermata) | 0.0214 | | |
| Fragments of Crustacea | (Crustacea) | 0.0022 | | |
| *Cymbium cymbium* (Linné) | (Gastropoda) | not in samples | | L |
| *Conus pulcher papilionaceus* Bruguière | (Gastropoda) | not in samples | | L |
| *Semifusus morio* (Linné) | (Gastropoda) | not in samples | | L |
| *Persicula miliaria* Linné | (Gastropoda) | not in samples | | O |
| *Cylichna grimaldii* Dautzenberg | (Gastropoda) | not in samples | | O |
| *Phacoides adansoni* (d'Orbigny) | (Bivalvia) | not in samples | | O |
| *Solen marginatus* (Pennant) | (Bivalvia) | not in samples | | O |
| *Dosinia exoleta* (Linné) | (Bivalvia) | not in samples | | O |
| *Diplodonta diaphana* Gmelin | (Bivalvia) | not in samples | | O |
| *Tagelus angulatus* (Sowerby) | (Bivalvia) | not in samples | | L, O |
| *Callinectes marginatus* | (Decapoda) | not in samples | | L, M |
| *Uca tangeri* Eydoux | (Decapoda) | not in samples | | M |
| *Carcinus mediterraneus* Czerniavsky | (Decapoda) | not in samples | | L, M |
| *Amphitrite edwardsii* | (Polychaeta) | not in samples | | O |
| *Eteone longa* (Fabricius) | (Polychaeta) | not in samples | | O |
| *Glycinde nordmanni* (Malmgren) | (Polychaeta) | not in samples | | O |
| *Hypsicomus* spec. | (Polychaeta) | not in samples | | O |
| *Sabellidae* spec. | (Polychaeta) | not in samples | | O |
| *Arabella mutans* (Chamberlin) | (Polychaeta) | not in samples | | O |
| *Cirrophorus* spec. | (Polychaeta) | not in samples | | O |
| *Tharyx dorsobranchialis* (Kirkegaard) | (Polychaeta) | not in samples | | O |
| *Caulleriella* aff. *bioculata* (Keferstein) | (Polychaeta) | not in samples | | O |

The total biomass of all species for the area of tidal flats we sampled, amounts to 17.0 g AFDW $m^{-2}$. One species, *Anadara senilis*, is responsible for 8.1 g (Wolff *et al.*, 1987) or 48% of the total biomass. All remaining species combined have a biomass of $8.9 \pm 1.0$ (s.e.) g $m^{-2}$. It appears that 5 species (including *A. senilis*) are responsible for 13.7 g or 81% of the total biomass, and 20 species for 16.1 g or about 95%. The remaining 84 taxa share the last 5%. The total number of individuals observed was 3967, the bivalve *A. senilis* excluded.

Table 4 presents the biomass per taxonomic group. Including the large bivalve *A. senilis* the bivalves are by far the most important group. Excluding this species, bivalves, gastropods, and polychaetes are of about equal importance. Next come oligochaetes, isopods, amphipods, echinoderms, and sipunculids. All other groups have negligible contributions to total biomass.

*Biomass per subarea*

Table 5 shows the biomass per subarea of the Banc d'Arguin, the bivalve *A. senilis* excluded. It appears that the more seaward subareas, *viz.* around Arel and around Niroumi have lower biomasses than the landward Baie de St. Jean and the area South of Iouik.

*Table 4.* Biomass (g AFDW $m^{-2}$) of the tidal flats of the Banc d'Arguin per taxonomic group. The data are derived from Table 3. Biomasses lower than 0.05 g AFDW $m^{-2}$ are given as 0.0 g.

| | | | |
|---|---|---|---|
| Coelenterata | | | 0.0 |
| Plathelminthes | | | 0.0 |
| Nemertini | | | 0.0 |
| Anadara senilis | 8.1 | | |
| Other Bivalvia | 2.9 | | |
| Total Bivalvia | | 11.0 | |
| Gastropoda | | 1.9 | |
| Total Mollusca | | | 12.9 |
| Polychaeta | | 3.9 | |
| Oligochaeta | | 0.1 | |
| Total Annelida | | | 4.0 |
| Sipunculida | | | 0.0 |
| Chilopoda | | 0.0 | |
| Insecta | | 0.0 | |
| Tanaidacea | 0.0 | | |
| Isopoda | 0.1 | | |
| Amphipoda | 0.0 | | |
| Other Crustacea | 0.0 | | |
| Total Crustacea | | 0.1 | |
| Total Arthropoda | | | 0.1 |
| Echinodermata | | | 0.0 |
| | | | 17.0 |

*Biomass per habitat type*

Table 6 shows the biomass per habitat type of the tidal flats of the Banc d'Arguin, the bivalve *A. senilis* excluded. The tidal levels denote the upper third of the tidal range (high tide), the middle third (half tide) and the lower third (low tide). Beaches have relatively low biomasses, seagrass beds average values, and mangrove areas and bare sandy flats seem to have relatively high biomasses; the differences are not significant, however.

Tidal level has no significant influence on biomass.

## Discussion

One of our objectives was to quantify the amount of food available for birds feeding on the tidal flats of the Banc d'Arguin. To do so we normally employed 0.6 mm sieves at 80 sampling stations; at 2 stations with coarse sediments the samples were sieved through a 1 mm sieve. Part of the biomass of characteristic benthic macrofaunal groups, such as molluscs, polychaetes, and amphipods, will have passed through our sieves (Reish, 1959; Nalepa & Robertson, 1981; Rees, 1984, Zwarts, 1988; Zwarts & Esselink, 1989). For a 1 mm sieve various authors mention percentages of 3–10% and even higher values do not seem unlikely. For individual species the biomass percentage lost may be even higher. The numbers of specimens thus lost can be much higher. This factor may also have been important for our 0.6 mm samples, even more so because the size spectrum of the individuals retained was relatively on the small side. Hence, it is conceivable that a relatively large, but unknown proportion of the benthic biomass taken by birds did not turn up in our samples. Moreover, an unknown part of the smallest specimens retained by our sieves will have been missed while sorting the samples (Rees, 1984; G. Moerland, unpubl. data). At the same time Zwarts *et al.* (1990) demonstrated that the smallest species of waders wintering at the Banc d'Arguin mainly feed on very small prey items, likely to pass a 0.6 mm sieve. Also Gaston (1992)

*Table 5.* Biomass ± s.e. (g AFDW $m^{-2}$) of benthic macrofauna of the tidal flats of the Banc d'Arguin per subarea (see Fig. 1). The bivalve *Anadare senilis* is not included.

| Subarea | Number of samples | Biomass | s.e. | Number of individuals |
|---|---|---|---|---|
| Around Isle of Arel | 17 * 5 | 7.2 | 1.5 | 658 |
| Around Isle of Niroumi | 15 * 5 | 6.9 | 1.5 | 806 |
| SE of Tidra | 11 * 5 | 8.4 | 1.6 | 207 |
| S of Iouik | 24 * 5 | 10.6 | 1.8 | 464 |
| Baie de Saint Jean | 5 * 5 | 16.7 | 9.8 | 525 |

demonstrated that another bird species (*Anas crecca*) was able to feed exclusively on tidal flat meiofauna. Hence, we have to conclude that our study does not give a complete picture of the food resources available for birds or any other predator of similar size feeding on the tidal flats. In addition, it should be remembered that benthic species occurring on or in tidal flats are not always physically available to predators because of behaviour, size, or burying depth.

Several authors have studied the effect of formalin preservation on AFDW values. Howmiller (1972) found that dry weight after 44 days in formalin 10% was reduced by 23% for freshwater tubificids and by 37% for chironomid larvae as compared to dry weight determined directly from fresh weight. Leuven *et al.* (1985) found no significant changes after four months preservation of freshwater gastropods in 10% formalin, but a small, but significant decrease (7%) after preservation in formalin 4%. Brey (1986) reported a decrease of about 22–23% for two marine bivalves during about 100 days of preservation in formalin 4%. Continued preservation in the same preservative for nearly another 100 days did not result in any further weight changes. Since our determination of AFDW was carried out about 1 year after original preservation, it is possible that our AFDW values are about 25% lower than we could have obtained when determining AFDW directly after sampling. Hence, it might be argued that all our biomass values should be increased by

*Table 6.* Biomass ± s.e. (g AFDW $m^{-2}$) per habitat type of the Banc d'Arguin tidal flats. The bivalve *Anadara senilis* is not included.

| Type of habitat | Tidal level | Number of stations | Biomass average | Standard error |
|---|---|---|---|---|
| Bare flats with *Anadara senilis* | Half tide | 3 | 11.9 | 2.8 |
| | Low tide | 1 | 4.2 | – |
| Mangroves | Half – low tide | 2 | 20.6 | – |
| Beach | High tide | 2 | 3.6 | 3.5 |
| | Low tide | 1 | 3.2 | – |
| Sandy flats | High tide | 6 | 7.5 | 3.2 |
| | Half tide | 3 | 10.2 | 1.3 |
| | Low tide | 3 | 17.3 | 5.9 |
| Seagrass beds | High tide | 1 | 7.0 | – |
| | Half tide | 30 | 9.0 | 1.4 |
| | Low tide | 14 | 7.3 | 1.4 |
| Miscellaneous | | 2 | 9.6 | – |

about 33%. However, we have refrained from this because of insufficient information on the effect of formalin on the species we sampled.

A number of conspicuous tidal flat species was too large or too mobile to turn up in our samples (Table 3). *Carcinus*, *Cymbium*, *Conus*, *Semifusus*, and *Tagelus* occurred in (very) low numbers ($<<1\ m^{-2}$) and can be neglected with regard to our biomass estimates. *Callinectes* and *Uca* are more common. The former species has a density of about 1 specimen per 1–10 $m^2$. The latter one probably has a much higher density, but is very hard to quantify due to its agility. Zwarts (pers. comm.) observed on average $10.2 \pm 1.6$ (s.e.) g AFDW $m^{-2}$ at a beach near Iouik. In our experience this particular beach is certainly not exceptional for its *Uca* densities. Also on the tidal flats *Uca* can be quite numerous. Hence, an average *Uca* biomass of the order of magnitude of 0.5–1.0 g AFDW $m^{-2}$ at the tidal flats of the Banc d'Arguin is not unlikely.

Piersma (1982) found for the benthic biomass of the tidal flats of the Banc d'Arguin 4.7 g AFDW $m^{-2}$ for the large bivalve *Anadara senilis* and 2.9 g AFDW $m^{-2}$ for all other species in January–February 1980. We arrive at 8.1 g AFDW $m^{-2}$ for *Anadara* and 8.9 g for all other species in February–April 1986. As already explained by Wolff *et al.* (1987) our *Anadara* value of 8.1 g pertains only to the area of flats we sampled; when recalculated for the entire area of tidal flats of the Banc d'Arguin its biomass will be about 5.5 g AFDW $m^{-2}$, which is close to the value by Piersma (1982). For all other species taken together Piersma (1982: Table 4.6) presents biomass values per habitat type similar to ours (Table 6) except for the habitat type of 'muddy *Zostera*'. For this type he finds only 2.08 g AFDW $m^{-2}$. The latter value, in combination with the estimate of the area of 'muddy Zostera' by Altenburg *et al.* (1982) (390 $km^2$ at a total tidal flat area of 536 km), explains how Piersma (1982) arrived at much lower value for average biomass of the entire tidal flat area. But why is his 'muddy *Zostera*' value so low? It appears that his 'muddy *Zostera*' samples are not randomly distributed over the tidal flats. They are strongly dominated by a series of samples from near-shore habitats, whereas samples from the vast areas of muddy seagrass beds West of the island of Tidra are nearly lacking. Moreover, large polychaetes hardly do occur in Piersma's 'muddy *Zostera*' samples, a phenomenon hard to explain.

Our samples covered about 70% of the tidal flats of the Banc d'Arguin (Fig. 2). Because of the satellite images investigated (Wolff & Smit, 1990) we believe the area not sampled to be rather similar to our subareas 'Arel' and 'Niroumi', which have biomasses, *Anadara* excluded, of 7.2 and 6.9 g AFDW $m^{-2}$. Assuming 7.0 g to hold true for the area not sampled we can calculate an estimate for the average biomass for the entire area of tidal flats of the Banc d'Arguin of 8.3 g AFDW $m^{-2}$. To this value about 5.5 g AFDW $m^{-2}$ has to be added for the large bivalve *A. senilis* (Wolff *et al.*, 1987).

It cannot be decided how representative these figures are for the yearly average of biomass. During our observations in February–April 1986 we did not see an obvious change of benthic biomass during this period. Samples collected in September 1988 have not been analyzed completely yet (Michaelis & Wolff, in prep.) but the first results do not point to a much higher biomass value. But even if benthic biomass in September will be higher, the yearly average of all species other than *Anadara* taken together probably will not be more than about 10 g AFDW $m^{-2}$.

Compared to other tidal flat areas the zoobenthic biomass of the Banc d'Arguin is lower than that in areas elsewhere along the East Atlantic seaboard (Zwarts, 1988; Wolff, 1991). Apart from explanations involving the nutrient status and the primary productivity of the flooding water masses, Wolff (1991) proposes that this is also due to the intense predation by birds and other predators. Zwarts *et al.* (1990), however, suggest that high bird predation at the Banc d'Arguin may be explained at least partly by low predation by fish, crabs, and shrimps. Comparison to other tropical tidal flat areas is hardly possible due to lack of data (Alongi, 1990).

Little is known about annual production/yearly average biomass (P/B) ratios for the benthic fauna of tropical seas in general (Longhurst & Pauly, 1987; Alongi, 1990) and of the Banc d'Arguin in particular. Wolff *et al.* (1987) arrived at a value of 0.02 $y^{-1}$ for the large bivalve *Anadara senilis* at the Banc d'Arguin, but they also concluded that this is an aberrant species. The other fauna largely consists of small species with probably short life-cycles as well as small individuals of large species (see also Piersma, 1982) and hence probably has fairly high P/B ratios (Robertson, 1979; Banse & Mosher, 1980). Ansell *et al.* (1978) compared macrofauna production of tropical and temperate beaches, and concluded that an equivalent biomass in the tropics produces at a rate of biomass turnover ten times faster than in temperate habitats. Edwards (1973), however, gives P/B values for intertidal macrofauna from Venezuela ranging from 0.3 to 2.8 $y^{-1}$. Alongi (1990), in a review of tropical soft-bottom benthos, lists P/B ratios ranging from 0.3 to 54.1. High P/B values apparently are possible for tropical benthic macrofauna, but they occur certainly not always and everywhere.

For the entire tidal flat region of the Banc d'Arguin the bivalve *A. senilis* has an average biomass of about 5.5 g AFDW $m^{-2}$ and, with a P/B ratio of 0.02 $y^{-1}$, an annual production of 0.11 g AFDW $m^{-2}$ $y^{-1}$. This is sufficient to support an annual predation of about 0.08 g AFDW $m^{-2}$ $y^{-1}$ by oystercatchers (Swennen, 1990). All other waders and several other predators are dependent on the remainder of the benthic fauna. Wolff & Smit (1990) calculated that birds are responsible for the removal of about 14 g AFDW $m^{-2}$ $y^{-1}$. There is no fishery for benthic tidal flat animals on the Banc d'Arguin. So, benthic consumption is at least 14 g AFDW $m^{-2}$ $y^{-1}$. Zwarts *et al.* (1990), however, found that about 40% of this consumption consists of prey sizes not retained in our sieves. Hence a consumption of about 8.5 g AFDW $m^{-2}$ $y^{-1}$ has to be supported by a benthic fauna with an average biomass for the entire tidal flat area of about 8.3 g AFDW $m^{-2}$. This requires a P/B ratio of about 1.0 $y^{-1}$. Wolff & Smit (1990) suggested that fish and invertebrates might take another 6 g AFDW $m^{-2}$ $y^{-1}$. If so this would raise the P/B ratio to about 1.7 $y^{-1}$. In view of the small size of most of the fauna, the tropical nature of the area, and the P/B values listed by Alongi (1990), these values are remarkably low. These low values even suggest that we strongly underestimated the part played by fish and invertebrate predators feeding on the benthic fauna of the tidal flats.

## Acknowledgements

We thank the Mauritanian authorities, in particular the Director of the Parc National du Banc d'Arguin Mr Hadya Amadou Kane, for their permission to carry out the investigations of the project Banc d'Arguin 1985–86 in the Parc National du Banc d'Arguin. We also thank the collaborators of the Parc National, particularly Messrs. El Hassane ould Mohammed el Abd, the late Ely ould Elemine, and Dr Pierre Campredon, for their cooperation in the field. Jan van de Kam and Jaap de Vlas assisted with our field sampling. The LANDSAT data were processed by Messrs. R. J. Bijlsma and E. J. Van Kootwijk. Dr G. C. Cadée, Dr. G. Eaton, Dr. G. Hartmann-Schröder, and Prof. dr C. den Hartog helped with some of our identifications.

Financial support for the project Banc d'Arguin 1985–86 was received from the Bachiene Foundation, the British Ornithological Union, the Commission of the European Communities, the Fund for Research for Nature Conservation ('FONA'), the Netherlands' Ministry of Agriculture and Fisheries, the National Geographic Society, the Society for the Protection of Nature Monuments ('Natuurmonumenten'), the Prince Bernhard Foundation, the Shell Internationale Research Maatschappij, and the Netherlands' State Forestry Service ('Staatsbosbeheer').

## References

Alongi, D. M., 1990. The ecology of tropical soft-bottom benthic ecosystems. Oceanogr. Mar. Biol. annu. Rev. 28: 381–496.

Altenburg, W., M. Engelmoer, R. Mes & T. Piersma, 1982. Wintering waders at the Banc d'Arguin, Mauritania. Report of the Netherlands Ornithological Expedition 1980. Stich-

ting Veth tot Steun aan Waddenonderzoek, Leiden, 283 pp.
Ansell, A. D., D. S. McLusky, A. Stirling & A. Trevallion, 1978. Production and energy flow in the macrobenthos of two sandy beaches in South West India. Proc. r. Soc. Edinburgh Sect. B 76: 269–296.
Banse, K. & S. Mosher, 1980. Adult body mass and annual production/biomass relationships of field populations. Ecol. Monogr. 50: 355–379.
Brey, T., 1986. Formalin and formaldehyde-depot chemicals: effects on dry weight and ash free dry weight of two marine bivalve species. Meeresforsch. 31: 52–57.
Edwards, R. R. C., 1973. Production ecology of two Caribbean marine ecosystems. II. Metabolism and energy flow. Estuar. coast. mar. Sci. 1: 319–333.
Ens, B. J., T. Piersma, W. J. Wolff & L. Zwarts, 1989. Report of the Dutch-Mauritanian project Banc d'Arguin 1985–86. Report WIWO 25/RIN-report 89/6, 336 pp.
Gaston, G. R., 1992. Green-winged teal ingest epibenthic meiofauna. Estuaries 15: 227–229.
Howmiller, R. P., 1972. Effects of preservatives on weights of some common macrobenthic invertebrates. Trans. amer. Fish. Soc. 101: 743–746.
Ivell, R., 1983. Technical note: the preparation of molluscan tissue for production estimates. J. moll. Stud. 49: 18–20.
Lappalainen, A. & P. Kangas, 1975. Littoral benthos of the northern Baltic Sea. II. Interrelationships of wet, dry and ashfree dry weights of macrofauna in the Tvärminne area. Int. Revue ges. Hydrobiol. 60: 297–312.
Leuven, R. S. E. W., T. C. M. Brock & H. A. M. van Druten, 1985. Effects of preservation on dry- and ash-free dry weight biomass of some common aquatic macro-invertebrates. Hydrobiologia 127: 151–159.
Longhurst, A. R. & D. Pauly, 1987. Ecology of tropical oceans. Academic Press, San Diego, 407 pp.
Michaelis, H. & W. J. Wolff, in prep. Benthic biomass of tidal flats before wintering waders arrive.
Nalepa, T. F. & A. Robertson, 1981. Screen mesh size affects estimates of macro- and meiobenthos abundance and biomass in the Great Lakes. Can. J. Fish. aquat. Sci. 38: 1027–1034.
Ould Dedah, Sidina, 1993. Wind, surface water temperature, surface salinity and pollution in the area of the Banc d'Arguin, Mauritania. Hydrobiologia 258: 9–19.
Piersma, T., 1982. Macrobenthic fauna of the intertidal flats. In: W. Altenburg, M. Engelmoer, R. Mes & T. Piersma (eds), Wintering waders on the Banc d'Arguin. Stichting Veth tot Steun aan Waddenonderzoek, Leiden: 49–66.
Rees, H. L., 1984. A note on mesh selection and sampling efficiency in benthic studies. Mar. Pollut. Bull. 15: 225–229.
Reish, D. J., 1959. A discussion of the importance of screen size in quantitative marine bottom samples. Ecology 40: 307–309.
Robertson, A. I., 1979. The relationship between annual production: biomass ratios and lifespans for marine macrobenthos. Oecologia (Berl.) 38: 193–202.
Sevrin-Reyssac, J., 1993. Hydrology and underwater climate of the Banc d'Arguin, Mauritania: a review. Hydrobiologia 258: 1–8.
Smit, C. J., A.-M. Blomert & L. Zwarts, 1989a. Meteorological measurements. In: B. J. Ens, T. Piersma, W. J. Wolff & L. Zwarts (eds), Report of the Dutch-Mauritanian project Banc d'Arguin 1985–86. Report WIWO 25/RIN-rapport 89/6: 16–42.
Smit, C. J., A.-M. Blomert, A. Meijboom, W. J. Wolff & L. Zwarts, 1989b. Hydrographical measurements. In: B. J. Ens, T. Piersma, W. J. Wolff & L. Zwarts (eds), Report of the Dutch-Mauritanian project Banc d'Arguin 1985–86. Report WIWO 25/RIN-rapport 89/6: 43–56.
Smit, C. J. & T. Piersma, 1989. Numbers, mid-winter distribution and migrations of wader populations using the East Atlantic flyway. In: H. Boyd & J.-Y. Pirot (eds), Flyways and reserve networks for water birds: 24–63. Internat. Waterfowl- and Wetland Res. Bureau Spec. Publ. 9, Slimbridge.
Smit, C. J. & W. J. Wolff, 1981. Birds of the Wadden Sea. Balkema, Rotterdam, 308 pp.
Swennen, C., 1990. Oystercatchers (Haematopus ostralegus) feeding on giant bloody cockles (Anadara senilis) on the Banc d'Arguin in Mauritania. Ardea 78: 53–62.
Wolff, W. J., 1991. The interaction of benthic macrofauna and birds in tidal flat estuaries: a comparison of the Banc d'Arguin, Mauritania, and some estuaries in The Netherlands. In: M. Elliott & J. P. Ducrotoy (eds), Proc. ECSA 19th Symp.: Estuaries and coasts: spatial and temporal intercomparisons. Caen, France, 4–8 Sept. 1989. Internat. Symp. Ser.: 299–306.
Wolff, W. J., A. Gueye, A. Meijboom, T. Piersma & M. A. Sall, 1987. Distribution, biomass, recruitment and productivity of *Anadara senilis* (L.) (Mollusca: Bivalvia) on the Banc d'Arguin, Mauritania. Neth. J. Sea Res. 21: 243–253.
Wolff, W. J. & C. J. Smit, 1990. The Banc d'Arguin, Mauritania, as an environment for coastal birds. Ardea 78: 17–38.
Zwarts, L., 1988. Numbers and distribution of coastal waders in Guinea-Bissau. Ardea 76: 42–55.
Zwarts, L., Blomert, B. J. Ens, R. Hupkes & T. M. van Spanje, 1990. Why do waders reach high feeding densities on the intertidal flats of the Banc d'Arguin, Mauritania? Ardea 78: 39–52.
Zwarts, L. & P. Esselink, 1989. Versatility of male curlews *Numenius arquata* preying upon *Nereis diversicolor*: deploying contrasting capture modes dependent on prey availability. Mar. Ecol. Prog. Ser. 56: 255–269.

*Hydrobiologia* **258**: 165–173, 1993.
*W. J. Wolff, J. van der Land, P. H. Nienhuis & P. A. W. J. de Wilde (eds),*
*Ecological Studies in the Coastal Waters of Mauritania.*



# The trawlfauna of the Mauritanian shelf (Northwest Africa): density, species composition, and biomass

G. C. A. Duineveld, M. S. S. Lavaleye & G. J. van Noort
*Netherlands Institute for Sea Research, P.O. Box 59, 1790 AB Den Burg, Texel, The Netherlands*

*Key words:* trawlfauna, Mauritanian shelf

## Abstract

In May 1988, epifaunal trawlsamples were collected at 17 stations on the Mauritanian shelf off the Banc d'Arguin. The species composition and biomass of the invertebrates and demersal fish were analysed. On the basis of species abundances, three station groups could be distinguished each corresponding to a subregion of the shelf: an inner shelf, a midshelf and an outer shelf station group. The average depth increased from 20 m for the inner to more than 100 m for the outer shelf stations. The biomass of mobile epifauna and fish was lowest on the inner shelf stations where the trawlfauna was dominated by large sessile filterfeeders. The trawlfaunas of the mid- and outer shelf were more similar in species composition. Demersal fish and crustaceans appeared to be more abundant on the outer shelf. The increase in density and biomass from inner to outer shelf is discussed in relation to the distribution of the infauna and the hydrographical conditions on the Mauritanian shelf. Being a demersal fishing area, density and biomass of the epifauna on the shelf are compared with that of the demersal fishing grounds in the southern North Sea. The result of this comparison shows that, except for echinoderms, all taxa attain higher densities and biomasses at the Mauritanian shelf.

## Introduction

The continental shelf of West Africa, between the Strait of Gibraltar and Cape Verde, is one of the most important fishery areas of the world (Bravo de-Laguna, 1982). This is related to hydrological conditions which give rise to a permanent or seasonal upwelling resulting in an enhanced primary production (Minas *et al.*, 1982). Not only have the pelagic resources, like fishes of the genera *Sardina* and *Sardinella*, attracted much fishing activity (Ansa-Emmim, 1982), but the demersal resources on the shelf are also subject to commercial exploitation. The Mauritanian shelf plays a vital role in this respect. It acts both as a nursery for some of the most important pelagic fish species (Ansa-Emmim, 1982; Boely, 1982) and as a source of exploitable stocks of fish, cephalopods and shrimps (Bravo de-Laguna, 1982).

The biological resources of the Mauritanian shelf have, therefore, been the subject of many scientific expeditions (Van der Knaap, 1985). During the three French 'Thalassa'-expeditions (Maurin & Bonnet, 1969; Bonnet *et al.*, 1971), extensive trawlsurveys were made on the Mauritanian shelf and slope between 10 and 800 meters in order to map the fishing grounds and estimate the yield of the trawlfishery. These efforts focussed mainly on large and commercial species. However, to obtain a better understanding of the parameters determining the quantity of benthos and bottomfish at the shelf, it is necessary to study the entire epibentic fauna as well as their potential food source, *i.e.* the macro- and meio-infauna living in the sediment.

One of the aims of the Dutch 'TYRO'-expedition to the Mauritanian shelf in May 1988, was to fill this gap by simultaneously making an inventory of the macro-, meio- and epibenthos of the area. The present paper deals with the taxonomic and quantitative composition of the trawl-samples taken on the shelf off the Banc d'Arguin at depths ranging from 20 to over 100 m.

## Material and methods

The positions of the trawlstations are shown in Fig. 1 and their depths are listed in Table 1. At each station one haul was made with a 5.5 m beam-trawl having a mesh-size of 5 × 5 mm in the cod end (see Creutzberg *et al.* (1987) for details). The beam was equipped with two tickler chains, the same number as used by the authors in the North Sea. To prevent clogging of the net by smaller epifauna and debris, and to keep the catch to a reasonable proportion, the duration of the hauls did not exceed 10 minutes. The area covered by the net, was derived from the initial and final position of the ship in combination with the towing speed (3 knots) and the width of the net. All quantities were converted to a standard area (10 000 $m^2$).

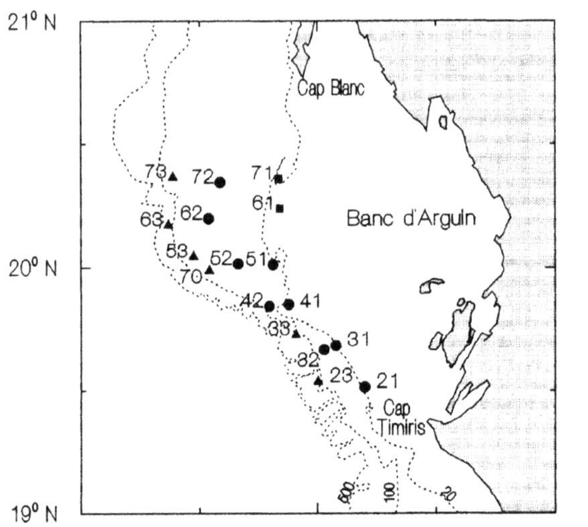


*Fig. 1.* Position of the trawlstations on the Mauritanian shelf. Symbols for station groups: ■ = inner shelf, ● = midshelf, and ▲ = outer shelf station group (cf. Table 1).

*Table 1.* Approximate depth at the trawlstations in three regions of the Mauritanian shelf. The positions of the stations are displayed in Fig. 1.

| Stat. | Outer shelf depth (m) | Stat. | Midshelf depth (m) | Stat. | Inner shelf depth (m) |
|---|---|---|---|---|---|
| 23 | 260 | 21 | 25 | 61 | 20 |
| 33 | 130 | 31 | 40 | 71 | 20 |
| 53 | 80 | 32 | 85 | | |
| 63 | 105 | 41 | 15 | | |
| 70 | 70 | 42 | 80 | | |
| 73 | 85 | 51 | 20 | | |
| | | 52 | 40 | | |
| | | 62 | 40 | | |
| | | 72 | 35 | | |

All large organisms from the catch were, if possible, identified to species level and weighed on board of the ship. Only the fish were individually measured. The smaller organisms were sorted on board and a volumetric subsample was preserved in formalin for later identification and weighing. Only those species which live on the sediment surface were considered. Infaunal species, such as burrowing bivalves and ophiuroids which were often caught in large quantities, were discarded.

The stations were classified on the basis of their epibenthic fauna, using the ordination programs TWINSPAN and DECORANA (Hill, 1979a, b). For this purpose, logarithmic abundance classes were used.

In conjunction with the present study, hydrographic data were collected at the trawlstations, including vertical profiles of temperature, salinity and percentage dissolved oxygen. Measurements were made with a Guildline 8705 CTD-probe equipped with an oxygen-sensor. The oxygen readings were calibrated by means of Winkler titrations of water samples.

## Results

In order to reduce the data to a synoptic and comprehensible form, the stations were subjected to a TWINSPAN classification on the basis of the species abundances. This showed that three station groups could be discriminated (Fig. 1), which differed by their distance from the shore as well as by their depth (Table 1). For reasons of convenience we have denominated the three groups as an inner shelf, a midshelf and an outer shelf station group. The inner shelf station group comprised only two stations, which were split from the other stations in the first division made by TWINSPAN. At the subsequent level a division was made between the mid- and outer shelf stations. Further division of these two groups showed that among the outer shelf stations, a separation could be made between the two deepest hauls (> 130 m) and the other stations. Similarly, in the midshelf group one station could be split off at this level, but for no obvious explanation in terms of abiotic factors. The arrangement of the stations along the two major ordination axes (DECORANA; Hill, 1979b) shows that the first axis contributes little to the separation between the midshelf and the outer shelf stations (Fig. 2). This implies that there is a clear discontinuity within the species set.

This discontinuity is illustrated in Table 2 by the average densities of the most abundant taxa in the three station groups. The two inner shelf stations are characterized by the low abundances of otherwise very common fish species, such as the dwarf sole *Microchirus boscanion* and the flatfish *Citharus lingulata*. But many crustacean species, and also gastropods (e.g. *Marginella*), were far less numerous or even absent at the inner shelf station. Notably high densities were found for the shrimp *Sycionia* and the firm tunicate *Pyura dura*. Present at the inner shelf station, but not found elsewhere, are two large bivalve species, *i.e. Chama* and *Glycymeris*, the echinoids *Psammechinus* and *Rotula* and the shrimp *Penaeopsis*. The main difference between the midshelf and outer shelf stations is that many species occurred in higher densities along the outer shelf. Species oc-

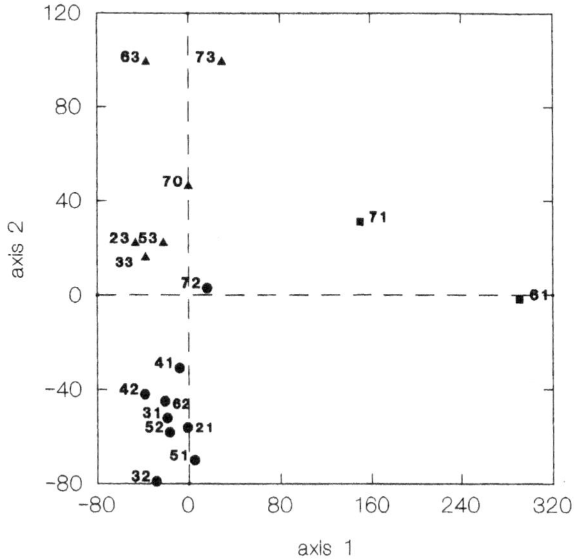


*Fig. 2.* Position of the stations along the first two ordination axis of DECORANA.

curring exclusively or with higher numbers at the mid-shelf stations are fish like *Platycephalus*, gobiids, large *Solea*, and shrimps belonging to the genus *Processa*. Species exclusively or predominantly found along the outer shelf are fish of the family Scorpaenidae, the echinoid *Centrostephanus*, crinoids, and the crustaceans *Plesionika*, *Pontocaris* and *Munida*.

In Table 3, the abundances of the separate species have been grouped into major taxonomical categories. These figures more clearly reveal the increasing abundance of mobile epifauna in an offshore direction, not only of invertebrates but of fish as well. Among the invertebrate groups, this gradient is most explicit with the crustaceans. Molluscs as a group attain their highest mean density on the midshelf which is entirely due to the numerous nassariid gastropods (125 000 ind. per 10 000 $m^2$) at one station (42) in this area. In the category of miscellaneous phyla, the numerically dominant tunicate *Pyura* reaches highest densities at the inner shelf stations.

The biomass of the major taxonomical groups in the three station groups is shown in Table 4. The biomass of fish, Crustacea, Echinodermata and Anthozoa increases from the inner to the outer shelf station group which agrees with the

*Table 2.* Abundance (ind. 10 000 m$^{-2}$) of most common epibenthic taxa in three subregions of the Mauritanian shelf. Figures are geometric means per stationgroup. Taxa at the family level have been printed in italics.

| | Shelf area: | Outer | Mid | Inner |
|---|---|---|---|---|
| Fish | *Scorpaenidae* | 36 | 5 | – |
| | *Triglidae* | 55 | 25 | 2 |
| | Arnoglossus spp. | 159 | 27 | 2 |
| | *Soleidae* small | 350 | 17 | 126 |
| | Citharus lingulata | 71 | 71 | – |
| | Platycephalus gruveli | – | 15 | – |
| | *Gobiidae* | 22 | 81 | – |
| | *Soleidae* large | 1 | 27 | 1 |
| | *Blenniidae* | 2 | 1 | 47 |
| | Ariosoma balearicum | 2 | 1 | 11 |
| | Diplecogaster | – | – | 9 |
| | Heteromycteris spec. | – | – | 54 |
| Crustaceans | Plesionika heterocarpus | 5211 | – | – |
| | Munida spp. | 218 | 2 | – |
| | *Inachinae* | 964 | 152 | 62 |
| | *Paguridae* | 5941 | 450 | 257 |
| | *Portunidae* | 164 | 38 | 28 |
| | *Parthenopidae* | 356 | 111 | 3 |
| | Pontocaris spec. | 368 | 2 | 2 |
| | *Leucosiidae* | 176 | 57 | 8 |
| | *Pisinae* | 129 | 9 | 5 |
| | *Dorippiidae* | 61 | 25 | 4 |
| | Scyllaris spp. | 85 | 77 | 1 |
| | Sycionia carinata | 4 | 11 | 1071 |
| | Penaeopsis spec. | – | – | 13 |
| Echinoderms | Centrostephanus longispinus | 10 | 1 | – |
| | Crinoids | 40 | 1 | – |
| | Ophiura spp. | 38 | 5 | – |
| | Ophiotrix spp. | 86 | 6 | 4 |
| | Luidia spp. | 4 | 8 | – |
| | Rotula spec. | – | – | 14 |
| | Psammechinus spec. | – | – | 134 |
| Molluscs | *Nassariidae* | 19 | 417 | – |
| | *Marginellidae* | 53 | 134 | 33 |
| | Cymbium spp. | – | 4 | 3 |
| | *Naticidae* | 33 | 19 | 110 |
| | Octopus | 4 | 3 | 26 |
| | *Conidae* | 2 | 11 | 20 |
| | Glycymeris spp. | – | – | 41 |
| | Crepidula porcellana | – | – | 227 |
| | Chama crenulata | – | – | 135 |
| Cnidarians | Anthozoa | 41 | 2 | 5 |
| | Epizoanthus | 5 | 1 | – |
| Miscellaneous | Porifera | 54 | 9 | 7 |
| | Pyura dura | – | 7 | 390 |

*Table 3.* Abundance (ind. m$^{-2}$) of principal taxonomic groups. Both values from the Mauritanian inner shelf ($N = 2$) are presented. Data for other areas are represented by geometric means; multiplication resp. division by the factor in brackets gives the 95% confidence limits. A. three subregions of the Mauritanian shelf. B. four subregions of the southern North Sea.

| **A** Mauritanian Shelf | Outer | Mid | Inner | |
|---|---|---|---|---|
| Fish | 0.235 (2.2) | 0.114 (2.5) | 0.043/0.043 | |
| Crustacea | 2.259 (2.6) | 0.395 (3.8) | 0.160/0.303 | |
| Echinodermata | 0.076 (11.) | 0.012 (5.5) | 0.019/0.035 | |
| Mollusca | 0.165 (3.0) | 0.241 (4.8) | 0.085/0.076 | |
| Anthozoa | 0.023 (11.) | 0.001 (4.5) | –/0.004 | |
| Miscellaneous | 0.006 (12.) | 0.004 (3.7) | 0.022/0.075 | |
| **B** Southern North Sea | N. of Dogger | Dogger | Oyster-Gr. | Coast |
| Fish | 0.026 (1.9) | 0.023 (1.8) | 0.027 (1.3) | 0.025 (2.7) |
| Crustacea | 0.046 (2.8) | 0.008 (2.3) | 0.026 (1.7) | 0.051 (2.3) |
| Echinodermata | 0.055 (2.3) | 0.078 (4.9) | 0.174 (2.4) | 0.060 (2.2) |
| Mollusca | 0.009 (1.8) | <0.001 (2.8) | 0.037 (2.7) | <0.001 (2.2) |
| Anthozoa | 0.001 (5.5) | <0.001 (2.1) | 0.002 (2.1) | <0.001 (1.6) |
| Miscellaneous | 0.006 (4.2) | 0.045 (3.7) | <0.001 (1.3) | – |

observed increase of their numerical density. The average weight of the fishes ($\Sigma$Biomass/$\Sigma$Numbers) at the midshelf stations seems to be higher than at the outer shelf stations whereas the fish at inner shelf stations have by far the smallest average weight. Because of the small size of the nassariid snails at the midshelf, the mean biomass of molluscs is relatively low. Higher mean

*Table 4.* Biomass (g Wet Weight m$^{-2}$) of principal taxonomic groups. Both values from the Mauritanian inner shelf ($N = 2$) are presented. Data for other areas are represented by geometric means; multiplication resp. division by the factor in brackets gives the 95% confidence limits. A. three subregions of the Mauritanian shelf. B. four subregions of the southern North Sea.

| **A** Mauritanian Shelf | Outer | Mid | Inner | |
|---|---|---|---|---|
| Fish | 1.907 (2.6) | 1.640 (2.6) | 0.181/0.215 | |
| Crustacea | 1.518 (2.9) | 0.294 (2.5) | 0.206/0.272 | |
| Echinodermata | 0.169 (15.) | 0.046 (7.8) | 0.076/0.143 | |
| Mollusca | 1.100 (2.4) | 0.483 (6.2) | 0.918/1.000 | |
| Anthozoa | 0.042 (4.4) | 0.002 (9.5) | –/0.009 | |
| Miscellaneous | 0.045 (6.5) | 0.274 (3.7) | 1.860/6.689 | |
| **B** Southern North Sea | N. of Dogger | Dogger | Oyster-Gr. | Coast |
| Fish | 0.927 (1.8) | 0.322 (1.9) | 0.518 (1.3) | 0.209 (4.7) |
| Crustacea | 0.029 (2.4) | 0.005 (2.6) | 0.028 (1.7) | 0.043 (2.2) |
| Echinodermata | 0.090 (2.8) | 0.212 (2.0) | 0.465 (1.5) | 0.359 (2.5) |
| Mollusca | 0.155 (1.9) | 0.002 (3.8) | 0.014 (2.6) | <0.001 (1.1) |
| Anthozoa | 0.007 (5.1) | 0.001 (2.7) | 0.026 (2.1) | 0.001 (1.9) |
| Miscellaneous | 0.070 (3.8) | 0.174 (3.6) | 0.001 (1.1) | – |

values were found at the outer shelf (different species, e.g. *Venus)* and at the inner shelf (*Cymbium*, *Chama*). The remaining miscellaneous phyla, which contribute little to the total density of the epifauna, become important when considering the total biomass. At the inner shelf stations this group constitutes 80% of the total biomass and consists almost entirely of the tunicate *Pyura dura*. This species by itself is responsible for the fact that there is no distinct trend in total epibenthic biomass among the station groups.

## Discussion

### *Relative richness and composition of the epifauna*

Because of the lack of comparable quantitative epifauna data from areas adjacent to the Mauritanian shelf, it is difficult to evaluate the relative richness of the trawlfauna with respect to its potential fishing yield and food conditions for demersal fish. To get some impression of its richness we have made a gross comparison with the trawlfauna from the southern North Sea (Duineveld & Van Noort, 1990). Although this area belongs to a different zoo-geographical province, this comparison is justified to some extent. The southern North Sea comprises a similar range in sediment-types and depths as the area we studied on the Mauritanian shelf. Moreover, primary production in the eutrophied southern North Sea (Zijlstra, 1988) matches the levels reported for the upwelling shelf off Cap Blanc, Mauritania (Huntsman & Barber, 1977), *viz.* 3 g C $m^{-2}$ $d^{-1}$. A comparison is furthermore validated by the fact that during both studies the same instruments and vessel were used. For the purpose of this comparison, the ashfree dry weight biomass data from the southern North Sea were converted into wet weight by means of the appropriate conversion factors given by Rumohr *et al.* (1987).

Total density and biomass of the major taxonomical groups in different parts of the southern North Sea are given in Table 3b and 4b. When comparing these data with those from Mauritania (Table 3a and 4a), it can be seen that the average density and biomass of most invertebrate taxa are higher in Mauritania. In the North Sea, only echinoderms attain a comparable or higher density and biomass. The species composition is, however, quite different; North Sea echinoderms are predatory starfish and mobile ophiuroids, whereas on the Mauritanian shelf, filter-feeding crinoids and *Ophiotrix* dominate. The relatively low values for fish at the inner shelf stations are comparable to the lowest value in the southern North Sea. At the outer- and midshelf stations, however, fish biomass and density exceed those found in any subregion of the southern North Sea. In both areas flatfishes constitute the major part of the bottom-dwelling fish fauna. With some of the invertebrate groups, such as crustaceans and molluscs, the differences between North Sea and Mauritania are even more pronounced. Striking are the high abundances of small predatory gastropods (Nassaridae, Marginellidae) on the shelf off Mauritania. This type of organism is very rare in trawlcatches from the southern North Sea where a usually small number of large predatory or sarcophageous Buccinidae is found.

Despite the inherent superficial character of the comparison between the two regions, we feel confident that we have shown that the epifauna on the Mauritanian shelf at least equals the epifauna in well-known demersal fishing areas of the North Sea in terms of numbers and biomass. Thiel (1982), however, states that in spite of the high density of benthos on the northwest African shelf, most larger organisms belong to filterfeeding groups such as Porifera, Actiniaria and Octocorallia, none of which are favoured by demersal fish as food. He argues that while the abundant infauna may constitute some food for fish, favourable fishing yields are not to be expected. Based on our comparison with the southern North Sea, we think that at least the mid- and outer shelf may offer much better feeding conditions for demersal fish than assumed by Thiel (1982). Our supposition is based on the extremely large numbers of small crustaceans in these regions, which are found nowhere in the southern North Sea in such high densities. The macro-infauna on the Mauritanian shelf also constitutes a better food source for demersal fish than in the North Sea. In this latter area, the major part of the infaunal biomass is made up of spatangoid echinoderms (Duineveld *et al.*, 1990), whereas on the shelf of the Banc d'Arguin polychaetes frequently dominate in terms of biomass (Duineveld *et al.*, 1993).

*Differences between the station groups*

As mentioned in chapter 3, we found a clear increase in density and biomass of fishes and of many invertebrate taxa from the shallow inner shelf stations to the deeper outer shelf stations. The results from the trawlsurveys with the 'Thalassa' off the Banc d'Arguin (Bonnet *et al.*, 1971), also show an elevated biomass of epifauna at the deeper stations along the shelf edge. These samples were collected with a net having an opening at least 10 times as large as that of our beamtrawl and a mesh size of 25 mm was used. Also Brulhet *et al.* (1974) report an increase of the total epibenthic biomass near the shelf edge (100 m depth) at 20 °N.

One should, however, consider this gradient with some reservation as fish as well as shrimps, are able to migrate, and our data only represent a one-time survey. Collignon (1971), for instance, demonstrated large seasonal variations in the abundance of selected fish species off the Atlantic coast near Morocco. It is, therefore, conceivable that over an annual cycle, differences between our station groups could become diminished, or even enlarged. However, given the fact that many of the epibenthic invertebrates we caught are poor migrators and that previous studies found a comparably increased biomass at the shelf edge, we feel there is sufficient ground to compare our epifauna distribution with that of its major food source *viz.* the infauna. This is of particular interest as the gradient in biomass of macro-infauna seems, to some extent, to be opposite to that of the epifauna (see Duineveld *et al.*, this volume). At the inner shelf stations, for instance, the biomass of many mobile epibenthic taxa was found to be minimal whereas filterfeeding tunicates (*Pyura*) and bivalves (e.g. *Chama*) reached their maximum biomass. The same large filterfeeders also account for the exceptionally high biomass (42 g afdw $m^{-2}$) of the macrofaunal coresamples taken at the inner shelf. The remaining biomass (5.7 g afdw $m^{-2}$) consisted mainly of numerous small interstitial annelids. Amphipods and isopods were remarkably scarce in samples from the inner shelf. The meio-infauna from the area yielded relatively low numbers of nematodes but elevated numbers of harpacticoid copepods and foraminiferans. Explanations for the low biomass of invertebrate predators in terms of prey availability, lead to the low abundances of small crustaceans and nematodes at the inner shelf. The shortage of these organisms is apparently not balanced by high densities of interstitial worms and copepods. Also, the cryptic lifehabit and small size of most macro-infauna probably affect food conditions for demersal fish. Hydrographic conditions may also play a role in the scarcity of epifaunal invertebrates at the inner shelf. The nature of the fauna as well as the presence of coarse sediment are indicative for strong currents (waves or tides) which could inhibit settlement of pelagic larvae of for instance crabs and hermit crabs.

At the shelf edge we find the reverse contrast between the biomass of epifauna and macro-infauna, *i.e.* a high epifauna and a moderate to low macro-infauna biomass. However, if the size fraction 0.4–1 mm is included, macro-infaunal density at the two shelf edge stations that were sampled, appears to be relatively high in comparison to stations on the shelf proper. The dominating taxa at the two stations were different: amphipods and isopods (12 000 ind. $m^{-2}$) being most numerous on the northern station and polychaetes (23 000 ind. $m^{-2}$) on the southern one. The polychaetes at the latter station are mainly surface-feeding spionids which form an easier prey for predators than the cryptic annelids at the inner shelf stations. The meio-infauna at the two shelf edge stations was quite different in that at the southern station all taxa had depressed densities whereas on the northern station densities were in the same range found on the shelf proper.

While we recognise the speculative nature of this discussion, it nevertheless serves to stress our point that detailed information on the infauna and their lifehabits should be part of discussions about the distribution of epifauna. It is evident that other factors govern the distribution of epifauna as well. In addition to the hydrographical conditions previously mentioned with respect to the inner shelf stations, we want to point out a particular feature of the upwelling water on the Mauritanian shelf, *viz.* its low content of dissolved oxygen (e.g. Weichart, 1974). During the present study the oxygen content near the bottom was found to decrease in an offshore direction (Fig. 3). Near the outer shelf, conditions could be characterised as being close to hypoxia (<25% saturation). A recent study on the behaviour of some epifaunal taxa under hypoxic conditions (Pihl *et al.*, 1991) showed that some taxa, in this case a small sole (*Trinectis*) and a mantis shrimp (*Squilla*), were able to withstand hypoxia better than others. According to the authors, this enabled these species to benefit from an increased food availability since many infaunal species come to the sediment surface under these conditions. Apart from this reaction of the infauna, the absence of larger predators that rapidly migrated from the hypoxic area, may form a secondary advantage for the resident species. We have no proof that the macro-infauna on the Mauritanian shelf reacts in a similar way to hypoxia. In fact, we did not observe any dead or moribund organisms in our coresamples. But it remains striking that small soles and crustaceans are far more numerous on the outer shelf with its adverse oxygen conditions. In this context, we should also point at the low oxygen values on part of the midshelf stations (Fig. 3). It is, however, likely

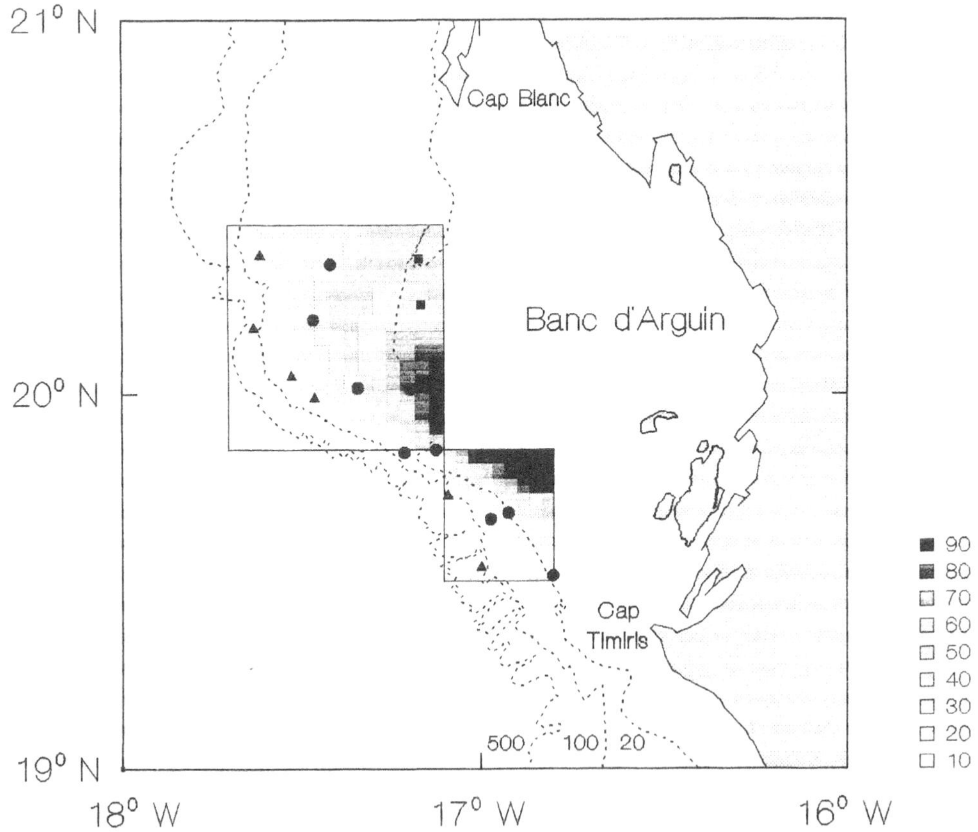


*Fig. 3.* The percentage oxygen saturation near the bottom on the Mauritanian shelf in May 1988. The positions of the trawlstations (cf. Fig. 1) have been indicated. The squares delimit the survey area.

that at the midshelf hypoxic periods are of shorter duration because any change in wind force and direction will first lead to a retreat of the upwelling from the more shallow places.

## Acknowledgements

We thank R. Witbaard, the technicians and the crew of the RV. 'TYRO' for their assistance during sampling. We acknowledge the S.O.Z. for financial support of the expedition.

## References

Ansa-Emmim, M., 1982. Fisheries in the CINECA region. Rapp. P.-v. Réun. Cons. int. Explor. Mer 180: 405–422.

Boely, T., 1982. Les ressources en poissons pélagiques des côtes ouest-africaines entre la Mauritanie et le fleuve Congo. Rapp. P.-v. Réun. Cons. int. Explor. Mer 180: 423–431.

Bonnet, M., J. Duclerc & P. Pichot, 1971. Nouvelle étude sur les fonds de pêche du Banc d'Arguin et de ses abords. Campagne de la 'Thalassa', janvier–février 1971. Science et Pêche, Bull. Inst. Pêches marit. 203: 1–15.

Brulhet, J., J. Maigret & Sy Moussa Arouna, 1974. Résultats de la campagne de chalutage expérimental autour du Banc d'Arguin. Bull. Lab. Pêches de Nouadhibou 3: 71–111.

Bravo de-Laguna, J., 1982. Distribution and abundance of demersal resources of the CINECA region. Rapp. P.-v. Réun. Cons. int. Explor. Mer 180: 432–446.

Collignon, J., 1971. Les poissons benthiques du plateau continental Marocain; 2eme note: le peuplement des sables cotiers. Bull. Inst. Pêches marit. 18: 37–69.

Creutzberg, F., G. C. A. Duineveld & G. J. van Noort, 1987. The effect of different numbers of tickler chains on beam-trawl catches. J. Cons. int. Explor. Mer 43: 159–168.

Duineveld, G. C. A. & G. J. van Noort, 1990. Geographical variation in the epifauna of the southern North Sea and adjacent regions. ICES paper C.M. 1990/Mini:9: 1–11.

Duineveld, G. C. A., P. A. W. J. de Wilde & A. Kok, 1990. A synopsis of the macrobenthic assemblages and benthic ETS activity in the Dutch Sector of the North Sea. Neth. J. Sea Res. 26: 125–138.

Duineveld, G. C. A., P. A. W. J. de Wilde, E. M. Berghuis & A. Kok, 1993. The benthic infauna and benthic respiration off the Banc d'Arguin (Mauritania, Northwest Africa). Hydrobiologia 258: 107–117.

Hill, M. O., 1979a. TWINSPAN – a FORTRAN program for arranging multivariate data in an ordered two-way table by classification of the individuals and their attributes. Ithaca, Cornell University, New York, 90 pp.

Hill, M. O., 1979b. DECORANA – a FORTRAN program for detrended correspondence analysis and reciprocal averaging. Ithaca, Cornell University, New York, 52 pp.

Huntsman, S. A. & R. T. Barber, 1977. Primary production off northwest Africa: the relationship to wind and nutrient conditions. Deep Sea Res. 24: 25–33.

Maurin, C. & M. Bonnet, 1969. Le chalutage au large des côtes nord-ouest africaines. Résultats des campagnes de la 'Thalassa'. Science et Pêche, Bull. Inst. Pêches marit. 177: 1–17.

Minas, H. J., L. A. Cospidoti & R. C. Dugdale, 1982. Nutrients and primary production in the upwelling region off Northwest Africa. Rapp. P.-v. Réun. Cons. int. Explor. Mer 180: 148–183.

Pihl, L., S. P. Baden & R. J. Diaz, 1991. Effects of periodic hypoxia on distribution of demersal fish and crustaceans. Mar. Biol. 108: 349–360.

Rumohr, H., T. Brey & S. Ankar, 1987. A compilation of biometric conversion factors for benthic invertebrates of the Baltic Sea. The Baltic Marine Biologists Publication 9: 1–56.

Thiel, H., 1982. Zoobenthos of the CINECA area and other upwelling regions. Rapp. P.v. Réun. Cons. int. Explor. Mer 180: 323–334.

Van der Knaap, M., 1985. Preliminary annotated inventory of the scientific expeditions and resource surveys carried out in the CECAF area. CECAF/TECH/85/64, Dakar: 50 pp.

Weichart, G., 1974. Meereschemische Untersuchungen im nordwestafrikanischen Auftriebsgebiet 1968. 'Meteor' Forsch.-Ergebnisse A 14: 33–70.

Zijlstra, J. J., 1988. The North Sea ecosystem. In H. Postma & J. J. Zijlstra (eds), Continental shelves. Elsevier Scientific Publishers, London: 231–277.

*Hydrobiologia* **258**: 175–183, 1993.
*W. J. Wolff, J. van der Land, P. H. Nienhuis & P. A. W. J. de Wilde (eds),*
*Ecological Studies in the Coastal Waters of Mauritania.*



# Food items of the grey mullet *Mugil cephalus* in the Banc d'Arguin area (Mauritania)

H. Michaelis
*Niedersächsisches Landesamt für Wasserwirtschaft, Forschungsstelle Küste, An der Mühle 5, D-2982 Norderney, Germany*

*Key words: Mugil cephalus*, sediment feeding, benthic diatoms, organic detritus, carbon content

## Abstract

The stomach and gut content of 22 *Mugil cephalus* from the Banc d'Arguin shallows (Mauritanian coast, West Africa) is examined. Quartz grains are the predominant material found in the stomach together with small portions of benthic diatoms and 'flakes', aggregates of fine-grained inorganic and organic particles. As the composition of this material is modified by a sorting procedure during feeding it is difficult to conclude, whether seagrass stands (muddy and mixed sediments with sand fraction) or sandy flats are preferred as feeding habitats. The animals studied did not or not relevantly utilize the non-diatom microphytes, though the environment of the Banc d'Arguin is rich in microbial mats (cyanobacteria), *Vaucheria* beds and epiphytic vegetation of seagrass leaves.

## Introduction

In the shallow water ecosystem of the Mauritanian Banc d'Arguin area mullets are an important element of the fish fauna (Maigret & Ly, 1986), and among the species occurring there, *Mugil cephalus* is predominating (Maigret & Ly, 1986). Due to its world-wide and abundant distribution in tropical and subtropical coastal waters, *M. cephalus* belongs to the most thoroughly investigated representatives of the Mugilidae. A common characteristic of this family is the highly specialised adaptation to feeding on surface sediments with the associated organic components including bacteria, cyanobacteria, microphytobenthos, macrophytic detritus and other organic particles, meiozoobenthos and small representatives of the macrofauna (Thomson, 1954, 1966; Odum, 1966, 1968, 1970; Hickling, 1970; Krefft & Michaelis, 1976). Uptake and ingestion are combined with a sorting and flushing procedure by which the too coarse material is rejected, the very fine material is released with the water leaving the gill slits, and the particles of the desired size range are retained by the gill apparatus for being passed through the digestive tract. Additionally, mullets graze on epiphytic microalgae covering hard bottoms and leaves of macrophytes and on trichomic algae and cyanobacteria forming mats on the sediment surface.

The present tentative study is based on a very limited material collected during a stay of one month only. It aims to contribute to the question which food sources *Mugil cephalus* exploits in the southern part of the Banc d'Arguin, a low-mesotidal environment with large intertidal flats divided by systems of shallow channels. According to Wolff & Smit (1990) about 80% of the intertidal areas are covered by seagrass (mainly *Zostera noltii)*, about 20% are made up of bare sandy flats inhabited by benthic diatoms (Sterrenburg & Sterrenburg, 1990) and some small areas are grown with a mangrove and with *Spartina*. In addition, there are extensive sebkha areas, a transition habitat between the tidal flats and the desert, where occasionally layers of cyanobacteria may occur. The subtidal seafloor is largely covered by stands of the seagrass *Cymodocea nodosa*.

## Material and methods

In September 1988 a benthic survey of the Banc d'Arguin eulittoral and supralittoral flats was carried out. A part of the benthos sampling was done with boats of the Imraguen fishermen from the village Iouik, who used the opportunities for catching grey mullets. From 6 of these catches 22 specimens of *Mugil cephalus* were collected and their intestines were preserved in 70% alcohol. All animals were caught during day and during low tide (about 3 hours before to 3 hours after low water). The total length of the animals ranged from 48.5 to 63 cm (16 measurements), their fresh weight from 1105 to 2162 g (15 measurements) and the length of the guts in preserved state varied from 180 to 225 cm (13 measurements). For microscopic analysis and for determination of organic carbon content of the food pulp samples were taken as denoted in Fig. 1: From all 22 animals out of the cardial part (I) and the pyloric part (II) of the stomach; samples out of the pyloric caeca (III), out of the gut just behind them (IV) and from the other parts of the gut (V–VIII) could only be taken from 13 well preserved animals.

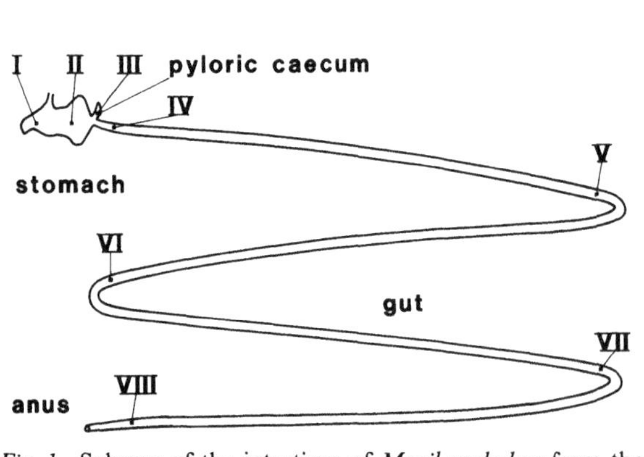


*Fig. 1.* Scheme of the intestines of *Mugil cephalus*; from the position I–VIII samples of the food pulp are examined.

The stomach content (cardial part) of the animals was analysed according to grain size distribution (one mixed sample of 20 animals, separate samples of 2 animals) and compared with the sediments of 29 sampling stations of the macrobenthic survey. With regard to sediment analysis the following methods were applied to all samples:

Grain size distribution: Sieving of the fractions $>0.06$ mm, areometer measurements of the fractions $<0.06$ mm.

Loss on ignition: Sediment dried at 105 °C was kept for one hour at 1000 °C. The difference in weight is given as percentage.

Organic matter: Calculated by substracting the weight of $CO_2$ supplied by $CaCO_3$ from the loss on ignition.

Organic carbon: Coloumetric measuring with the C-analyzer 'Coloumat 702-LI' after dry combustion in oxygen flow.

## Results

### *State of filling*

The intestines of all animals investigated showed the same state of filling: The cardial part of the stomach was only half-filled, the elastic caecum being almost unexpanded (Fig. 6d). In the gut, food material was equally distributed over the whole length.

### *Microscopic and grain size analysis*

Microscopic examination revealed, that all animals had sandy sediment in the cardial parts of their stomachs with quartz grains representing by far the biggest portion (Fig. 2a). The grain size distribution of the cardial stomach contents (Fig. 3) shows fine sand being the main fraction and medium sand or coarse silt being present in

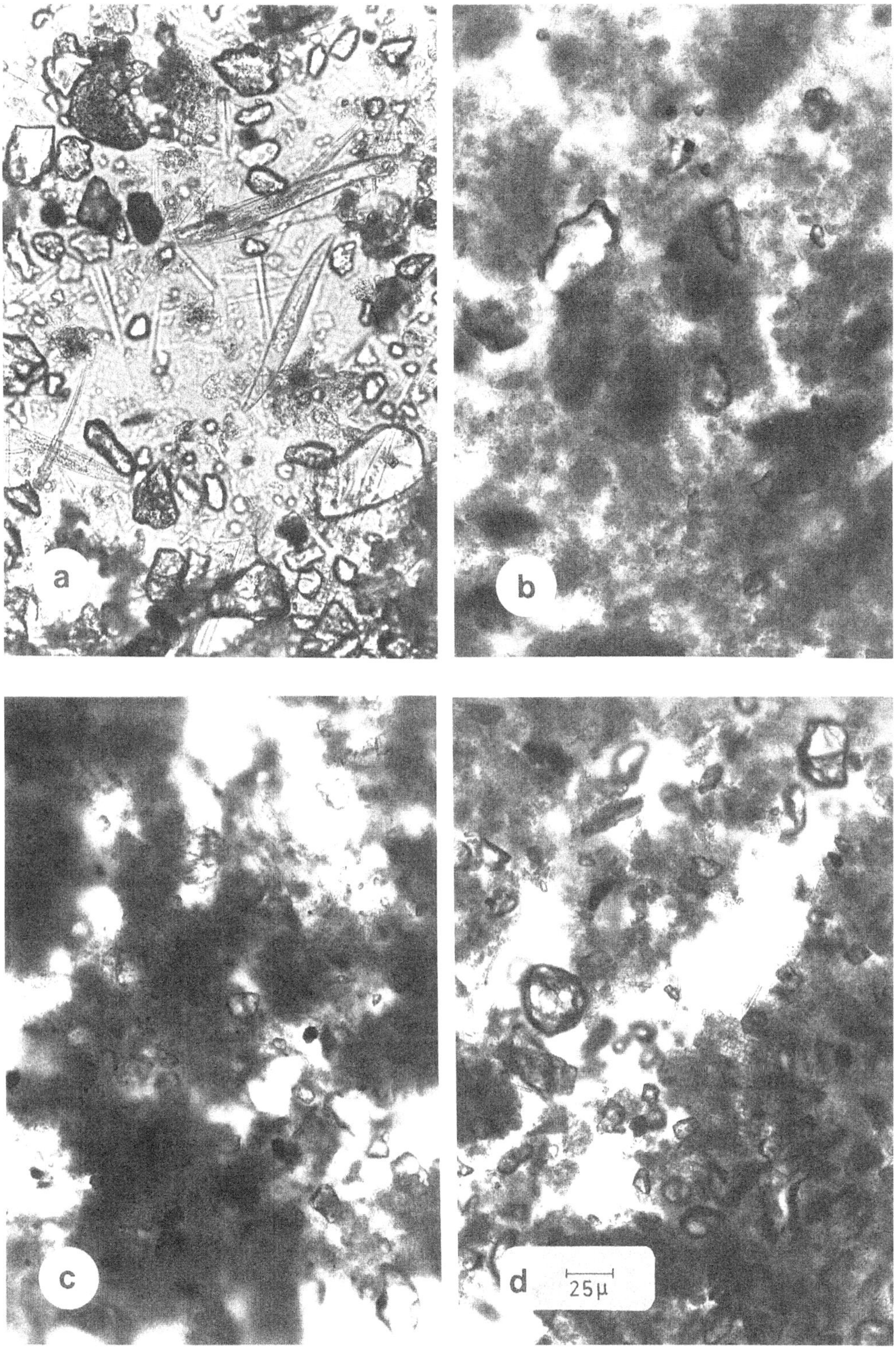


*Fig. 2.* Microscopic views of the food pulp from different parts of the intestines. a: Content of the cardial part of the stomach with quartz grains, diatoms and 'flakes'. b: Content of the pyloric caeca with secretion and admixed food material. c: Content of the anterior part of the gut; secretion material is prevailing. d: Content of the posterior part of the gut.

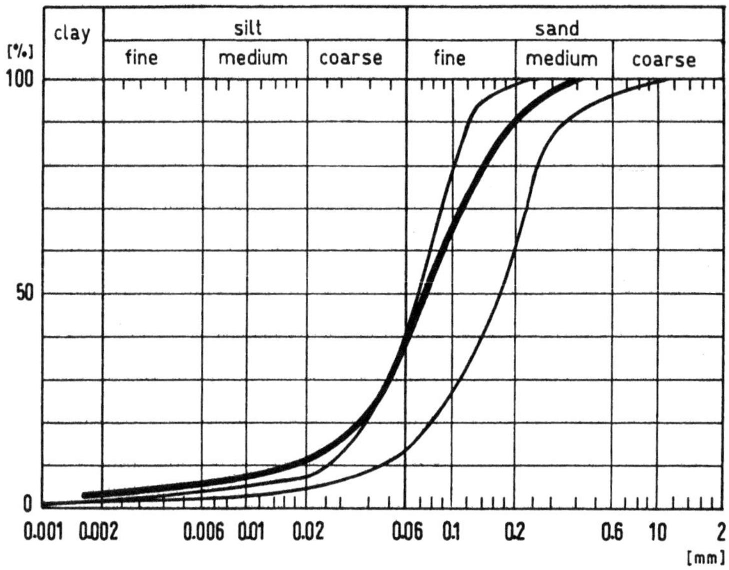


*Fig. 3.* Grain size distribution of food material from the cardial part of the stomach. Thick line: Mixed sample of 20 animals; thin lines: 2 separate animals.

certain amounts. Fine and medium silt are poorly represented and clay is almost absent.

In addition to the quartz grains, there were regularly three kinds of other materials present in the food pulp:

- Benthic diatoms; in the cardial part of the stomach they were found unbroken, mainly species of the genera *Synedra*, *Nitzschia*, *Gyrosigma*, *Pleurosigma*, *Amphora*, and others (Fig. 2a).
- Flakes of a probably organic matrix to which fine organic and mineral particles were agglomerated (Fig. 2a).
- Pieces of decaying higher plant material, probably seagrass leaves; the amount was always unimportant.

Not regularly and never in relevant amounts the following other items were found in the ingested material: dinoflagellates, trichomic algae, foraminifers, nematodes, and, rarely, harpacticoids and other small crustaceans.

One step further, in the muscular, pyloric part of the stomach (II), the material is kneaded and ground. Therefore the diatoms were found with cracked shells and their plasma was flown out.

Passing over from the stomach to the gut, the food pulp receives the secretion of the pyloric caeca (III), into which small amounts of food material may intrude (Fig. 2b). This secretion visibly remained the dominating constituent in the anterior quarters of the gut (IV und V, Fig. 2c). In the middle and posterior parts of the gut (VI and VII, Fig. 2d) the secretion appeared to be reduced and was no longer discernable from other fine-grained materials.

### *Carbon content*

The organic carbon was analysed as a parameter of the potentially available energy in the food material. Figure 4 represents the changes of the carbon content which the food pulp underwent during the passage. The freshly collected sediment in the cardial part of the stomach contained carbon values of less than 3%, on average 2%. Behind the pylorus of the stomach, where the secretion of the pyloric caeca is added, the carbon increased essentially to 12% on average with a maximum of 30%. On the way through the gut there is a gradual decrease. However, in the ultimate part the

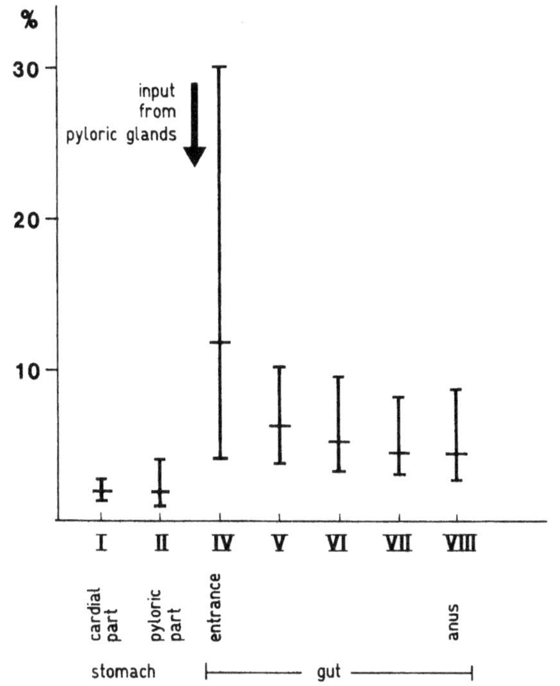


*Fig. 4.* Organic carbon content of the food pulp (% of dry weight) in different parts of the intestines; concerning locations I–VIII compare Fig. 1.

average value was still about 4%, which is the double amount of the carbon found in the stomach.

## Discussion

This study on food and feeding of *Mugil cephalus* in the Banc d'Arguin shallows provides only preliminary information, because the material examined is very limited in number (22 animals) and covers only a short time of the year (September). In the Banc d'Arguin waters seagrass beds with muddy and mixed sediments (Fig. 5) are the most widely occurring habitat, which is richest in energy in terms of organic matter and carbon (average 2.17%, Table 1), and which therefore might be suspected of being a main feeding ground of *Mugil cephalus*. Additionally, there are extensive stands of microphytes easily available: The epiphyton of seagrass leaves, cyanobacteria forming microbial mats, and *Vaucheria* beds; the latter trichomic xanthophycean algae form mats of considerable size and thickness in *Zostera* stands of lower density.

In view of the rich potential food sources mentioned above, it is a somewhat surprising result of this study that all mullet specimens examined contained exclusively sandy material in their stomachs, as similarly stated by Maigret & Ly (1986) also with regard to *M. cephalus* of the Banc d'Arguin. There is no 'preponderance of very fine sediment particles' as found by Odum (1968) in juvenile animals of 5 to 25 cm standard length in American waters. In the Banc d'Arguin material one difference can be found between the stomach content and the sediments of the sandy habitats: The carbon content of the stomach material (average: 2%) is 10 to 15 times higher than in the sandy flats (average: 0.2%) or in the sebkha (0.14%). Wolff & Smit (1990) give an average of 0.9% organic matter for the sandy flats, corresponding to roughly 0.4% C. As it is known, that sediments ingested in the field and stomach material have a different composition due to the sorting and flushing procedure of the Mugilidae (Günther, 1880; Nikolskii, 1957; Thomson, 1966; Odum, 1968; Hickling, 1970, and others) the following explanations appear conceivable:

- Feeding in sandy habitats; the small fraction of isolated fine particles would be largely removed while – together with sand and coarse silt – the diatoms and 'flakes' (aggregates of mineral and organic fine particles) would be retained, causing a considerable enrichment of organic carbon.
- Feeding in seagrass beds with muddy and mixed sediments; in this case the coarser material of decaying leaves has to be rejected and the large fine-grained sediment fractions have to be removed by the flushing technique. As the sediments of the seagrass beds are generally badly sorted with always relatively big portions of the sand fractions (Fig. 5), the retained food material might consist of sand plus coarse silt together with diatoms and flakes. A carbon enrichment does not occur in this case, the contents in the feeding habitat (in average

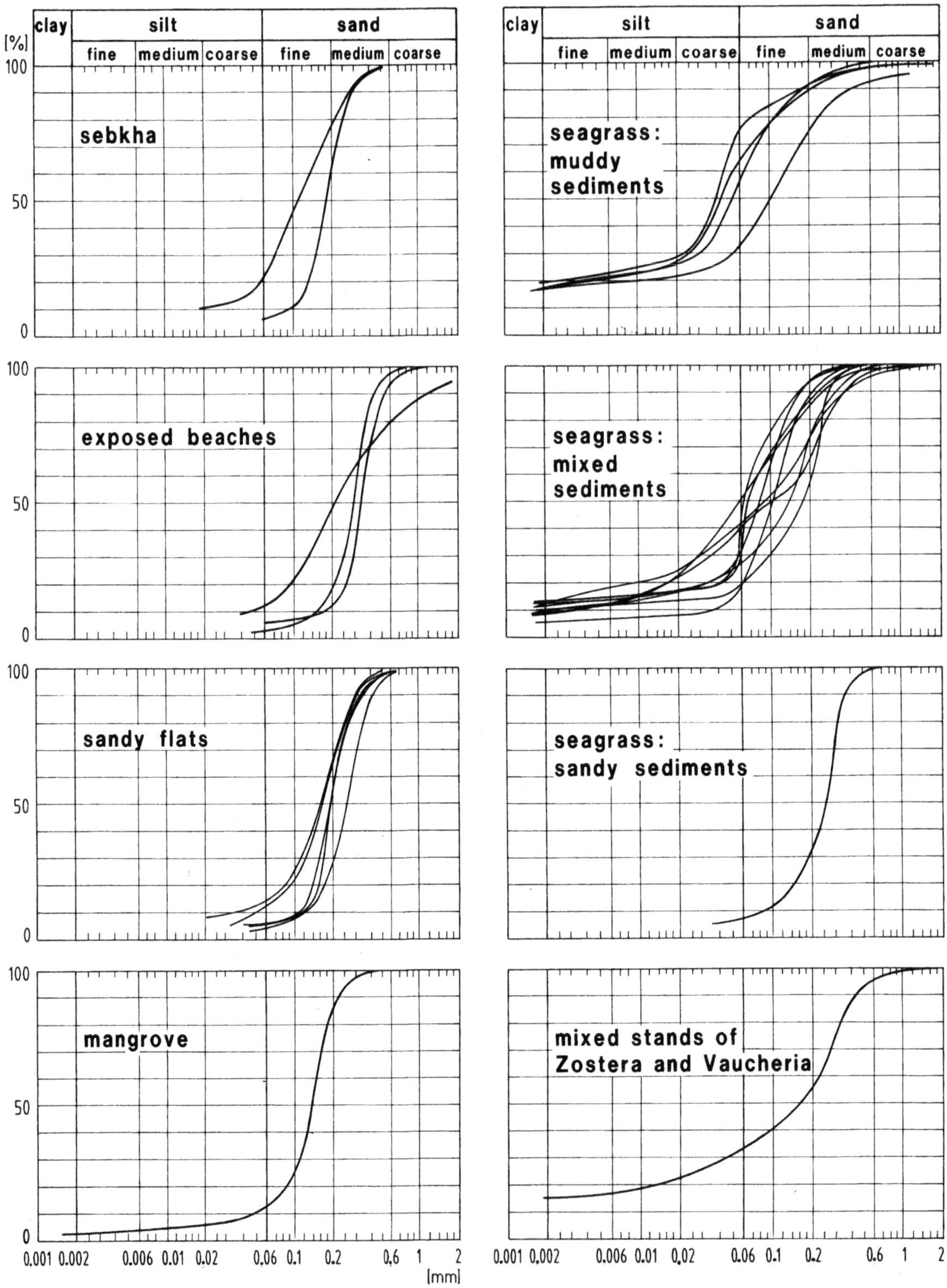


*Fig. 5.* Grain size distribution of sediments from 29 macrobenthic sampling stations in the different intertidal habitats of the Banc d'Arguin (only 26 curves, as 3 belong to identical pairs).

*Table 1.* Organic matter and carbon content in sediments of the Banc d'Arguin intertidal area (sampling stations of a macrobenthic survey in September 1988).

| Habitat | Station nr. | Loss on ignition | Organic matter % of dry weight | Organic carbon | Organic C average ± SD |
|---|---|---|---|---|---|
| Sebkha | 1 | 0.99 | 0.99 | 0.05 | 0.14 ± 0.08 |
| | 2 | 7.36 | 7.36 | 0.22 | |
| | 3 | 3.94 | 3.80 | 0.14 | |
| Sandy flats | 4 | 0.97 | 0.81 | 0.14 | 0.20 ± 0.19 |
| | 5 | 2.28 | 2.21 | 0.07 | |
| | 6 | 1.97 | 1.65 | 0.12 | |
| | 7 | 1.06 | 0.98 | 0.03 | |
| | 8 | 3.13 | 2.36 | 0.28 | |
| | 9 | 3.86 | 1.54 | 0.55 | |
| Exposed beaches | 10 | 7.13 | 5.89 | 0.12 | 0.05 ± 0.06 |
| | 11 | 1.54 | 1.08 | 0.03 | |
| | 12 | 0.91 | 0.56 | 0.01 | |
| Mangrove | 13 | 1.47 | 1.47 | 0.16 | |
| Stands of seagrasses, muddy and mixed sediments | 14 | 10.56 | 10.01 | 2.12 | 2.17 ± 0.78 |
| | 15 | 15.82 | 15.82 | 3.74 | |
| | 16 | 13.76 | 9.77 | 2.33 | |
| | 17 | 10.75 | 9.03 | 2.25 | |
| | 18 | 7.94 | 7.04 | 1.82 | |
| | 19 | 11.10 | 10.77 | 2.77 | |
| | 20 | 8.39 | 8.05 | 1.98 | |
| | 21 | 6.57 | 5.48 | 1.62 | |
| | 22 | 8.63 | 7.41 | 1.99 | |
| | 23 | 3.68 | 2.95 | 0.49 | |
| | 24 | 12.47 | 12.19 | 3.01 | |
| | 25 | 11.64 | 10.72 | 2.92 | |
| | 26 | 9.82 | 8.64 | 1.88 | |
| | 27 | 4.82 | 4.40 | 1.52 | |
| Stands of seagrasses, sandy sediments | 28 | 1.65 | 1.44 | 0.33 | |
| *Zostera/Vaucheria* | 29 | 8.87 | 8.41 | 1.92 | |

2,17%, Table 1; according to Wolff & Smit, 1990, also about 2%) and in the stomach (in average 2%) are similar.

No information is available from the sediments of the sublittoral habitats, which probably will be intensely utilized as feeding grounds.

As to the carbon measurements, they should be expected to show lower values at the end of the passage than at the beginning in the stomach. However, Fig. 4 seems to suggest, that the high input of secretion by the pyloric caeca is not completely reabsorbed. This remains an unsatisfactory observation which requires closer examination, as the animals have to supply their energy budget permanently with new carbon.

The material studied allows some conclusions concerning the rhythm of feeding activities: The incomplete filling of the stomach and the equal distribution of the food pulp over the gut do not support the presence of any tidal rhythm in the food uptake. Odum (1970) in Florida and Georgia waters found that *M. cephalus* was feeding almost permanently at day and night, but the stomachs contained more material during high water than during low water. In the Banc d'Arguin waters *Mugil cephalus* probably takes up food continuously during the whole day, which is corroborated by the steady flow of material through the gut. This means that during low water, when the intertidal flats are exposed, feeding is continued in the subtidal parts of the seafloor.

This behaviour is quite different from that of a mullet species, *Crenimugil labrosus*, living in European waters including the Wadden Sea. As described by Hickling (1970), feeding is interrupted by digestion phases, and Krefft & Michaelis (1976) showed that these phases are closely related to the tidal cycle. The fishes invade the intertidal area with the rising tide, gather on the mud flats covered by diatom layers, and scrape off the top layer with their sharp lower jaws. During this phase the cardial part of the stomach is filled completely until it is expanded to the uttermost (Fig. 6a). At the same time the food material is passed on to the pylorical part of the stomach and to the gut. During low water the fishes stay in the channels and spend the time with digestion. Then the stomach becomes totally empty (Fig. 6b, c) and the anterior parts of the gut too.

It seems remarkable that in the limited number of *Mugil cephalus* collected on the Banc d'Arguin the non-diatom microphytes are not or scarcely represented as food materials. In one single place the characteristic feeding marks of mullets – not necessarily *Mugil cephalus* – were observed in cyanobacteria-layers in the sebkha. Schwarz *et al.* (1975) describe feeding marks of herbivorous fishes in intertidal 'stromatolites' formed by cyanobacteria. According to the figures given (Schwarz *et al.*, 1975, Fig. 5c, 6c) one type of these marks might be derived from mullets. In the *Vaucheria* beds of the Banc d'Arguin no traces of any consumption by mullets could be observed.

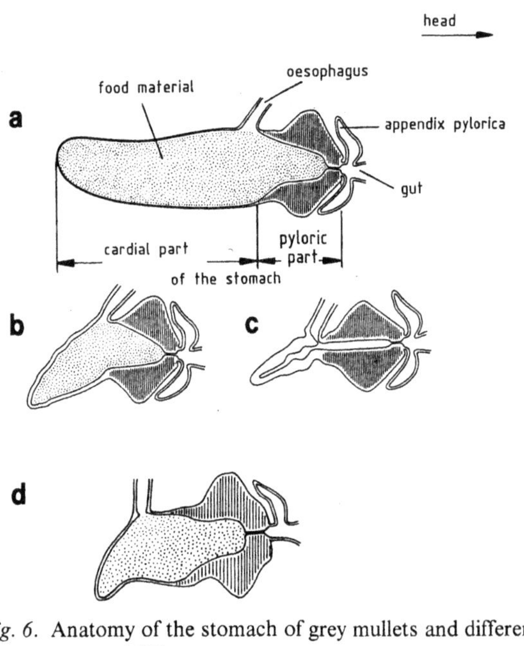


*Fig. 6.* Anatomy of the stomach of grey mullets and different states of filling.
a, b, c: Stomach of *Crenimugil labrosus* completely filled (a), half-filled (b) and empty (c). From Krefft & Michaelis (1976).
d: Stomach of *Mugil cephalus*; all animals of this study were encountered with a cardial part of the stomach only half-filled.

## Acknowledgement

The financial support of this study received from the Deutsche Forschungsgemeinschaft, Bonn-Bad Godesberg, and the Umweltstiftung WWF-Deutschland, Frankfurt, is gratefully acknowledged.

## References

Günther, A. C. L. G., 1880. An introduction to the study of fishes. Edinburgh, 720 pp.

Hickling, C. F., 1970. A Contribution to the Natural History of the English Grey Mullets (Pisces, Mugilidae). J. mar. biol. Ass. U.K. 50: 609–633.

Krefft, G. & H. Michaelis, 1976. Die Meeräsche im niedersächsischen Wattenmeer. Natur u. Museum 106: 23–29.

Maigret, J. & B. Ly, 1986. Poissons de Mer de Mauritanie. Centre National de Recherche Oeceanographique et de Pêche. Compiegne, France, 213 pp.

Nikolskii, G. W., 1957. Spezielle Fischkunde (translated from

Russian). VEB Deutscher Verlag der Wissenschaften, Berlin, 632 p.
Odum, W. E., 1966. The food and feeding of the striped mullet, *Mugil cephalus*, in relation to the environment. M.S. Thesis, Institute of Marine Sciences, University of Miami. Miami, Florida, 118 p.
Odum, W. E., 1968. The ecological significance of fine particle selection by the striped mullet *Mugil cephalus*. Limnol. Oceanogr. 13: 92–98.
Odum, W. E., 1970. Utilization of the direct grazing and plant detritus food chain by the striped mullet *Mugil cephalus*. In Steele, H. J. (ed.): Marine Food Chains: 1–43. Oliver & Boyd, Edinburgh.
Schwartz, H. U., G. Einsele & D. Herm, 1975. Quartz-sandy, grazing-contoured stromatolites from coastal embayments of Mauritania, West-Africa. Sedimentology 22: 539–561.
Sterrenburg, F. A. S. & F. J. G. Sterrenburg, 1990. An Outline of the Marine Littoral Diatom Biocoenosis of the Banc d'Arguin, Mauritania, West Africa. Botanica Marina 33: 459–465.
Thomson, J. M., 1954. The organs of feeding and the food of some Australian mullet. Aust. J. mar. Freshwat. Res. 5: 469–484.
Thomson, J. M., 1966. The Grey Mullets. Oceanogr. Mar. Biol. annu. Rev. 4: 301–335.
Wolff, W. J. & C. Smit, 1990. The Banc d'Arguin, Mauritania, as an environment for coastal birds. Ardea 78: 17–38.

*Hydrobiologia* **258**: 185–196, 1993.
*W. J. Wolff, J. van der Land, P. H. Nienhuis & P. A. W. J. de Wilde (eds),*
*Ecological Studies in the Coastal Waters of Mauritania.*




# The distribution and abundance of young fish in the Banc d'Arguin, Mauritania

Zwanette Jager
*DLO-Institute for Forestry and Nature Research, Texel, The Netherlands;*
*Present address: Tidal Waters Division, P.O. Box 207, 9750 AE Haren, The Netherlands*

*Key words:* Banc d'Arguin, fish, Mauritania, nursery

## Abstract

The importance of the Banc d'Arguin (Mauritania) for juvenile fish has been investigated in September 1988. Fish have been sampled with a 2 m beam trawl at 50 locations in the coastal zone between Cap Sainte Anne and Cap Timiris. Water depth varied between 1 and 16 metres. Temperature and salinity at each station have been measured, as well as the weight of the catch and total length of the individual fish. The largest catches – in numbers as well as in weight – were obtained around the shallows of Tidra and Arel. This region has a large diversity in species, and contains large numbers of juvenile fish and cephalopods. The deeper areas seem to be much poorer in species, but average length and weight of the sampled fish are higher than around Tidra and Arel.

## Introduction

The Banc d'Arguin (Fig. 1) is an area of about 9000 km$^2$, consisting of sandbanks and mudflats covered with seagrasses. Depth generally does not exceed 4 m, except in the eastern part where depths down to 16 metres occur. Water temperatures and salinities can reach quite extreme values; there is no inflow of fresh water from the continent (Wolff & Smit, 1990). Furthermore, the area is situated in a transition zone, which is twice a year swept by the passage of the northern subtropical hydrodynamic front. This causes large variability of a number of characteristics like temperature, salinity, nutrient concentrations, productivity etc. throughout the year. These variations are reflected by changes in catchability, composition and abundance of fish stocks, and cause fish migrations parallel to the coast over long distances (Troadec & Garcia, 1980). Hence, the Banc d'Arguin fish fauna will show temperate, sub-tropical as well as tropical elements.

Unfortunately, little is known of the distribution, composition and densities of the fish fauna in the shallow coastal area of the Banc d'Arguin. The trawling surveys that were carried out until present were restricted to the continental shelf and the deeper northeastern part of the Banc d'Arguin. Very few studies exist of the shallow area around Tidra (a.o. Campredon & Schrieken, 1989). The outer edge of the Banc, shallower than 20 m, has not been covered at all.

The potential function of the Banc d'Arguin as a nursery for larval and juvenile fish has hardly been investigated until now. This can partly be explained by the difficult access to this uncharted area, where only shallow-draught vessels can navigate. However, for the management of the Mauritanian marine living resources, it is important to know the relationship between the fish stocks present on the continental shelf, and those living on the Banc d'Arguin.

This inventory survey was carried out in September 1988 on the Banc d'Arguin, covering the

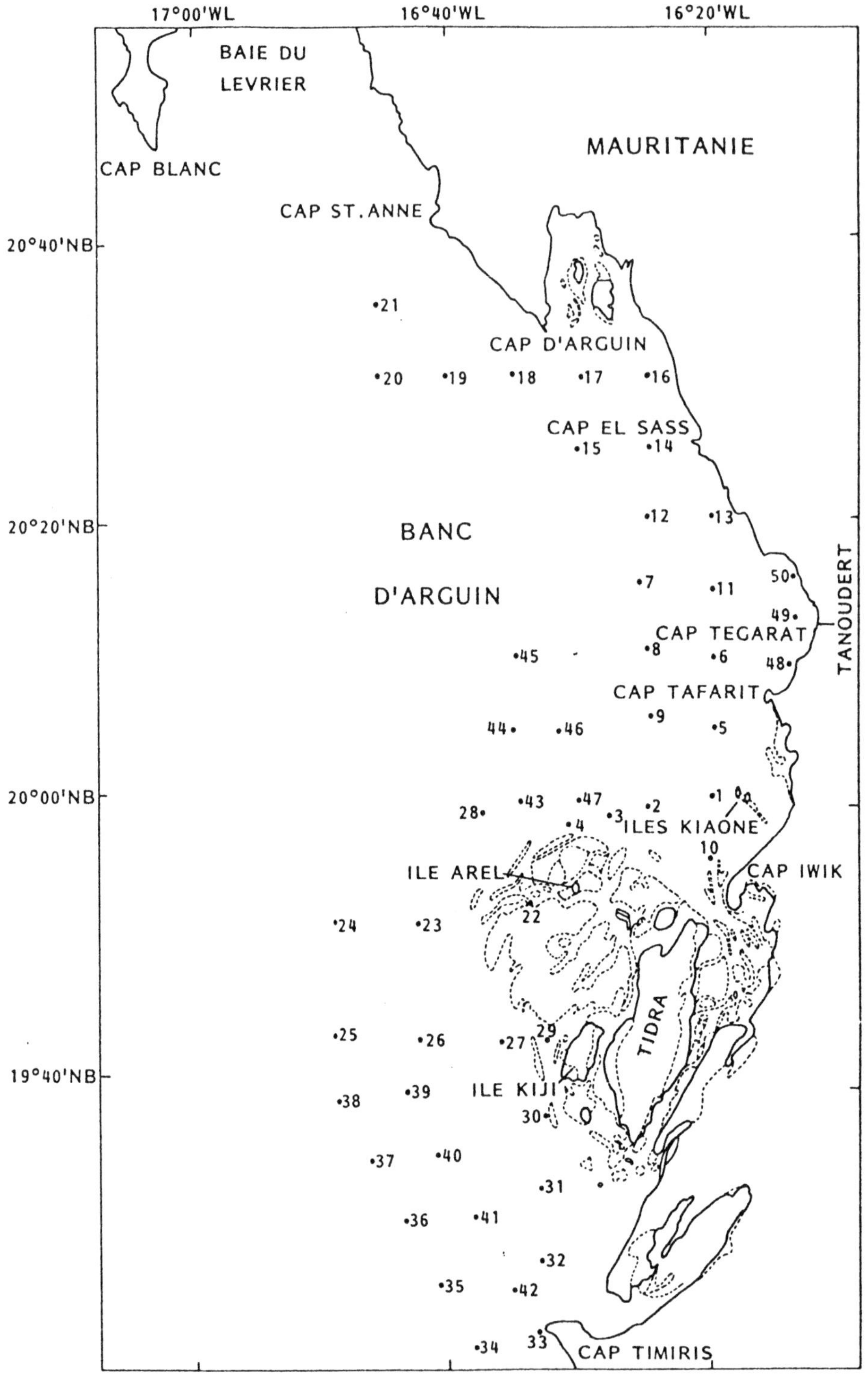


*Fig. 1.* Sample stations on the Banc d'Arguin, Mauritania, in September 1988. (Numbers indicate the sequence of sampling.)

area between Cap Sainte Anne and Cap Timiris (Fig. 1). Main purpose was to estimate the importance of the Banc d'Arguin for juvenile fish. To achieve this, an attempt was made to sample as many locations as possible.

## Materials and methods

A two-metre beam trawl was chosen for sampling. This trawl with a mesh size of 4.7 cm at the front part and 1.7 cm at the codend is efficient in catching small fish living close to the bottom and enables a semi-quantitative comparison of the locations sampled. At some locations an extra weight of 2 kg was put on the beam of the net to be able to fish in deeper water. Generally, one haul lasted ten minutes, except in the area southwest of Tidra where hauls of five minutes were made due to the abundant vegetation bycatch filling the trawl. Haul duration was measured from the moment that the net touched the bottom until the moment we started lifting the net. The length of the pulling rope was always about four times the water depth. With a trawling speed of about two knots and haul durations of 5 or 10 minutes, 600 or 1200 $m^2$ were covered in one haul. Unfortunately, a meter wheel for exact determination of the distance fished was not available. Considering also that the speed indicator of the ship was often not working and that there was no information about current velocities, only rough estimates of haul length could be obtained. At each location depth, temperature and salinity were measured. Depth was measured by means of an echo-sounder; salinities were determined from watersamples afterwards in the laboratory of the Forschungsstelle Küste, Norderney, Germany. The captured fish were identified, and the total lengths of the individuals were measured. Fifty locations were sampled. Due to lack of accurate navigation means, the positions given in Fig. 1 have to be considered as a rough indication of the places sampled.

It was attempted to cluster the locations by number of species and individuals. Clustering was performed with different methods, a.o. Ward's method and Euclidian Squared Distance as (dis-)similarity measurement. The most important and consistent clusters are described.

## Results

### *Abiotic factors*

The locations sampled varied in depth between 1 and 16 m (Fig. 2). As expected, the area between Cap El Sass and Iles Kiaone is the deeper part (10–16 m). Towards Cap Timiris the 10 m-isobath is situated relatively close to the coast.

Temperatures varied between 25 and 30 °C, salinities between 38 and 43‰ (Table 1). The lower temperatures were found in the northern part, the higher temperatures in the vicinity of Ile Tidra. Salinities were highest near the coast and decreased in westward direction.

Temperature and salinity were positively correlated ($r^2 = 0.31$; $p < 0.01$). Depth and temperature were negatively correlated ($r^2 = 0.22$; $p < 0.01$); however, depth and salinity were not significantly correlated ($r^2 = 0.05$; $p > 0.05$).

### *Biota*

In the catches, 45 fish species belonging to 22 families were present (Table 2). Catch weight reached a maximum of almost 8 kg per haul at location 48. The number of species ranged from 1 to 16 per location, the number of individuals in the catches varied between 1 and about 650 (Table 1).

A significant negative correlation existed between the natural logarithm of the number of individuals and depth, as well as between the number of species and depth. The natural logarithm of the mean length of the individuals and depth were positively correlated, whereas the natural logarithm of the mean length was not significantly correlated with temperature or salinity (Table 3).

Most of the locations shallower than 6 m (northwest of Tidra, near Tanoudert and near Timiris) were characterised by a bycatch of sea-

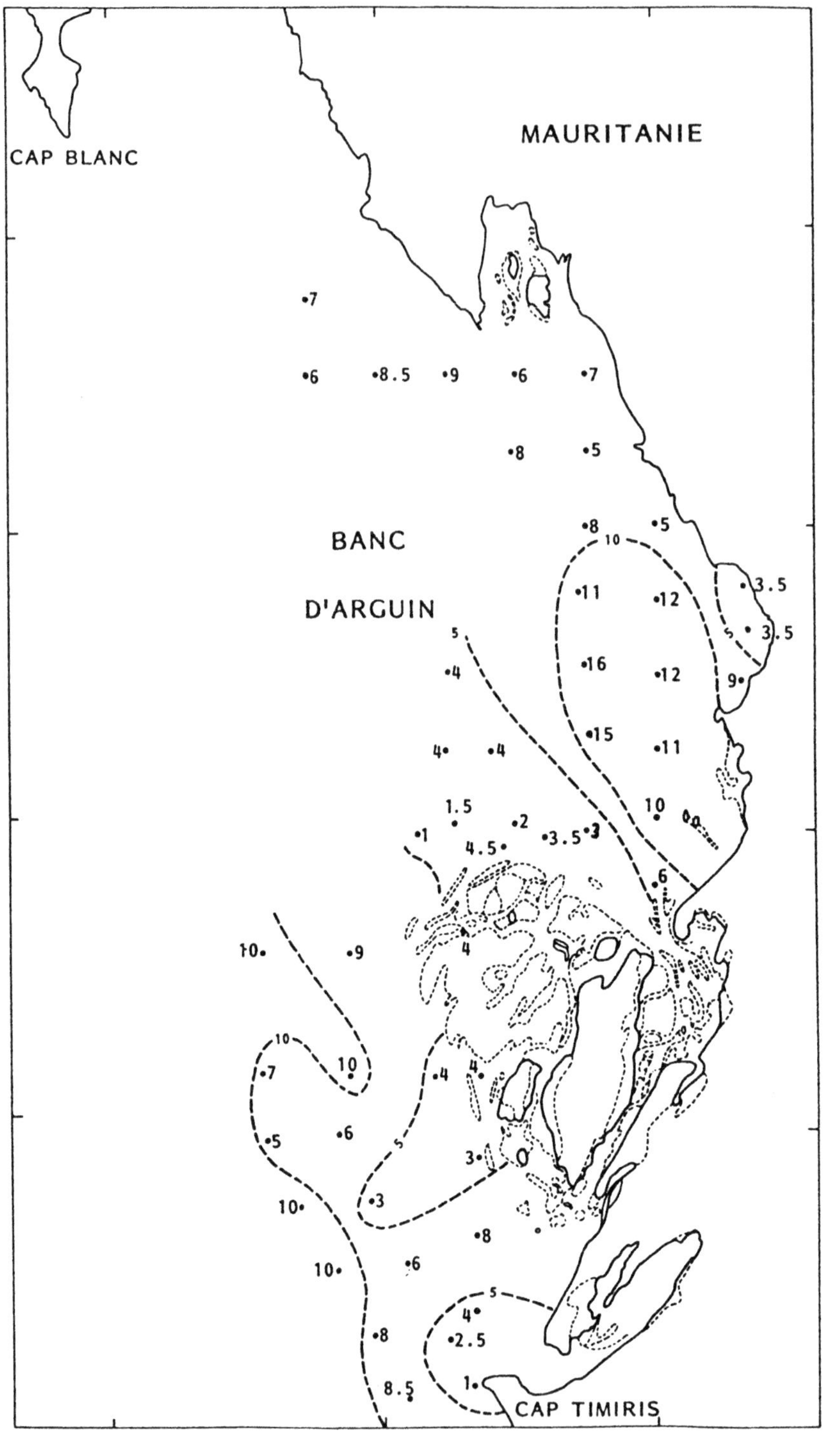


*Fig. 2.* Depth (in metres) of the sampled stations on the Banc d'Arguin in September 1988.

*Table 1.* Abiotic parameters and catch characteristics of the sampled locations on the Banc d'Arguin in September 1988.

| Location | Depth (m) | Temperature (°C) | Salinity (‰S) | Weight [W] (g) | Species number [Nsp] | Number of fish [nind] | Mean length (cm) |
|---|---|---|---|---|---|---|---|
| 1 | 10 | | 41.32 | 2000 | 3 | 5 | 28.0 |
| 2 | 3 | | 41.32 | 2969 | 10 | 375 | 5.9 |
| 3 | 3.5 | | 41.95 | 4205 | 16 | 646 | 5.8 |
| 4 | 4.5 | | 40.99 | 2909 | 11 | 423 | 6.2 |
| 5 | 11 | 26.6 | 40.99 | 0 | 0 | 0 | |
| 6 | 12 | 26.3 | 40.99 | 0 | 0 | 0 | |
| 7 | 11 | 26.8 | 40.99 | 0 | 0 | 0 | |
| 8 | 16 | 27.0 | 40.99 | 492 | 5 | 12 | 10.8 |
| 9 | 14 | 26.9 | 41.32 | 38 | 2 | 7 | 8.7 |
| 10 | 6 | | | 28 | 5 | 12 | 3.8 |
| 11 | 12 | 27.8 | 40.99 | 346 | 4 | 6 | 14.5 |
| 12 | 8 | 27.7 | 41.62 | 3265 | 3 | 4 | 24.8 |
| 13 | 5 | 28.5 | 41.62 | 795 | 4 | 33 | 7.8 |
| 14 | 6.5 | 28.2 | 42.60 | 559 | 3 | 3 | 19.7 |
| 15 | 8 | 27.5 | 40.99 | 516 | 2 | 4 | 10.3 |
| 16 | 7 | 26.9 | 41.95 | 249 | 7 | 29 | 7.0 |
| 17 | 6 | 25.8 | 40.67 | 256 | 4 | 5 | 10.8 |
| 18 | 9 | 25.7 | 39.38 | 14 | 1 | 4 | 4.0 |
| 19 | 9 | 25.4 | 38.10 | 0 | 0 | 0 | |
| 20 | 6 | 25.2 | 38.10 | 0 | 0 | 0 | |
| 21 | 7 | 24.9 | 38.43 | 127 | 3 | 3 | 12.7 |
| 22 | 6 | 27.9 | 41.32 | 1482 | 7 | 97 | 8.6 |
| 23 | 9 | 26.8 | 39.38 | 90 | 2 | 4 | 10.0 |
| 24 | 10 | 26.5 | 39. 38 | 225 | 4 | 12 | 12.8 |
| 25 | 7 | 26.9 | 39.38 | 16 | 1 | 2 | 3.0 |
| 26 | 10 | 27.0 | 40.03 | 286 | 3 | 10 | 13.0 |
| 27 | 4 | 28.0 | 40.99 | 423 | 9 | 58 | 7.5 |
| 28 | 1 | 28.3 | 40.99 | 1111 | 11 | 169 | 5.5 |
| 29 | 4 | 28.5 | 41.95 | 2050 | 7 | 22 | 10.6 |
| 30 | 3 | 29.3 | 40.99 | 1370 | 11 | 97 | 6.3 |
| 31 | 8 | 30.0 | 41.32 | 82 | 3 | 6 | 9.0 |
| 32 | 4 | 28.9 | 40.36 | 480 | 11 | 24 | 10.4 |
| 33 | 1 | 28.8 | 41.32 | 1522 | 10 | 146 | 6.3 |
| 34 | 9 | 27.6 | 38.43 | 10 | 3 | 3 | 12.3 |
| 35 | 8 | 27.6 | 39.38 | 0 | 0 | 0 | |
| 36 | 10 | 27.6 | 39.08 | 0 | 0 | 0 | |
| 37 | 10 | 28.0 | 39.71 | 1050 | 2 | 2 | 28.0 |
| 38 | 5 | 28.0 | 39.71 | 125 | 2 | 6 | 7.7 |
| 39 | 6 | 28.5 | 40.67 | 96 | 3 | 14 | 6.4 |
| 40 | 3 | 28.5 | 40.67 | 935 | 9 | 43 | 8.9 |
| 41 | 6 | 28.9 | 42.92 | 2100 | 6 | 9 | 11.0 |
| 42 | 3 | 29.4 | 43.23 | 410 | 15 | 50 | 7.0 |
| 43 | 2 | 28.2 | 41.32 | 4071 | 13 | 490 | 6.2 |
| 44 | 4 | 27.9 | 40.36 | 1077 | 7 | 42 | 8.0 |
| 45 | 4 | 27.5 | 38.75 | 1116 | 6 | 116 | 6.5 |
| 46 | 4 | 27.8 | 39.71 | 1500 | 1 | 1 | 47.0 |
| 47 | 2 | 28.8 | 40.99 | 582 | 8 | 112 | 6.1 |
| 48 | 9 | 27.4 | 41.32 | 7822 | 7 | 121 | 11.7 |
| 49 | 3.5 | 26.9 | 41.62 | 1879 | 12 | 45 | 9.2 |
| 50 | 3.5 | 27.4 | 41.62 | 80 | 2 | 3 | 10.7 |
| Avg | 6.8 | 27.6 | 40.66 | 1015.2 | 6.0 | 76.16 | 10.94 |
| Std | 3.4 | 1.1 | 1.2 | 1449.6 | 4.0 | 142.5 | 8.0 |

*Table 2.* Fish species, present in the catches on the Banc d'Arguin in September 1988. Additionally, four unidentified species were present in small numbers.

| | |
|---|---|
| Discobathidae | *Zanobathus schoeleinii* |
| Myliobathidae | *Rhinoptera marginata* |
| Tetraodontidae | *Ephippion guttiferum* |
| | *Sphaeroides spengleri* |
| Ariidae | *Arius heudeloti* |
| Syngnathidae | *Syngnathus* sp. |
| Polynemidae | *Galeoides decadactylus* |
| Serranidae | *Serranus scriba* |
| | *Promicrops esonue* |
| | *Epinephelus aeneus* |
| | *Epinephelus alexandrinus* |
| | *Epinephelus guaza* |
| Pomadasyidae | *Pomadasys incisus* |
| Sciaenidae | *Argyrosomus regius* |
| | *Dicentrarchus punctatus* |
| | *Pseudotolithus brachygnatus* |
| | *Pseudotolithus senegalensis* |
| | *Sciaena umbra* |
| Sparidae | *Diplodus bellottii* |
| | *Diplodus sargus* |
| | *Pagellus acarne* |
| | *Sparus auratus* |
| | *Sparus auriga* |
| | *Spondyliosoma cantharus* |
| Ephippidae | *Chaetodipterus goreensis* |
| Labridae | *Crenilabrus bailloni* |
| | *Xyrichtys* sp.? |
| Blenniidae | *Malacoctenus* sp. *nov.?* |
| | *Parablennius incognitus?* |
| Monacanthidae | *Stephanolepis hispidus* |
| Gobiidae | |
| Batrachoidae | *Batrachoides didactylus* |
| Scorpaenidae | *Scorpaena angolensis* |
| Dactylopteridae | *Cephalacanthus volitans* |
| Psettodidae | *Psettodes belcheri* |
| Bothidae | *Syacium micrurum* |
| Soleidae | *Synaptura lusitanica* |
| | *Pegusa triophthalma* |
| | *Solea senegalensis* |
| | *Dicologoglossa cuneata* |
| Cynoglossidae | *Cynoglossus* sp. |

grasses or green algae. At most of the locations deeper than 8 m west of Tidra, red algae were present as bycatch. At the other sites, a large variety of bycatch showed up, such as corals, sponges, shells, crabs, hermit crabs and worms. Locations 32 and 42 showed a high diversity (11 and 15 species respectively), and were special due to the occurrence of *Chaetodipterus goreensis* (Cuvier, 1831), a guinean species only present on the Banc d'Arguin in the warm season (Sevrin-Reyssac, 1983).

Juvenile *Sepia* sp. (Cephalopoda) were caught in some quantity (7 to 27 per haul) at the relatively shallow (2–3 m) locations 4, 28, 30 and 43. Of some species, almost exclusively adult individuals were caught, *viz. Ephippion guttiferum* (Bennett, 1831), *Diplodus sargus* (Linnaeus, 1758), *Syngnathus* sp., and the rays *Zanobathus schoeleini* (Muller & Henle, 1841) and *Rhinoptera marginata* (E. Geoffroy Saint-Hillaire, 1817).

### *Cluster analysis*

Cluster analyses, based on Table 4, resulted in the following clusters (Fig. 3). The largest cluster comprises all sites with a very small catch, ranging in depth from 3.5 to 16 m (7 m on average). The second cluster (2) consists of the locations 2, 3, 4 and 43 (depth 2–4.5 m, mean length of the fish *ca* 6.0 cm). These are very rich sites, with more than 350 individuals of at least 10 species per haul, and abundant presence and dominance of *Stephanolepis hispidus* (Linnaeus, 1766) and *D. bellottii* (Steindachner, 1882). The length-frequency distribution (Fig. 4) shows that the individuals range from 4 to 6 cm mainly, whereas the maximum attainable length is 25 cm for *S. hispidus*, and 15 cm for *D. bellottii* (Maigret & Ly, 1986). This means that of these two species, principally the juveniles were caught. Besides, large numbers of *Crenilabrus bailloni* (Valenciennes, 1839) and moderate numbers of *Sphaeroides spengleri* (Bloch, 1792) and gobiids were present. Of these species, a wider length-range occurred. Fish smaller than 4 cm were not fully retained by the mesh of the net.

The third cluster (3), which consists of the locations 28, 30, 33 and 47 (depth 1–3 m, mean length of the fish *ca* 6.0 cm) with between 97 and 168 individuals of about 10 species per haul, is characterised by a quite large abundance of *S. hispidus*, and the presence of several *C. bailloni*. Also, *D. bellottii* and *S. spengleri* are present. The

*Table 3.* Correlations between abiotic and biotic parameters.

| Correlation matrix | No. of species | Ln (no. of individuals) | Ln (mean length) |
|---|---|---|---|
| Depth | $r^2 = -0.36$*** | $r^2 = -0.30$*** | $r^2 = 0.21$** |
| Temperature | $r^2 = 0.20$** | $r^2 = 0.15$* | $r^2 = 0.05$ |
| Salinity | $r^2 = 0.20$** | $r^2 = 0.11$* | $r^2 = 0.03$ |

*** = Very highly significant; $p < 0.001$.
** = Highly significant; $p < 0.01$.
* = Significant; $p < 0.05$.

lengths of *S. hispidus* and *D. bellottii* are about 5 cm.

The fourth cluster (4) is composed of the sites 27, 40, 44 and 49 (depth 3–4 m, mean length of the fish *ca* 8.4 cm). This cluster is characterised by 40 to 60 animals per site; mostly, these sites have moderate (10–40) numbers of *S. hispidus* and *D. bellottii*.

*Table 4.* Numbers of individuals per haul of 30 species at 50 locations on the Banc d'Arguin (Sept. 1988).

| Species | Code loc. | 1 | 2 | 3 | 4 | 8 | 9 | 10 | 11 | 12 | 13 | 14 | 15 | 16 | 17 | 18 | 21 | 22 | 23 | 24 | 25 | 26 | 27 |
|---|---|---|---|---|---|---|---|---|---|---|---|---|---|---|---|---|---|---|---|---|---|---|---|
| *Rays* | 1 | 1 | 0 | 0 | 0 | 0 | 0 | 0 | 0 | 1 | 0 | 0 | 0 | 0 | 0 | 0 | 0 | 0 | 0 | 0 | 0 | 0 | 0 |
| *Ephippion guttiferum* | 2 | 0 | 0 | 0 | 0 | 0 | 0 | 0 | 0 | 1 | 1 | 0 | 0 | 0 | 0 | 0 | 0 | 0 | 0 | 0 | 0 | 0 | 0 |
| *Sphaeroides spengleri* | 3 | 0 | 4 | 5 | 18 | 0 | 0 | 2 | 0 | 0 | 0 | 0 | 0 | 0 | 1 | 4 | 1 | 68 | 3 | 6 | 0 | 4 | 1 |
| *Arius heudeloti* | 4 | 0 | 0 | 0 | 0 | 0 | 0 | 0 | 0 | 0 | 0 | 0 | 0 | 1 | 0 | 0 | 0 | 0 | 0 | 0 | 0 | 0 | 0 |
| *Syngnatus typhle* | 5 | 0 | 0 | 3 | 0 | 0 | 0 | 0 | 0 | 0 | 0 | 0 | 0 | 0 | 0 | 0 | 0 | 0 | 0 | 0 | 0 | 0 | 0 |
| *Galeoides decadactylus* | 6 | 0 | 0 | 0 | 0 | 1 | 0 | 0 | 1 | 0 | 0 | 0 | 0 | 1 | 0 | 0 | 0 | 0 | 0 | 0 | 0 | 0 | 0 |
| *Serranus scriba* | 7 | 0 | 1 | 7 | 8 | 0 | 0 | 0 | 0 | 0 | 0 | 0 | 0 | 0 | 0 | 0 | 0 | 0 | 0 | 1 | 0 | 0 | 1 |
| *Epinephelus* spp. | 8 | 0 | 2 | 2 | 0 | 0 | 0 | 0 | 0 | 0 | 0 | 0 | 0 | 0 | 0 | 0 | 0 | 0 | 0 | 1 | 0 | 0 | 0 |
| *Pomadasys incisus* | 9 | 0 | 0 | 0 | 0 | 0 | 0 | 0 | 0 | 0 | 0 | 1 | 0 | 0 | 0 | 0 | 0 | 0 | 0 | 0 | 0 | 0 | 0 |
| *Pseudotolithus* spp. | 10 | 0 | 0 | 0 | 0 | 3 | 0 | 0 | 0 | 0 | 0 | 0 | 0 | 10 | 0 | 0 | 0 | 0 | 0 | 0 | 0 | 1 | 0 |
| *Sciaena umbra* | 11 | 0 | 1 | 1 | 0 | 0 | 0 | 0 | 0 | 0 | 0 | 0 | 0 | 0 | 0 | 0 | 0 | 0 | 0 | 0 | 0 | 0 | 1 |
| *Diplodus bellottii* | 12 | 0 | 186 | 64 | 93 | 0 | 0 | 0 | 2 | 0 | 24 | 0 | 0 | 12 | 0 | 0 | 0 | 21 | 0 | 0 | 0 | 0 | 26 |
| *Diplodus sargus* | 13 | 0 | 1 | 1 | 0 | 0 | 0 | 0 | 0 | 0 | 0 | 0 | 0 | 0 | 0 | 0 | 0 | 0 | 0 | 0 | 0 | 0 | 0 |
| *Sparus* spp. | 14 | 0 | 0 | 0 | 0 | 0 | 0 | 0 | 0 | 0 | 0 | 0 | 0 | 0 | 0 | 0 | 0 | 0 | 0 | 0 | 0 | 0 | 6 |
| *Spondyliosoma cantharus* | 15 | 0 | 10 | 6 | 15 | 0 | 0 | 0 | 0 | 0 | 0 | 0 | 0 | 0 | 0 | 0 | 0 | 2 | 0 | 0 | 0 | 0 | 0 |
| *Chaetodipterus goreensis* | 16 | 0 | 0 | 0 | 0 | 0 | 0 | 0 | 0 | 0 | 0 | 0 | 0 | 0 | 0 | 0 | 0 | 0 | 0 | 0 | 0 | 0 | 0 |
| *Crenilabrus bailloni* | 17 | 0 | 0 | 42 | 30 | 0 | 0 | 0 | 0 | 0 | 0 | 0 | 0 | 0 | 0 | 0 | 0 | 0 | 0 | 0 | 0 | 0 | 3 |
| *Stephanolepis hispidus* | 18 | 2 | 166 | 492 | 244 | 0 | 0 | 2 | 0 | 0 | 7 | 0 | 0 | 0 | 0 | 0 | 2 | 3 | 1 | 3 | 0 | 4 | 18 |
| *Gobius* spp. | 19 | 0 | 0 | 13 | 2 | 0 | 0 | 0 | 0 | 0 | 1 | 0 | 0 | 0 | 0 | 0 | 0 | 0 | 0 | 0 | 0 | 0 | 1 |
| *Batrachoides didactylus* | 20 | 0 | 0 | 2 | 2 | 0 | 0 | 0 | 0 | 0 | 0 | 0 | 0 | 0 | 0 | 0 | 0 | 0 | 0 | 0 | 0 | 0 | 0 |
| *Scorpaena angolensis* | 21 | 0 | 0 | 0 | 0 | 0 | 0 | 0 | 0 | 0 | 0 | 0 | 0 | 0 | 0 | 0 | 0 | 0 | 0 | 1 | 0 | 0 | 0 |
| *Cephalacanthus volitans* | 22 | 0 | 0 | 0 | 0 | 0 | 0 | 0 | 0 | 0 | 0 | 0 | 0 | 0 | 0 | 0 | 0 | 0 | 0 | 0 | 0 | 0 | 0 |
| *Psettodes belcheri* | 23 | 2 | 0 | 0 | 0 | 0 | 0 | 0 | 0 | 0 | 0 | 1 | 1 | 0 | 0 | 0 | 0 | 0 | 0 | 0 | 0 | 0 | 0 |
| *Syaium micrurum* | 24 | 0 | 0 | 0 | 0 | 0 | 0 | 0 | 0 | 0 | 0 | 0 | 0 | 0 | 0 | 0 | 0 | 0 | 0 | 0 | 0 | 1 | 0 |
| *Synaptura lusitanica* | 25 | 0 | 0 | 0 | 0 | 0 | 0 | 0 | 0 | 0 | 0 | 0 | 0 | 0 | 0 | 0 | 0 | 1 | 0 | 0 | 0 | 0 | 0 |
| *Pegusa triophtalma* | 26 | 0 | 0 | 0 | 0 | 0 | 0 | 1 | 0 | 0 | 0 | 0 | 0 | 0 | 0 | 0 | 0 | 0 | 0 | 0 | 0 | 0 | 0 |
| *Solea senegalensis* | 27 | 0 | 0 | 1 | 1 | 0 | 0 | 0 | 0 | 0 | 0 | 0 | 0 | 0 | 0 | 0 | 0 | 0 | 0 | 0 | 0 | 0 | 0 |
| *Cynoglossus* spp. | 28 | 0 | 1 | 0 | 0 | 2 | 2 | 0 | 2 | 0 | 0 | 0 | 0 | 2 | 1 | 0 | 0 | 0 | 0 | 0 | 0 | 0 | 0 |
| *Sepia* spp. | 29 | 0 | 0 | 4 | 7 | 2 | 0 | 2 | 0 | 2 | 0 | 0 | 3 | 3 | 1 | 0 | 0 | 1 | 0 | 0 | 2 | 0 | 1 |
| *Penaeus* spp. | 30 | 0 | 0 | 0 | 1 | 4 | 5 | 5 | 1 | 0 | 0 | 1 | 0 | 0 | 2 | 0 | 0 | 0 | 0 | 0 | 0 | 0 | 0 |
| | Total | 5 | 372 | 643 | 421 | 12 | 7 | 12 | 6 | 4 | 33 | 3 | 4 | 29 | 5 | 4 | 3 | 96 | 4 | 12 | 2 | 10 | 58 |

*Table 4.* (continued).

| Species | 28 | 29 | 30 | 31 | 32 | 33 | 34 | 37 | 38 | 39 | 40 | 41 | 42 | 43 | 44 | 45 | 46 | 47 | 48 | 49 | 50 |
|---|---|---|---|---|---|---|---|---|---|---|---|---|---|---|---|---|---|---|---|---|---|
| *Rays* | 0 | 0 | 0 | 0 | 0 | 0 | 0 | 1 | 0 | 0 | 0 | 0 | 0 | 0 | 0 | 0 | 0 | 0 | 2 | 0 | 0 |
| *Ephippion guttiferum* | 0 | 3 | 0 | 0 | 0 | 0 | 0 | 0 | 0 | 0 | 1 | 1 | 0 | 0 | 0 | 0 | 1 | 0 | 1 | 1 | 0 |
| *Sphaeroides spengleri* | 34 | 0 | 0 | 0 | 05 | 01 | 0 | 0 | 0 | 0 | 0 | 0 | 9 | 1 | 1 | 2 | 0 | 5 | 0 | 1 | 0 |
| *Arius heudeloti* | 0 | 0 | 0 | 0 | 0 | 0 | 0 | 0 | 0 | 0 | 0 | 0 | 0 | 0 | 0 | 0 | 0 | 0 | 4 | 0 | 0 |
| *Syngnathus typhle* | 4 | 0 | 0 | 0 | 0 | 0 | 0 | 0 | 0 | 0 | 0 | 0 | 0 | 1 | 0 | 0 | 0 | 0 | 0 | 0 | 0 |
| *Galeoides decadactylus* | 0 | 0 | 0 | 0 | 0 | 0 | 0 | 0 | 0 | 0 | 0 | 0 | 0 | 0 | 0 | 0 | 0 | 0 | 103 | 2 | 0 |
| *Serranus scriba* | 1 | 1 | 0 | 0 | 0 | 2 | 0 | 0 | 0 | 0 | 3 | 0 | 5 | 17 | 0 | 0 | 0 | 0 | 0 | 1 | 0 |
| *Epinephelus* spp. | 0 | 0 | 1 | 0 | 0 | 0 | 0 | 0 | 0 | 0 | 2 | 0 | 1 | 0 | 0 | 0 | 0 | 0 | 0 | 0 | 0 |
| *Pomadasys incisus* | 0 | 0 | 0 | 0 | 1 | 0 | 0 | 0 | 0 | 0 | 0 | 1 | 1 | 0 | 0 | 0 | 0 | 0 | 2 | 1 | 0 |
| *Pseudotolithus* spp. | 0 | 0 | 0 | 0 | 0 | 0 | 0 | 0 | 0 | 0 | 0 | 0 | 0 | 0 | 0 | 0 | 0 | 0 | 4 | 0 | 0 |
| *Sciaena umbra* | 0 | 0 | 0 | 0 | 0 | 0 | 0 | 0 | 0 | 0 | 0 | 0 | 2 | 0 | 0 | 0 | 0 | 0 | 0 | 1 | 0 |
| *Diplodus bellottii* | 3 | 2 | 4 | 0 | 1 | 21 | 0 | 0 | 0 | 11 | 10 | 2 | 10 | 71 | 7 | 93 | 0 | 14 | 5 | 18 | 3 |
| *Diplodus sargus* | 0 | 0 | 1 | 0 | 0 | 0 | 0 | 0 | 0 | 0 | 0 | 0 | 0 | 1 | 6 | 0 | 0 | 1 | 0 | 2 | 0 |
| *Sparus* spp. | 0 | 0 | 0 | 0 | 2 | 0 | 0 | 0 | 0 | 0 | 0 | 0 | 0 | 1 | 0 | 0 | 0 | 0 | 0 | 0 | 0 |
| *Spondyliosoma cantharus* | 2 | 0 | 3 | 0 | 2 | 22 | 0 | 0 | 0 | 0 | 0 | 0 | 7 | 5 | 0 | 7 | 0 | 4 | 0 | 2 | 0 |
| *Chaetodipterus goreensis* | 0 | 0 | 0 | 0 | 1 | 0 | 0 | 0 | 0 | 0 | 0 | 0 | 1 | 0 | 0 | 0 | 0 | 0 | 0 | 0 | 0 |
| *Crenilabrus bailloni* | 1 | 1 | 10 | 2 | 0 | 3 | 0 | 0 | 0 | 0 | 6 | 0 | 2 | 57 | 1 | 0 | 0 | 4 | 0 | 0 | 0 |
| *Stephanolepis hispidus* | 82 | 13 | 57 | 3 | 3 | 76 | 0 | 0 | 5 | 2 | 17 | 0 | 5 | 298 | 25 | 12 | 0 | 79 | 0 | 14 | 0 |
| *Gobius* spp. | 13 | 0 | 9 | 0 | 0 | 0 | 0 | 0 | 0 | 0 | 0 | 0 | 3 | 19 | 0 | 0 | 0 | 2 | 0 | 0 | 0 |
| *Batrachoides didactylus* | 0 | 0 | 2 | 0 | 1 | 0 | 0 | 0 | 0 | 0 | 2 | 0 | 0 | 2 | 0 | 0 | 0 | 0 | 0 | 0 | 0 |
| *Scorpaena angolensis* | 0 | 0 | 0 | 0 | 0 | 0 | 0 | 0 | 0 | 0 | 0 | 0 | 0 | 0 | 0 | 0 | 0 | 0 | 0 | 0 | 0 |
| *Cephalacanthus volitans* | 0 | 0 | 0 | 1 | 2 | 0 | 0 | 0 | 0 | 0 | 0 | 1 | 0 | 0 | 0 | 0 | 0 | 0 | 0 | 0 | 0 |
| *Psettodes belcheri* | 0 | 0 | 0 | 0 | 1 | 1 | 0 | 0 | 0 | 0 | 0 | 0 | 1 | 0 | 0 | 0 | 0 | 0 | 0 | 0 | 0 |
| *Syacium micrurum* | 0 | 0 | 0 | 0 | 0 | 0 | 0 | 0 | 0 | 0 | 0 | 0 | 0 | 0 | 0 | 0 | 0 | 0 | 0 | 0 | 0 |
| *Synaptura lusitanica* | 0 | 1 | 0 | 0 | 0 | 0 | 1 | 0 | 0 | 0 | 0 | 0 | 0 | 0 | 0 | 0 | 0 | 0 | 0 | 0 | 0 |
| *Pegusa triophtalma* | 0 | 0 | 0 | 0 | 4 | 0 | 0 | 0 | 0 | 0 | 0 | 1 | 1 | 0 | 0 | 0 | 0 | 0 | 0 | 0 | 0 |
| *Solea senegalensis* | 0 | 0 | 0 | 0 | 0 | 2 | 0 | 1 | 0 | 0 | 0 | 0 | 0 | 0 | 0 | 1 | 0 | 0 | 0 | 0 | 0 |
| *Cynoglossus* spp. | 0 | 0 | 0 | 0 | 0 | 0 | 0 | 0 | 0 | 0 | 0 | 0 | 0 | 0 | 0 | 0 | 0 | 0 | 0 | 0 | 0 |
| *Sepia* spp. | 27 | 1 | 9 | 0 | 1 | 4 | 1 | 0 | 1 | 1 | 1 | 3 | 1 | 15 | 0 | 1 | 0 | 2 | 0 | 1 | 0 |
| *Penaeus* spp. | 1 | 0 | 1 | 0 | 0 | 4 | 1 | 0 | 0 | 0 | 1 | 0 | 1 | 2 | 1 | 0 | 0 | 0 | 0 | 1 | 0 |
| *Total* | 168 | 22 | 97 | 6 | 24 | 146 | 3 | 2 | 6 | 14 | 43 | 9 | 50 | 490 | 41 | 116 | 1 | 111 | 121 | 45 | 3 |

Finally, a fifth cluster (5) is formed by one single location, *viz.* 48. This site is unique by the presence of a large number of juvenile *Galeoides decadactylus* (Bloch, 1795), a semi-pelagic schooling species.

## Discussion and conclusions

### *Abiotic factors*

The observed temperatures and salinities in the area agree well with the descriptions given by Peters (1976) and Sevrin-Reyssac (1983): upwelling water flows into the bay south of Cap Blanc and is warmed up quickly by solar radiation, causing a negative correlation between temperature and depth. The water flowing southward over the Banc d'Arguin gains a high salinity by evaporation. This is in agreement with the observed positive correlation between temperature and salinity. The absence of a significant correlation between salinity and depth may indicate an intensive exchange of water between deeper and shallow areas at the time of the survey.

### *Biota*

The collected number of 45 fish species is high compared to the study of Campredon &

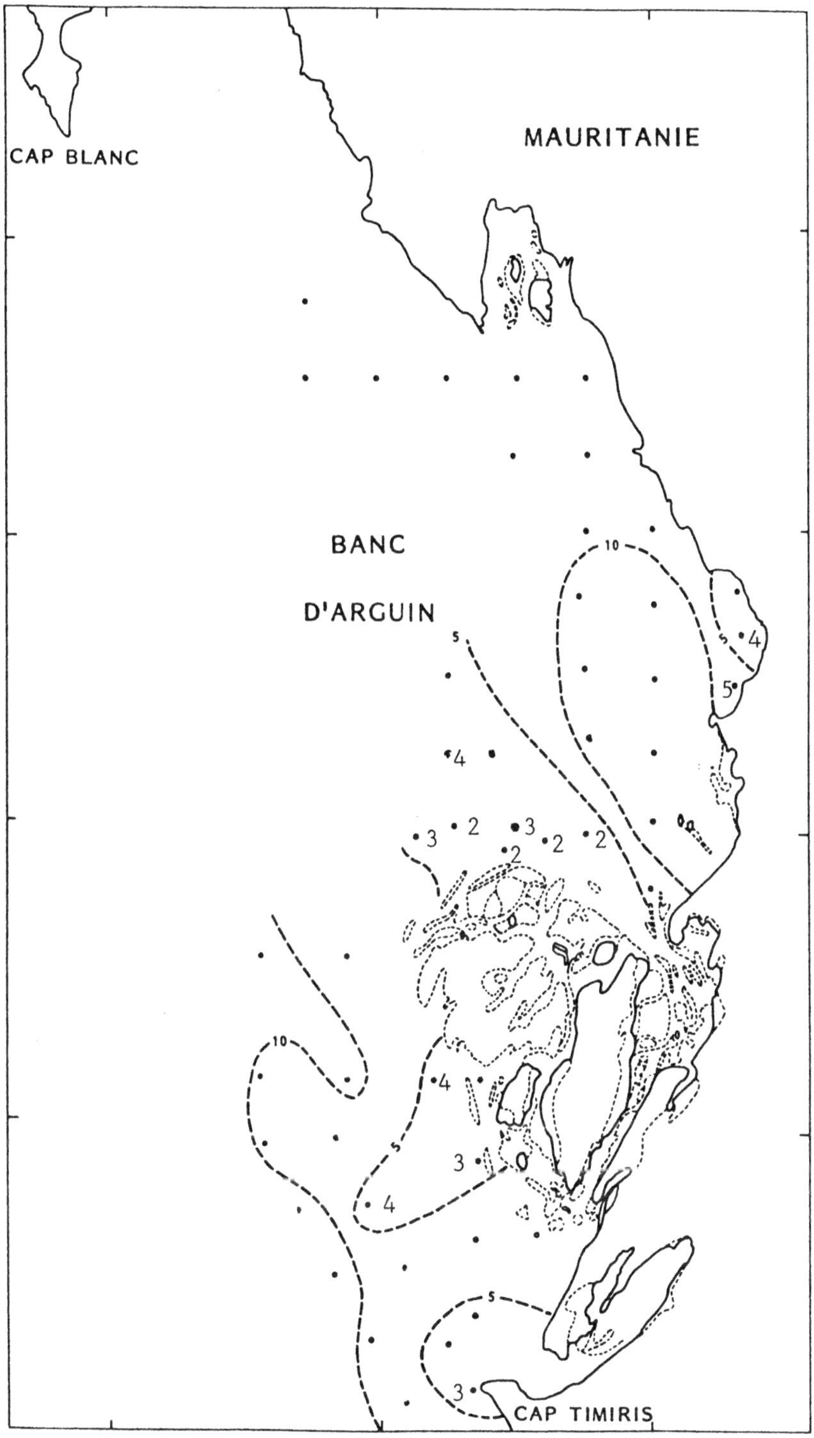


*Fig. 3.* Distribution of the station clusters on the Banc d'Arguin in September 1988. The figures denote the cluster to which the stations belong, stations without figure belong to cluster 1. The dashed lines denote the 5 and 10 m isobaths.

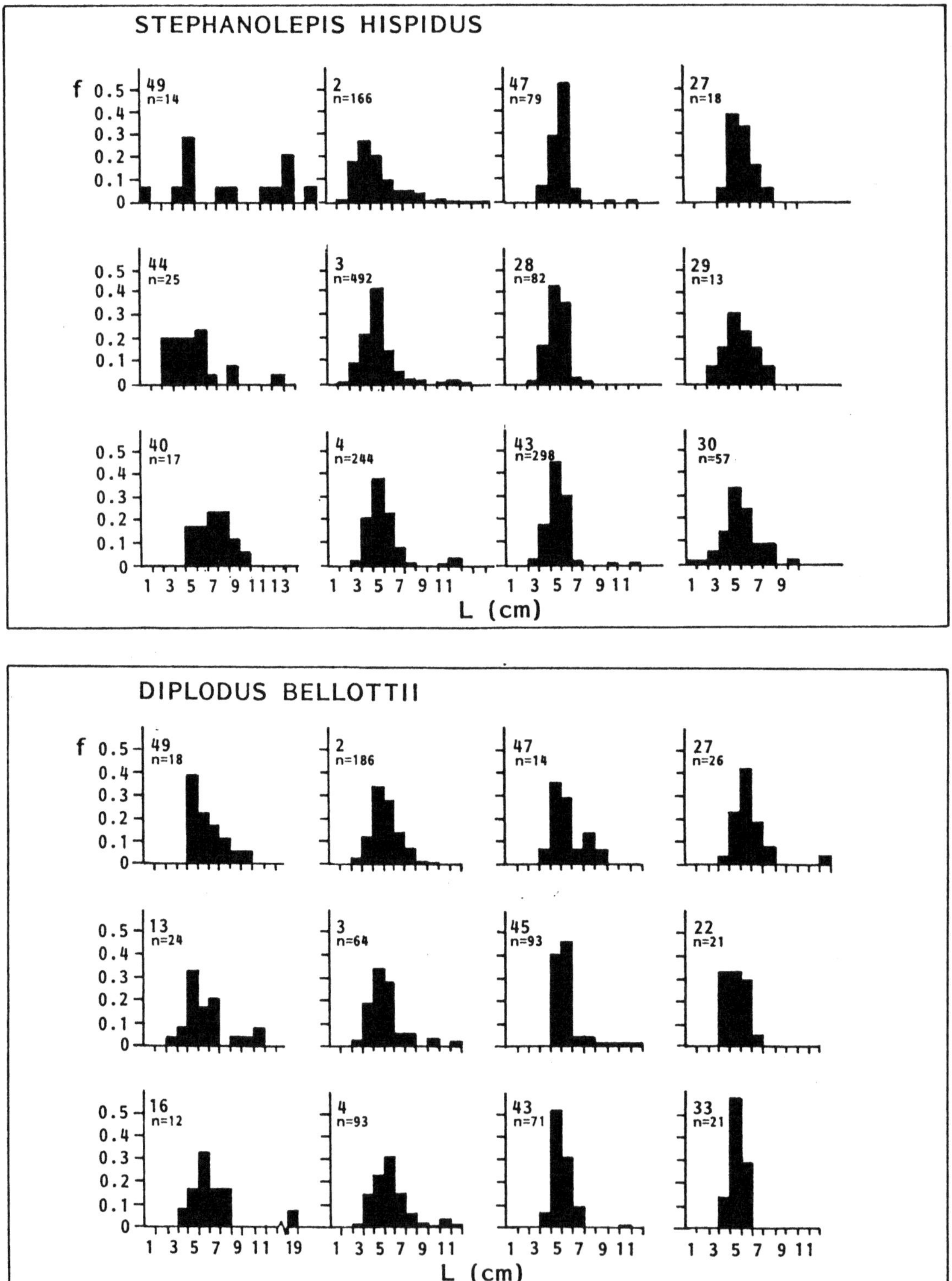


*Fig. 4.* Length-frequency distributions of *Stephanolepis hispidus* (top) and *Diplodus bellottii* (bottom) on the Banc d'Arguin in September 1988. Only stations with more than 10 individuals of the species are shown.

Schrieken (1989) who collected 19 fish species, and 10 species of prawns. The higher number found in this study is probably caused by the larger area covered in the survey of September 1988, including a wider depth range.

The observed numerical dominance of juvenile *S. hispidus*, *D. bellottii* and gobiids in the area north of Tidra in this study, is in good agreement with the results of Campredon & Schrieken (1989). The most striking differences between our results in September 1988 and those of Campredon & Schrieken (October 1985) are that the latter caught higher numbers of prawns, much smaller numbers of *S. hispidus*, and no *D. bellottii*. Remarkable is also their record in October 1985 of 13–18 *Ephippion guttiferum* on seagrass beds, whereas we only caught single individuals of this species on depths between 3 and 9 m. These differences are hard to explain.

Most of the relationships between number of species, *Ln* (number of individuals) and *Ln* (mean length) on one hand, and depth, temperature and salinity on the other hand were statistically significant. But highest significance was found for the correlations with depth. This might indicate that the shallow parts of the Banc d'Arguin are important as a habitat for large numbers of small fish. Of these small individuals, more than 50% are juveniles of *S. hispidus* and *D. bellottii*. Moreover, the locations shallower than 5 m appear to be very rich in species. The smaller number of species at the deeper locations may have been caused by a lower gear-efficiency in deeper water. The absence of small individuals in the catches at the deeper locations can not be explained by this factor, since these fish are expected to be efficiently captured by a 2 m beam trawl (cf. Kuipers, 1975). In shallow water (1–2 m), the observed large quantities of fish may even be underestimated because of reduced gear-efficiency due to disturbance effects of the boat; this effect is more important for larger fish (Kuipers, 1975). Furthermore, experiments by Gray & Bell (1986) showed that on seagrass beds, 30% less species and 85% less individuals were obtained by trawling during day-time, compared to poisoning by rotenon. Although specific data on the efficiency of the beam trawl used in this study for the fish species present on the Banc d'Arguin are lacking, it is concluded that depth is an important explanatory factor for the observed differences in abundance of young fish. Temperature and salinity apparently play a less important role in structuring the fish biocenosis.

*Cluster analysis*

The first cluster is a very diverse group of locations, which have their low catches in common. The second and third cluster do not differ substantially, except that cluster 2 has a slightly higher species diversity and a higher abundance of *D. bellottii* than cluster 3. All locations in cluster 2 and 3 have depths shallower than 5 m. The mean length of the fish has an average of 6.0 cm in these clusters.

Related with depth is the presence of seagrasses. It is known that on the Banc d'Arguin up to a depth of about 5 m, enough light can penetrate the water column to allow growth of a seagrass vegetation. Seagrasses were shown to be present northwest of Tidra, and near Cap Tafarit (Wolff & Smit, 1990; Nienhuis, 1990). From our own observations, seagrass appeared to be present as bycatch at the locations northwest of Tidra, and near Tanoudert.

The presence of seagrasses may be an important explanation for the observed species diversity, and for the abundance of juvenile fish at the locations shallower than 5 m, as reflected in the clusters 2 and 3. Unfortunately, the vegetation bycatch in the trawl was not recorded accurately enough to be able to test this hypothesis.

*S. hispidus* is known to be a direct consumer of seagrass (Sevrin-Reyssac, 1983). Other species may use the vegetation for shelter.

The species composition at site 48 ressembles the 'sciaenid community' as characterised by Longhurst & Pauly (1986), occurring in tropical waters on soft bottoms above the thermocline, consisting of *Pomadasys* sp., *Galeoides decadactylus*, *Pseudotolithus typus*, *Arius* sp., and *Cynoglossus goreensis*.

An important part of the Banc d'Arguin could not be sampled during this survey. Due to the timing of the survey, this study offers a picture of the demersal nectonic populations by the end of the warm season. This paper suggests the function of the Banc d'Arguin as a nursery for some fish species, especially *S. hispidus* and *D. bellottii*, and for *Sepia* species. More study is required to assess the importance of the Banc d'Arguin in the lifecycle of other fish species, those of commercial value included.

## Acknowledgements

Thanks are due to the Mauritanian authorities who gave permission to carry out this study. I thank Messrs. J. van Dijk, T. M. van Spanje, D. Yahya, and the crew of the Bentili I for assistance with field work. I am very grateful to T. C. Iles for his statistical support, and to K. Essink for his advice concerning the manuscript. The help of many others was very much appreciated. The study was funded by the Netherlands Directorate-General for International Cooperation, the World Wide Fund for Nature (Netherlands), the Netherlands Marine Science Foundation and the former Research Institute for Nature Management.

## References

Campredon, P. & B. Schrieken, 1989. Fishes and shrimps of the tidal flats. In: B. J. Ens, T. Piersma, W. J. Wolff & L. Zwarts (eds), Report of the Dutch-Mauritanian project Banc d' Arguin 1985–1986. RIN Rapport 89/6, Texel: 222–227.

Gray, C. A. & J. D. Bell, 1986. Consequences of two common techniques for sampling vagile macrofauna associated with the seagrass Zostera capricorni. Mar. Ecol. Progr. Ser. 28: 43–48.

Kuipers, B., 1975. On the efficiency of a two-metre beam trawl for juvenile plaice (Pleuronectes platessa). Neth. J. Sea Res. 9: 69–85.

Maigret, J. & B. Ly, 1986. Les poissons de mer de Mauritanie. Sciences Nat., Compiègne.

Nienhuis, P. H., F. van Lent & J. M. Verschuure, 1990. Distribution of seagrasses in Banc d'Arguin, Mauritania, N.W. Africa. Delta Institute for Hydrobiological Research Progress Report 1989. K.N.A.W. Amsterdam, 1990.

Peters, H., 1976. The spreading of the water masses of the Banc d'Arguin in the upwelling area off the northern Mauritanian coast. 'Meteor' Forschungsergebnisse, Reihe A, 18: 78–100.

Sevrin-Reyssac, J., 1983a. Quelques particularités de la chaine alimentaire marine dans la région des îles du Banc d' Arguin (Parc National du Banc d'Arguin, Mauritanie). Bull. Centre Nat. Rech. Océanogr. Pêches 11: 41–52.

Sevrin-Reyssac, J. 1983b. Affinité biogéographique de la région des îles du Banc d'Arguin (Parc National du Banc d'Arguin, Mauritanie). Bull. Centre Nat. Rech. Océanogr. Pêches 11: 53–56.

Troadec, J.-P. & S. Garcia, 1980. The fish resources of the Eastern Central Atlantic. Part one: the resources of the Gulf of Guinea from Angola to Mauritania. FAO Fisheries Technical Paper No. 186.1. Rome, 1980.

Wolff, W. J. & C. J. Smit, 1990. The Banc d' Arguin, Mauritania, as an environment for coastal birds. Ardea 78: 17–38.

*Hydrobiologia* **258**: 197–210, 1993.
*W. J. Wolff, J. van der Land, P. H. Nienhuis & P. A. W. J. de Wilde (eds), Ecological Studies in the Coastal Waters of Mauritania.*



# Seabirds in the shelf edge waters bordering the Banc d'Arguin, Mauritania, in May

M. F. Leopold
*Netherlands Institute for Sea Research (NIOZ), P.O. Box 59, 1790 AB Den Burg, Texel, The Netherlands*

*Key words:* seabirds, Storm-petrels, upwelling, atlantic, Mauritania, Banc d'Arguin

## Abstract

Seabird distribution was mapped over the shelf off the Banc d'Arguin, Mauritania, in May 1988 at the end of the upwelling season. Storm-petrels (Oceanitidae), were the most numerous with an average density of 14.5 $km^{-2}$. They were most abundant over water-patches rich in primary production or zooplankton, but the simultaneous presence of trawlers in these areas made it impossible to relate seabird density directly to hydrography. The trawlers attracted the seabirds and their waste served as an important source of food. The most numerous species in the area was the Wilson's Storm-petrel *Oceanites oceanicus*. Northern-hemisphere migrants, including summering sub-adults also used the area in numbers, and they too, profited from the fisheries. The Royal Tern *Sterna maxima* was the only local breeding seabird reaching the shelf slope area.

## Introduction

Seabirds show a numerical response in relation to food availability. Highly productive areas at sea, such as upwelling systems and oceanographical fronts can support large numbers of seabirds, whereas the unproductive areas in tropical oceans are often devoid of seabirds. Distributions of seabirds were first mapped, and considered against the background of oceanographical parameters by Jespersen (1924) and Murphy (1936). Since then, many ornithologists have been engaged in counting seabirds at sea, and today a great deal of information is available on the marine distribution and ecology of these animals (reviews in Brown, 1980, 1988). Still, the magnitude of the seas and oceans inevitably makes that in some areas the avifauna has been poorly studied.

The Atlantic waters off northwest Africa have received comparatively little attention from ornithologists. The area is dominated by a band of rich upwelling water, and although breeding seabirds are rather scarce in northwest Africa, large numbers of seabirds may be expected to use these waters on migration and during the winter. Some information on seabirds in the area has been collected by observers on passing freighters and research vessels (Bourne, 1963; Brown, 1979; Moerbeek, 1982; Cadée, 1981; Hazevoet, 1985; Camphuysen, 1988). Most observers spent only brief periods of time in the area, and were not engaged in systematic surveys. The only systematic survey of seabirds at sea in the area is that by Brown (1979) in the waters off Senegal, in February/March 1976.

A seabird-survey was made during a biological and hydrographic cruise in the waters off the Banc d'Arguin, Mauritania in May 1988. Other workers on board collected data relevant to the seabird work: Loktionov on hydrography, Berghuis *et al.* on primary production and coworkers of Kuipers on zooplankton (see elsewhere in this volume). In this paper a description of the avifauna is given in terms of species, age composition and densities. Two questions concerning the seabirds at sea are discussed:

1. Do local breeders of the Banc d'Arguin go beyond the sheltered Banc on their foraging trips?
2. Do the seabirds show a numerical response to hydrographical parameters and/or zooplankton distribution, or is seabird distribution governed by other factors in this region?

## Study area

### *Hydrography and plankton*

The study area is restricted to the waters over the shelf break off Mauritania, northwest Africa, between Cap Blanc (*ca* 20° 45′ N) and Cap Timiris (*ca* 19° 20′ N). Water depths ranged from less than 20 m to over 1000 m. In the south the shelf is comparatively narrow, and traversed by several deep trenches that connect the shallow Banc d'Arguin with the deep ocean (Fig. 1).

A description of the oceanographic properties of the waters off northwest Africa is given by Mittelstaedt, (1991); Van Camp *et al.*, (1991) and Loktionov (1993). The northern trade winds transport relatively cold water to the area. This current follows the coast until Cap Blanc, where it becomes detached from the continent, and gradually turns west. Coastal upwelling, driven by the Coriolis force is observed year round in the waters north of Cap Blanc, but south of the Cap (in the study area) upwelling is seasonal, occurring mainly in winter and spring. Here, the cold subsurface waters cannot reach the surface during the summer, due to the spreading of a surface layer of warm tropical water. The local oceanography is complicated by the presence of the shallow Banc d'Arguin. The water over this rather

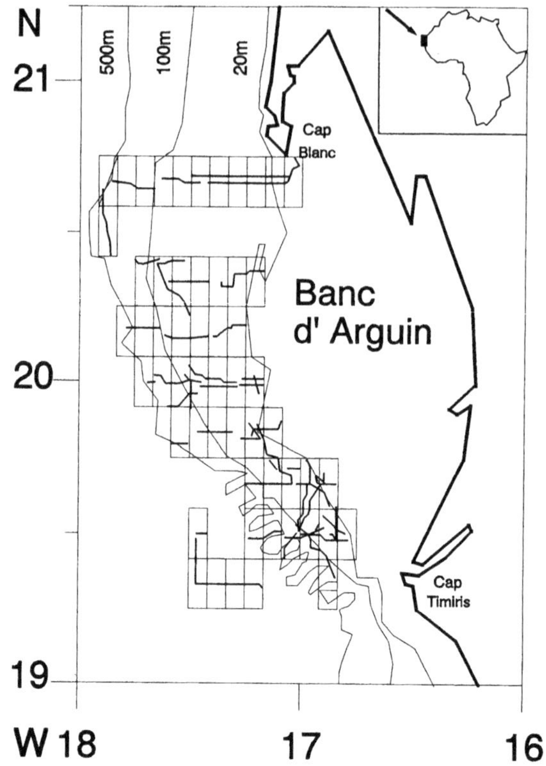


*Fig. 1.* The study area, divided into rectangles of 10 × 5 geographical minutes. The cruise-tracks during which seabirds were counted are indicated as thick lines within these rectangles. The 20 m, 100 m, and 500 m isobaths are presented as thin lines.

sheltered, shallow part of the continental shelf is warmed by the sun and persistently 7–8 °C warmer than the cool upwelled water offshore. Evaporation is high on the Banc d'Arguin, and the water is highly saline as a result. This 'Banc Water' occasionally reaches such a high density, that it sinks off the Banc, mainly through the deep trenches across the continental shelf near Cap Timiris (Peters, 1976).

The present survey took place at the end of the upwelling season. Only one patch of freshly upwelled water was found, in the southeast of the study area. Here, surface temperatures were relatively low and a high biomass of the copepod *Calanoides carcinatus* was found. This copepod is regarded as an indicator species for upwelling. This upwelling patch also contained high numbers of euphausids, fish-larvae and larvae of crustaceans. No other fresh patches of upwelling were found in the area, and other zooplankton species dominated in other parts of the study area. Total zooplankton biomass was highest in the upwelling patch, and in general, inshore of the 40 m depth line (Kuipers *et al.*, 1993). The study area was thus a transition zone in several properties. A strong depth-gradient was present, as were several different water-masses and zooplankton communities (Loktionov, 1993; Kuipers *et al.*, this vol.).

*Fisheries*

Seabirds are affected by human fishing activities in different ways. A fishing industry aimed at small fish or crustaceans may be an important competitor for food, but on the other hand, many species benefit by feeding on discards and offal (Furness, 1982; Abrams, 1983, 1985; Furness & Ainley, 1984; Ryan & Moloney, 1988). During the survey, large fishing vessels were regularly encountered. They were mainly Russian, Rumanian and Korean trawlers fishing for pelagic roundfish for fishmeal. Most were fishing along the shelf edge. Fisheries statistics show that the region supports important, international fisheries for pelagic, shoaling fish species (Carangidae, Clupeidae and Scombridae), cephalopods (M'Bareck Ould Soueilem, 1986; Bravo de Laguna, 1989) and shrimp (Kaczynski, 1989).

*Weather conditions*

During the survey, the wind was always northerly, ranging from NW to NE. Windspeed ranged from 1 to 14 m $s^{-1}$, and was 4 to 6 Beaufort (7 to 14 m $s^{-1}$) for 90% of the time. The magnitude of the swell differed from less than 1 metre at the innermost transects to about 4 metres at some of the most oceanic transects. No major high or low pressure zones passed the study area during the survey. Barometric pressure ranged from 990 to 1000 hPa in the hours spent counting. There was no precipitation, and visibility was good for most of the time, although occasionally clouds of very fine desert sand reduced visibility to about 2 km.

## Methods

Seabirds were counted over the shelf edge waters from the RV 'Tyro'. This ship has a length of 84 m, and outdoor observation posts on the wings of the bridge, at 10 m above sea level. Apart from the last day of the survey, during the approach of Cap Blanc, land was never in sight. The ship was primarily engaged in other marine research, and the seabird observer could not influence the sailing schedules.

The survey started at Cap Blanc at 10 May. The ship worked its way south, following transects in an east-west direction, which were 10 Nautical Miles (18.5 km) apart. The latitude of Cap Timiris was reached on 12 May. The ship remained south of 19° 40′ N for 10 days, whereafter it worked its way north again. until arrival at Cap Blanc at 29 May.

Seabirds were counted whenever the ship was steaming between stations during the daylight hours (Fig. 1). Birds were counted using the standard 'snapshot strip-transect' method (Tasker *et al.*, 1984). Binocular scanning was added to the procedure to make sure that all birds (especially the very small storm-petrels) present within the counting-strip were discovered (cf. Ainley & Boekelheide, 1983; Follestad, 1990). Ship speeds varied between 5 and 12 knots.

The counts were broken up into 10 minute-bouts, but a bout was terminated whenever the ship changed its course or stopped at a sampling station. For each individual bout, positions were taken at the start and end, in degrees, minutes and hundredths of minutes, using the ship's SATNAV-equipment. The observed area per counting bout was calculated from the linear distance travelled multiplied by the strip width (300 m).

The survey area was divided into rectangles, measuring 10 geographical minutes latitude by 5 minutes longitude. This grid size was chosen in order to get a good resolution over the depth gradient, and because the principal survey tracks were 10 NM apart in the north-south direction (Fig. 1). For each counting bout, the central position was calculated as the mean of the starting and end position. According to its central position, each count was placed in the corresponding rectangle. In order to calculate bird densities, the results of all counts were pooled per rectangle. For each rectangle, this procedure gives us the total surface area observed, the number of birds of a given species seen in that area, and hence the density of birds per square kilometre. Data were collected in 55 rectangles. Rectangles in which a given species was only seen outside the counting strips are given the lowest density-class in maps presenting the distribution of that species. For species of which in the majority of the rectangles no birds were counted within the strips, distribution is mapped in terms of total numbers seen per kilometre traveled.

## Results

### *Species accounts*

#### *Shearwaters*

Four species of shearwaters were seen, of which the Cory's Shearwater *Calonectris diomedea* was the most numerous (Table 1). Even so, densities of Cory's Shearwater were low: 0.08 $km^{-2}$ on average. As in most larger species, more birds were seen outside the counting strips than within. Cory's Shearwaters were seen throughout the study area, but few were seen in the transects over deeper water in the northwest and southwest of the study area. Seven rarer shearwaters, all of the genus *Puffinus*, were seen during the survey (Table 1; Fig. 2).

Age and subspecific status of the Cory's Shearwaters could not be discerned with certainty. In May, adult birds are engaged in their 'pre-laying

*Table 1.* Seabirds seen during the survey. Total numbers seen are divided into those encountered within the counting strip, (using the 'snapshot' counting method of Tasker *et al.*, 1984), and those outside these strips. Only the first category was used for density estimates. Bird-names after Harrison (1983).

| English name | Scientific name | Inside | Outside |
|---|---|---|---|
| Cory's Shearwater | Calonectris diomedea | 20 | 83 |
| Great Shearwater | Puffinus gravis | 0 | 1 |
| Sooty Shearwater | Puffinus griseus | 2 | 2 |
| Manx Shearwater | Puffinus puffinus | 1 | 1 |
| Storm-Petrels | Oceanitidae | 3519 | 1766 |
| Northern Gannet | Sula bassana | 5 | 62 |
| Pomarine Skua | Stercorarius pomarinus | 36 | 59 |
| Arctic Skua | Stercorarius parasiticus | 39 | 56 |
| Skua spec. | S. pomarinus/parasiticus | 0 | 16 |
| Long-tailed Skua | Stercorarius longicaudus | 19 | 43 |
| Great Skua | Catharacta skua | 8 | 13 |
| Sabine's Gull | Larus sabini | 237 | 421 |
| Audouin's Gull | Larus audouinii | 0 | 1 |
| Lesser Black-backed Gull | Larus fuscus | 10 | 22 |
| Herring Gull | Larus argentatus | 0 | 2 |
| Gull-billed Tern | Sterna nilotica | 1 | 0 |
| Caspian Tern | Sterna caspia | 2 | 8 |
| Royal Tern | Sterna maxima | 80 | 140 |
| Sandwich Tern | Sterna sandvicensis | 125 | 259 |
| Commic Tern | S. hirundo/paradisaea | 645 | 976 |
| Bridled Tern | Sterna anaethetus | 3 | 5 |
| Black Tern | Chlidonias niger | 97 | 251 |

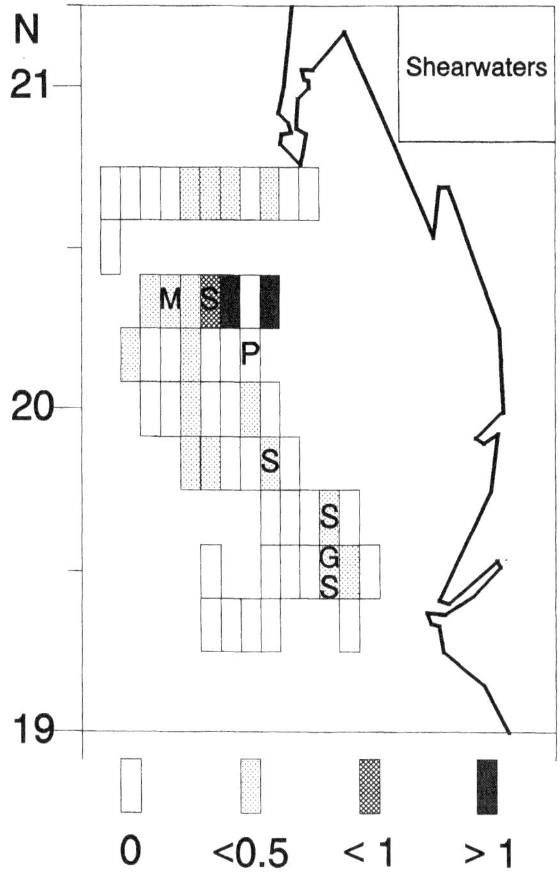


*Fig. 2.* Distribution of Cory's Shearwater (shading) in numbers seen per kilometre sailed. Observations of other shearwaters are marked with lettering. G: Greater Shearwater; S: Sooty Shearwater; P: Manx Shearwater, subspecies *puffinus*; M: Manx Shearwater, subspecies *mauretanicus*.

exodus' (Moerbeek, 1982), and the distance to the nearest breeding colonies (> 500 km) may not have been prohibitively large for breeding birds of the region. The dark-billed subspecies *edwarsii* of the Cape Verdes was not seen: all birds observed at close range had yellow bills, so either belonged to the Atlantic subspecies *borealis* or the Mediteranian subspecies *diomedea*. Most of the Cory's Shearwaters were seen flying north, a phenomenon also noted by Brown (1979) in February/March off Senegal.

*Storm-petrels*

Storm-petrels were the most numerous birds in the study area (Table 1). Wilson's Storm-petrels *Oceanites oceanicus* comprised some 90%, and British Storm-petrels *Hydrobates pelagicus* 10% of 591 individuals identified to species. Most of the Wilson's Storm-petrels were in heavy wing moult, whereas wing moult was not seen in the British Storm-petrel (cf. Cramp & Simmons, 1977). Madeiran Storm-Petrels *Oceanodroma castro*, which may reach the area in this time of the year, or Leach's Storm-petrels *O. leucorhoa*, which winter here (Brown, 1979; Lamarche, 1987), were not seen.

Storm-petrels were seen in nearly all rectangles of the survey. In two separate areas over the slope major concentrations were found: around latitudes 20° 10′ and 19° 40′ North (Fig. 3). In both areas, hundreds of Storm-petrels were seen in several 10 minute counts. They were found sitting

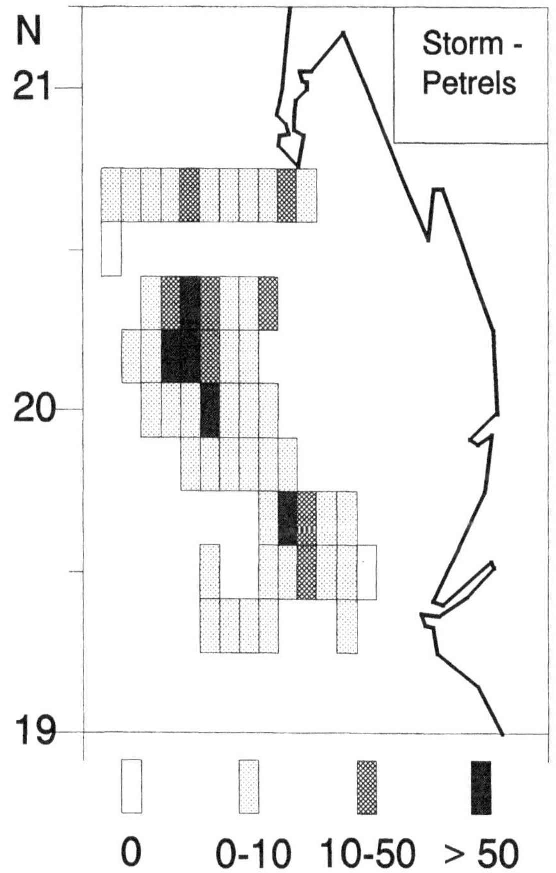


*Fig. 3.* Distribution of Storm-petrels, in 4 density classes (birds $km^{-2}$).

in dense flocks on the water (largest group estimated at 250 birds), just 'swarming' in great numbers over the surface, and associated with trawlers. The trawlers discarded an oily substance, probably fish-oil and tiny particles of fish-meat during fish-meal preprocessing at sea. Storm-petrels gathered in large numbers around these slicks to forage (largest groups 500 and 650 birds). The birds were seen to take small particles from the slicks, mainly at the edges. Storm-petrels are known to take oil globules (Mayo, 1948; Cramp & Simmons, 1977; Brown *et al.*, 1982) in addition to their natural prey, zooplankton and fish. In the stomach of a Wilson's Storm-petrel caught on board, both copepod and fish remains were found, the latter being two tiny (<0.5 mm) clupeoid otoliths. On two occasions Wilson's Storm-petrels were observed to feed on carrion: groups of 5 to 10 birds were seen to tear at a dead conspecific and at a large dead fish. In stomachs of both Wilson's and British Storm-petrels caught in the area one month later, J. C. den Hartog (Nat. Mus. of Nat. Hist., Leiden, pers. comm.) found mainly cypris larvae of Cirripedia (Crustacea). Other items found in these stomachs were small plastic particles and one insect: a weevil (Coleoptera, Curculionidae) in a Wilson's Storm-petrel. Storm-petrels in the area thus appeared to feed opportunistically, taking a variety of food, both natural and supplied by man.

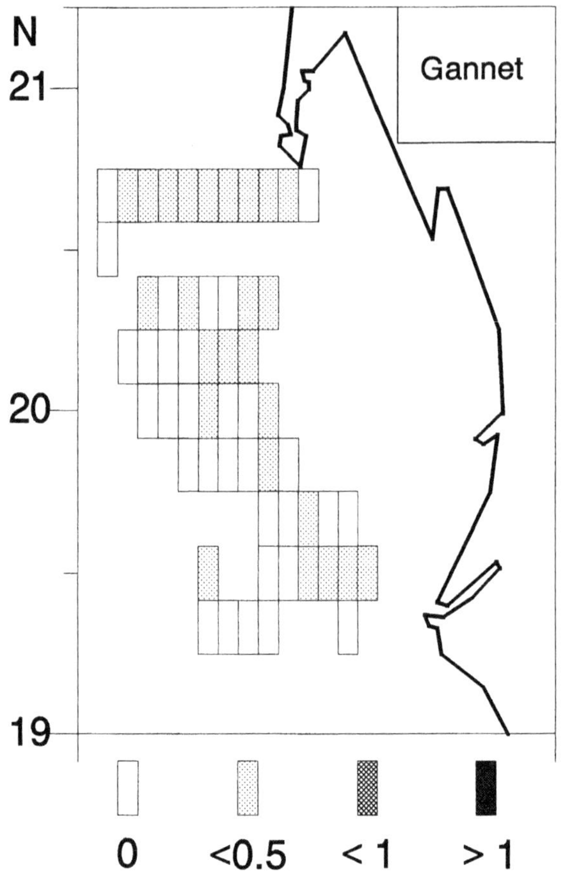


*Fig. 4.* Distribution of Northern Gannet. Shading represents numbers of birds seen per kilometre sailed.

*Northern Gannet*

Gannets *Sula bassana* were seen throughout the area (Fig. 4), but densities were low (on average 0.02 km$^{-2}$). Mauritania is close to the southern end of their wintering range, and immatures predominate here. Some second calendar year birds are known to spend the summer in the region (Nelson, 1978). In Mauritania, the first birds arrive in October, and from December to February several thousands may be seen off Cap Blanc and Cap Timiris (Lamarche, 1987).

In Gannets six plumage types are distinguished, from 1 (first year) to 6 (fully adult: Nelson in Blake *et al.*, 1984). As expected, most birds present in May 1988 were immatures. Many of the birds seen were in poor condition. Of 12 birds in type 4 to 6 plumages, only one type 6 bird had no obvious problems. Of the remaining eleven, five had plastics, nylon ropes, or netting around their bill, head, body and/or feet; five were oiled, and four had very worn plumages, so some birds even suffered from a combination of these. The condition of the type 1 to 3 birds seemed somewhat better, but out of 55 birds, eight type 2 and two type 1 birds were entangled in plastics or nylon, one type 2 bird was oiled, three type 2 birds had very worn plumages, and one type 1 bird had a hole in one wing, so in total, one in four of the younger birds also had obvious problems. It appears therefore, that although some younger birds may summer in the area voluntarily, most older birds should have left in May. Those remaining in the area were in a condition which was too poor to leave, as were a considerable number of the younger birds. Contamination with human waste (netting, oil) seems a major threat for Gannets in the area.

*Skuas*

During the survey all four northern hemisphere Skua-species were regularly encountered, especially when fishing vessels were in sight. Taken together, the skuas had an average density of 0.4 $km^{-2}$. Arctic *Stercorarius parasiticus* and Pomarine Skuas *S. pomarinus* were the most numerous. Some immature Pomarine/Arctic Skuas were sometimes difficult to separate, especially at larger distances. Within the counting strip however, this was only a minor problem. Skuas occurred throughout the study area, but in general were less common in the deeper parts of the study area (Fig. 5). In the most shallow parts, Arctic Skuas were the most numerous species (cf. Lamarche, 1987), especially in the central part of the study area, where the Pomarine was relatively rare.

The Pomarine Skua is the only skua that winters in great numbers in the upwelling waters off West Africa. As many as 500 have been seen around a fishing fleet in the area (Urban *et al.*, 1986). Arctic and Long-tailed *S. longicaudus* Skuas generally winter further south, off Namibia and South Africa, and Great Skuas *Catharacta skua* further north, especially the adult birds (Furness, 1987). Spring migration mainly takes place from March to May (Lambert, 1980; Furness, 1987), so large numbers of Arctic and Long-tailed Skuas must have passed the study area shortly before the time of the survey. However, almost 90% of the Long-tailed Skuas were in full adult plumage, indicating that their migration was still in full swing. Of the Pomarine and Arctic Skuas, some 60% were immatures (in this figure 6 unidentified 'immature' Pomarine/Arctic Skuas are included). The distribution of plumage types in the three smaller species (Table 2) was as expected (Furness, 1987), with a majority of light birds in all three species, and dark and intermediate birds only seen in Pomarine and Arctic Skuas.

Great Skuas were aged as adult or immature. Two birds in adult plumage were seen, and 10 in immature plumages. Age determination by plumage is difficult in Great Skuas, however (Harrison, 1983; Furness, 1987), and the two 'adults' may well have been older immatures.

*Gulls*

Four species of gulls were seen, of which the Sabine's Gull *Larus sabini* was the most common (Table 1). Sabine's Gulls winter further south of the study area, mainly in the upwelling off southern Africa. During spring migration, these birds follow the shelf edge and the Senegal/Canary upwellings. The peak of this migration is in the first half of May (Cramp, 1983; Urban *et al.*, 1986). Off Mauritania, spring migration has mainly been documented for April/May (Lamarche, 1987).

During the present survey Sabine's Gulls were seen mainly over the shelf, between the 20 and 100 m depth lines (Fig. 6). Groups of several tens were seen around fishing vessels. Most Sabine's Gulls seen at close range had whitish foreheads, so probably were first summer birds (Grant, 1982). Only 55 out of 242 birds were in full summer plumage. This indicates that the majority was not fully adult, which explains why these birds were still present in the area in May.

Lesser Black-backed Gulls *L. fuscus* are known to winter with 15–25000 birds at the Banc d'Arguin (Lamarche, 1987; Urban *et al.*, 1986), but are considered to remain inshore (Brown, 1979). By the time of the survey, most appeared to have left the area. Only a few immatures remained: 31 first summer birds, and one second summer bird were seen, mainly close to Cap Blanc. The other gulls seen, two Herring Gulls *L. argentatus* and one Audouin's Gull *L. audouinii*, were also in first summer plumage and seen near Cap Blanc.

*Terns*

Terns are numerous throughout the year on the Banc d'Arguin. Thousands of pairs breed on the islands of the National Park (Campredon, 1987) and over 100000 may be present on migration and during winter (von Westernhagen, 1970;

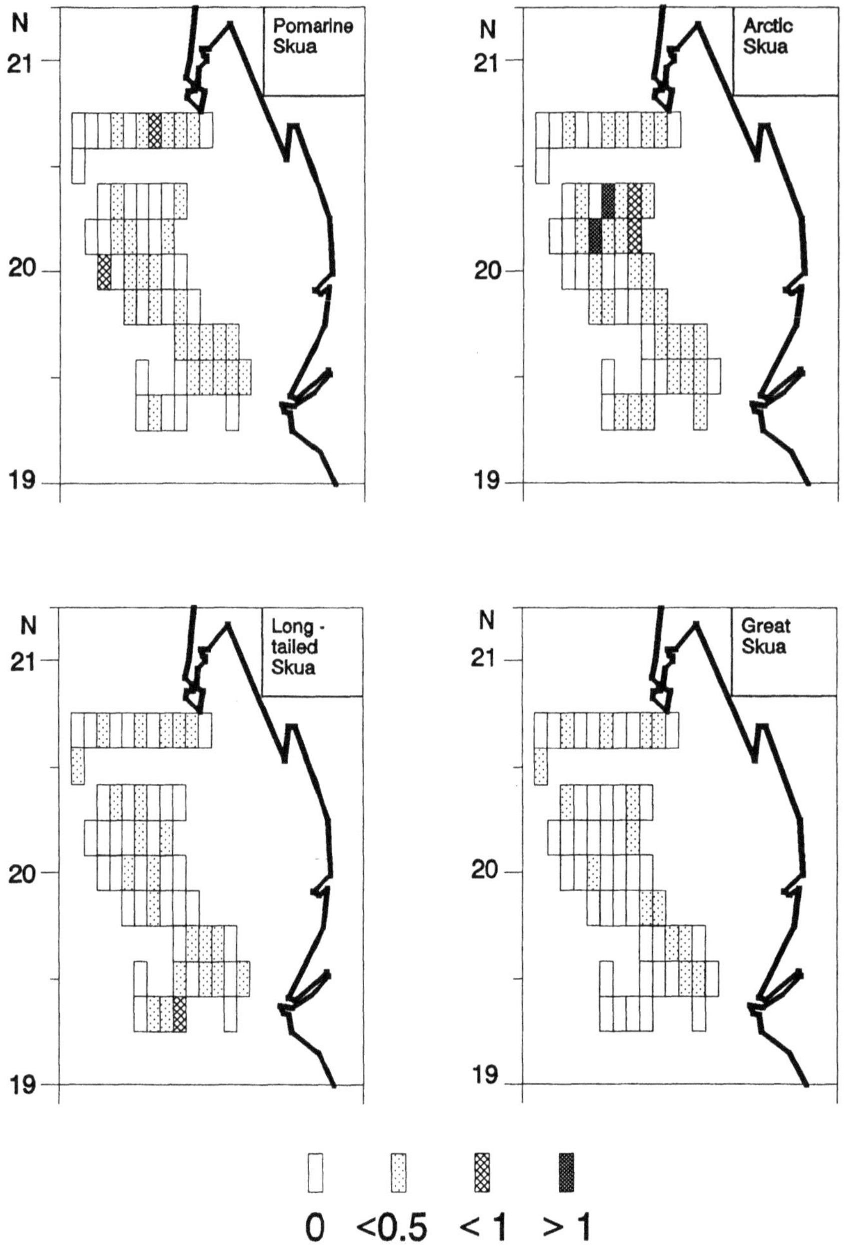


*Fig. 5.* Distribution of the four Skua-species seen during the survey. Shading represents numbers of birds seen per kilometre sailed.

Gandrille & Trotignon, 1973; Duhautois *et al.*, 1974; Trotignon *et al.*, 1980; van Dijk *et al.*, 1988). Most numerous are Black Terns *Chlidonias niger* (100 000 estimated in autumn), Sandwich Terns *Sterna sandvicensis* (30–40 000 wintering), and Royal Terns *S. maxima* (14 000 in summer).

Royal Terns breed on small islands on the Banc d'Arguin, mainly on the Ilet des Flamants in the North, and small islands around the Isle of Tidra in the centre. The total population is estimated at 5630 pairs. At the time of the survey, most pairs had eggs (Campredon, 1987). Campredon (pers. comm.) estimated from the lengths of foraging trips, that Royal Terns may venture as far as 50 kilometre away from the colonies. He considers this species the only one capable of going this far.

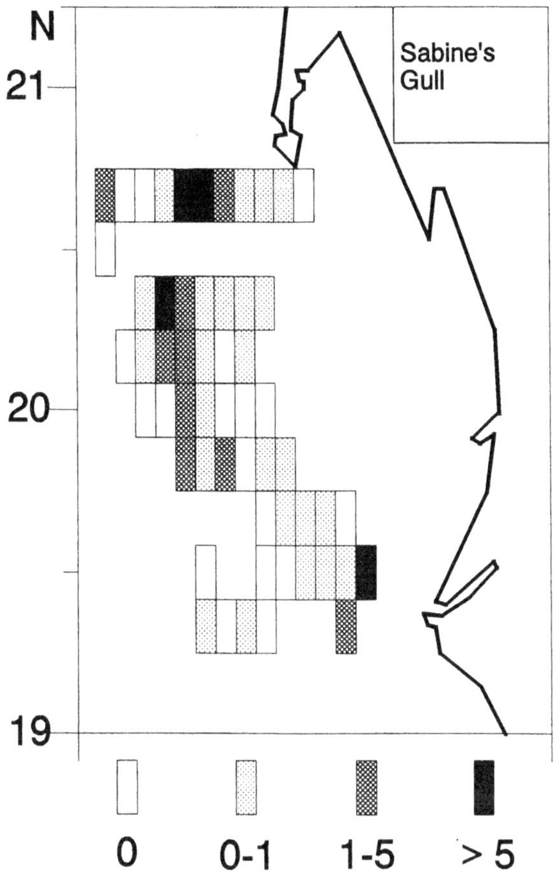


*Fig. 6.* Distribution of Sabine's Gull, in 4 density classes (birds $km^{-2}$).

*Table 2.* Age and plumage types in the *Stercorarius*-skuas observed.

| | Pomarine | Arctic | Long-tailed |
|---|---|---|---|
| Adult | 21 | 27 | 35 |
| Subadult | 22 | 15 | 3 |
| First summer | 8 | 14 | 2 |
| Unknown | 44 | 39 | 22 |
| Light | 66 | 52 | 42 |
| Intermediate | 1 | 4 | 0 |
| Dark | 7 | 18 | 0 |
| Unknown | 21 | 21 | 20 |

A distance of 50 km would bring the shelf slope within reach of the breeding birds. Campredon's estimate is supported by the present survey: Royal Terns were seen in numbers, and mainly in two areas, each relatively close to the breeding islands. The radius of the Royal Terns even seems a little larger than 50 km (Fig. 7).

Of 102 Royal Terns the plumage was noted. The majority (67) were in full adult breeding plumage, 20 were seemingly adult too, but had white or mottled foreheads, so probably were non-breeding, and 15 were obvious immatures. Two adults were seen flying towards the land with a fish in their bills: one just south of Cap Blanc (20° 42′ N, 17° 03′ W), and the other one off Tidra (19° 39′ N, 16° 56′ W). These observations further support the suggestion that the shelf slope is within reach of Royal Terns breeding at the Banc d'Arguin. However, the majority of the Royal Terns were not flying directly to or away from land. Most were seen flying into the wind, *i.e* in northerly directions.

Three other species of terns were also commonly seen during the survey (Table 1). With the exception of some Common Terns *Sterna hirundo*, these species do not breed on the Banc d'Arguin. Numbers of Common Terns were also too large to be all of Mauritanian origin. The hundreds of Sandwich, Common and Black Terns seen during the survey must have been either on migration, or they were summering in the area, and thus probably sub-adults. Common and Arctic Terns *Sterna paradisaea* are often difficult to separate, especially the immatures. The majority of these terns were identified as *S. hirundo*, but the positive identification of one Arctic (and one Roseate Tern *S. dougallii*) casts some doubt on these identifications, and this group is further treated as 'Commic' Terns. The Commic Terns were the most numerous terns. There was a constant stream of Commics, flying north, into the wind. Commic Terns were present in the highest densities in the shallow part of the study area (Fig. 7). Local concentrations resulted from birds being attracted to fishing vessels. Plumages were noted for 218 Commics: 29 were Common Terns in adult summer plumage, 109 were second summer Commics, and 60 were first summer Commics. Hence only a minority were adult birds.

Sandwich Terns showed a similar distribution pattern, behaviour and age composition, but occurred slightly closer inshore (Fig. 7). Again, the

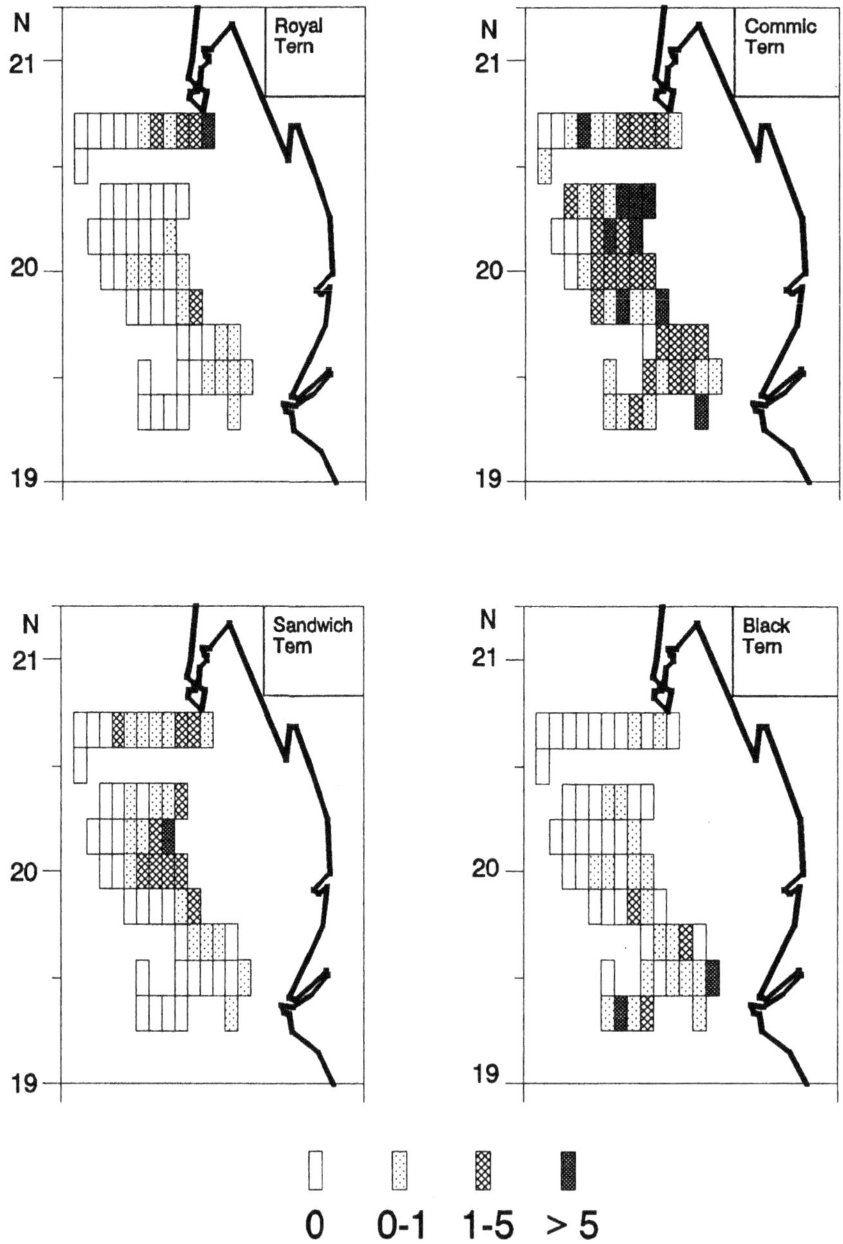


*Fig. 7.* Distribution of the four most common tern-species, in 4 density classes (birds $km^{-2}$).

majority of the birds (87% of 63 birds) was immature. Black Terns occurred both inshore and offshore, but were most numerous in the south of the study area. The majority of the Black Terns was in adult summer plumage (122 out of 173 birds). Unlike the other two species, Black Terns are known to migrate still in large numbers along west Africa in May (Urban *et al.*, 1986), and a considerable proportion of the Black Terns seen may have been breeders of northern Europe.

Other breeding species of the Banc d'Arguin, the Gull-billed *S. nilotica*, Caspian *S. caspia* and Bridled Tern *S. anaethetus* were very rare in the study area (Table 1), and were only seen in inshore parts. Caspian Terns were seen in the vicinity of Cap Blanc and near Cap Timiris. Bridled Terns were only seen over very shallow water near the Banc: between 19°40′ and 19°42′ N and 16°55′ and 17°01′ W. The Gull-billed Tern was seen in the same area.

*Flying directions*

Flying directions were noted for birds seemingly flying on a definite course. Birds approaching fishing vessels or the research vessel were excluded, as were the storm-petrels, since these mostly flew in a zigzag-fashion. Data were collected for Cory's Shearwater, Sabine's Gull, and four Tern-species. The vast majority of these birds were flying into northerly directions (Fig. 8). Three explanations may be offered for this phenomenon: 1, they were either migrating, so came from a pool of birds further south; 2, they were compensating for the drift resulting from the northerlies; or 3, they flew into the wind in order to forage most efficiently. Spring-migration supposedly was the reason for many birds to fly north, but drift compensation and the need to fly into the wind while foraging must have been important for birds summering in the area or breeding at the Banc d'Arguin.

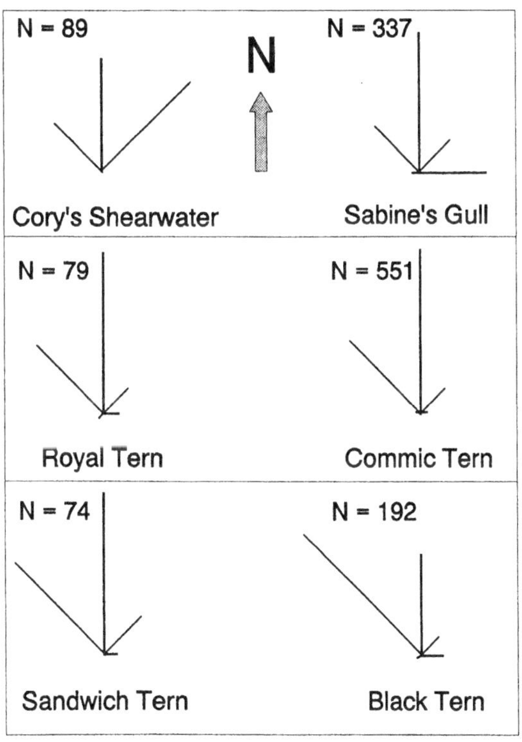


*Fig. 8.* Percentual distribution of flying directions of Cory's Shearwater, Sabine's Gull, and the four most common tern-species. In each species, the sum of the lines represents 100%.

*Total numbers seen and biomass*

In total, 46 hours were spent on watch, during which time 242 $km^2$ of strip-area was observed. Within this surface area, 4849 seabirds were noted (20 $km^{-2}$). Storm-petrels were numerically the most important group, with an average density of 14.5 $km^{-2}$. Outside the strips, 4187 more seabirds were seen (Table 1). On average, 32.74 birds were seen per 10 minute watch. This is about 5 times as many as seen by Brown in the waters off Senegal. In terms of biomass, the seabirds amounted to 7 kg $km^{-2}$ (using body masses from Brown *et al.*, (1982) and Urban *et al.*, (1986)). Brown (1979) uses a measure of kg/10 minutes to compare seabird biomass in different upwelling areas. For the Senegal upwelling in February/March, he arrives at a figure of 6.1. In the Mauritanian studied here, this figure is 11.1, so birds off Mauritania were more numerous, but on average smaller than those in the Senegal upwelling.

## Discussion

The Mauritanian slope area was found to hold several categories of seabirds:

1. local breeders (Royal Tern);
2. birds migrating to breeding colonies at higher latitudes (skuas, Black Tern, possibly some Sabine's Gulls);
3. immature, northern hemisphere seabirds spending (part of the) summer in the area (Gannet, most gulls and terns, part of the skuas, and probably most of the Cory's Shearwaters and British Storm-petrels);

4. The Wilson's Storm-petrels, breeding birds of the Antarctic, using the area to moult and spend the southern winter.

The survey showed that the vast majority of the *ca* 50 000 piscivorous seabirds breeding at the Banc d'Arguin (Campredon, 1987), do not leave the Banc on foraging trips. None, or very few of the Great and Longtailed Cormorants *Phalacrocorax carbo* and *P. africanus*, White Pelicans *Pelecanus onocrotalus*, Slender-billed Gulls *Larus genei*, or Gull-billed and Caspian Terns all with over 1000 breeding pairs, were seen in the study area. Only Royal Terns apparently go to the slope zone in numbers. In May these birds were brooding eggs, and the off-duty birds apparently had sufficient time to venture this far from their nests. During a second cruise of the RV Tyro in the area one month later, Royal Terns were still frequently seen (data provided by Smeenk and Addink, pers. comm.), showing that even when having chicks, Royal Terns reach the slope.

During this second cruise of the Tyro on the Mauritanian slope, the ship was almost continuously engaged in fishing activities. This attracted large numbers of seabirds, making only qualitative observations on seabirds possible. Hundreds of Commic, Sandwich and Black Terns were seen in June, and Cory's Shearwaters, British Storm-petrels (and Wilson's) were also regularly seen. Gannets, Skuas and Sabine's Gulls were relatively less common than in May, indicating that many of the latter category were leaving the area in May, while most terns, British Storm-petrels and Cory's Shearwaters were summering there. The Mauritanian waters appear to be an important area for these species during their first summer(s). This is analogous to the *ca* 200 000 waders spending their first summer on the tidal flats of the Banc d'Arguin (van Dijk *et al.*, 1990). The fishing industry in the area obviously supplies large amounts of food for these birds, and may be an important feature of the area in the non-upwelling season.

Studying the relation between the local hydrography and the distribution patterns of the seabirds off the Banc d'Arguin proved very difficult. The trawlers had a profound effect on the distribution of the seabirds, masking any effects of hydrographical nature. Most of the larger birds seen during this survey were strongly attracted to the trawlers in the area, rendering any analysis of distribution patterns in relation to hydrography senseless. Moreover, large numbers of birds were probably migrating through the area at the time of the survey, and these birds too, may not have been linked to local hydrographical circumstances.

Storm-petrels, the most numerous birds in the study area, are often found to be concentrated in productive waters, such as upwellings and convergences (Brown, 1988), and their distribution pattern off the Banc d'Arguin may thus also be notably linked to hydrographical parameters, as well as being influenced by the trawlers. Indeed, a large concentration of storm-petrels was found in the patch of upwelling water in the southeast of the study area, which was rich in zooplankton. However, a second, equally large concentration of storm-petrels was found in the central part of the study area, where no upwelling had occurred for some time (Kuipers *et al.*, this volume). In both areas a large number of trawlers was fishing and as a result, food patches in the form of oily slicks were amply available. In June, Smeenk & Addink (pers. comm.) found storm-petrel concentrations of similar magnitude in the area. These were all situated along the slope, where fishing still was intense.

The situation off Mauritania appears similar in this respect to that described for the southern Benguela upwelling region, where trawlers persistently operate along the shelf, attracting large numbers of offal-eating seabirds (Abrams, 1983; 1985). Storm-petrels, as well as the larger seabirds in the study area, benefit from human fisheries in the area, by foraging on food-stuff that would otherwise be unavailable to them. During the non-upwelling season, the large-scale fishing operations may be of overriding importance, since zooplankton biomass and availability of small fish is likely to be lower than in the upwelling season.

During this survey, outwelling through the deep trenches in the south (Loktionov, 1993), as proposed by Peters (1976), was not found, but Banc Water left the Banc d'Arguin at the surface in the central part of the study area. In this tongue of Banc Water outflow, very high values for primary production were measured (Berghuis *et al.*, this volume). Hence, two hydrographically different areas were found to be enriched within the study area: the patch of upwelling water in the south-east, and an area in the centre, where warm, highly productive water left the Banc d'Arguin. Both areas supported large concentrations of storm-petrels, but also substantial fishing fleets, making it impossible to estimate the relative contribution of either factor for storm-petrel distribution.

It may be concluded, that not only the waders (Ens *et al.*, 1990) and the piscivorous breeding birds benefit from this Ocean-Banc d'Arguin system, but also seabirds over the slope. Human fishing activities serve as an intermediate, turning fish into food available for the seabirds.

## Acknowledgements

I would like to thank Rob Witbaard for assistance during the observations, and Chris Smeenk and Marjan Addink for the additional observations in June. Koos den Hartog allowed me to use his stomach analyses of storm-petrels. Rob Dapper and Katja Philippart provided expert help with computer-analysis and presentation of the results. Kees Camphuysen, Cor Smit and Eric Marteijn are thanked for making available to me some of the 'grey' literature. Wim Wolff, Jan Andries van Franeker and an anonymous referee read and improved the manuscript. This study was part of the Dutch 1988 Mauritania expedition, financed by the Netherlands Marine Science Foundation.

## References

Abrams, R. W., 1983. Pelagic seabirds and trawl-fisheries in the southern Benguela Current region. Mar. Ecol. Prog. Ser. 11: 151–156.

Abrams, R. W., 1985. Pelagic seabird community structure in the southern Benguela region: changes in response to man's activities? Biol. Conserv. 32: 33–49.

Ainley, D. G. & R. J. Boekelheide, 1983. An ecological comparison of oceanic seabird communities of the south Pacific Ocean. Studies in Avian Biol. 8: 2–23.

Berghuis, E. M., G. C. A. Duineveld & J. Hegeman, 1993. Primary production and distribution of phytopigments in the watercolumn and sediments on the upwelling shelf off the Mauritanian coast (northwest Africa). Hydrobiologia 258: 81–93.

Blake, B. F., M. L. Tasker, P. Hope Jones, T. J. Dixon, R. Mitchell & D. R. Langslow, 1984. Seabird distribution in the North Sea. Nature Conservancy Council, Huntingdon.

Bourne, W. R. P., 1963. A review of oceanic studies of the biology of seabirds. Proc. 13 Intl. Ornithol. Congr.: 831–854.

Bravo de Laguna, J., 1989. Managing an international multispecies fishery: the saharan trawl fishery of cephalopods. In J. F. Cady (ed.), Marine invertebrate fisheries: their assessment and management. J. Wiley & Sons, NY, 591–612.

Brown, L. H., E. K. Urban & K. Newman, 1982. The Birds of Africa, Vol. 1. Academic Press, London, 521 pp.

Brown, R. G. B., 1979. Seabirds of the Senegal upwelling and adjacent waters. Ibis 121: 283–292.

Brown, R. G. B., 1980. Seabirds as marine animals. In J. Burger, B. L. Olla & H. E. Winn (eds), Behavior of marine animals, Plenum Press, NY, 4: 1–39.

Brown, R. G. B., 1988. The influence of oceanographic anomalies on the distributions of Storm-petrels (Hydrobatidae) in Nova Scotian waters. Colonial Waterbirds 11: 1–8.

Cadée, G. C., 1981. Seabird observations between Rotterdam and the equatorial Atlantic. Ardea 69: 211–216.

Camphuysen, C. J., 1988. Zeevogelwaarnemingen aan boord van MS Plancius, van Nederland tot de Kaapverdische Eilanden, winter 1985/86. Sula 2: 37–46.

Campredon, P., 1987. La reproduction des oiseaux d'eau sur le Parc National du Banc d'Arguin (Mauritanie) en 1984–1985. Alauda 55: 187–210.

Cramp, S. (ed.), 1983. The Birds of the Western Palearctic, Vol. 3. Oxford University Press, 913 pp.

Cramp, S. & K. E. L. Simmons (eds), 1977. The Birds of the Western Palearctic, Vol. 1. Oxford University Press, 722 pp.

Duhautois, M. C., F. Charmoy, D. Reyjal & J. Trotignon, 1974. Seconde prospection post-estivale au Banc d'Arguin (Mauritanie). Alauda 42: 313–332.

Ens, B. J., T. Piersma, W. J. Wolff & L. Zwarts (eds), 1990. Homeward Bound: Problems waders face when migrating from the Banc d'Arguin, Mauritania, to their northern breeding grounds in spring. Ardea 78: 1–364.

Follestad, A., 1990. The pelagic distribution of Little Auk *Alle alle* in relation to a frontal system off central Norway, March/April 1988. Polar Res. 8: 23–28.

Furness, R. W., 1982. Competition between fisheries and seabird communities. Adv. mar. Biol. 20: 225–307.

Furness, R. W., 1987. The Skuas. Poyser, Calton, 363 pp.

Furness, R. W. & D. Ainley, 1984. Threats to seabird populations presented by commercial fisheries. In J. P. Croxall, P. G. H. Evans & R. W. Schreiber (eds), Status and conservation of the world's seabirds. ICBP-Technical Publication, Norwich, 701–708.

Gandrille, G. & J. Trotignon, 1973. Prospection post-estivale au Banc d'Arguin (Mauritanie). Alauda 41: 129–159.

Grant, P. J., 1982. Gulls – a guide to identification. Poyser, Calton, 280 pp.

Harrison, P. 1983. Seabirds – an identification guide. Croom Helm, Beckenham, 448 pp.

Hazevoet, C. J., 1985. Bird records from Mauritania in December 1984. Dutch Birding 7: 27–28.

Jespersen, P., 1924. The frequency of birds over the high Atlantic Ocean. Nature 114: 281–283.

Kaczynski, V. M., 1989. Foreign fishing fleets in the subSaharan West African EEZ. Marine Policy 13: 2–15.

Kuipers, B. R., H. J. Witte & S. R. Gonzalez, 1993. Zooplankton distribution in the coastal upwelling system along the Banc d'Arguin, Mauritania. Hydrobiologia 258: 133–149.

Lamarche, B., 1987. Liste commentée des oiseaux de Mauritanie. Nouakchott, 202 pp.

Lambert, K., 1980. Ein Uberwinterungsgebiet der Falkenraubmöwe, *Stercorarius longicaudus* Vieill. 1819, vor Südwest-mid Südafrika entdeckt. Beitr. Vogelkde. 26: 199–212.

Loktionov, Y., 1993. Hydrographical observations west of the Banc d'Arguin in May 1988. Hydrobiologia 258: 21–32.

Mayo, A. L. W., 1948. Birds seen in the tropical Atlantic and western approaches to Gibraltar. Ibis 90: 22–25.

M'Bareck Ould Soueilem, M., 1986. Statistiques de la pêche pelagique en Mauritanie. In: E. Josse & S. Garcia (eds), Description et évaluation des ressources halieutiques de la ZEE Mauritanienne. Rapp. Working Group CNROP/FAO/ORSTOM, Nouadhibou, Mauritania, Sept. 1985. CECAF/ECAF Series, FAO, Rome, pp. 194–204.

Mittelstaedt, E., 1991. The ocean boundary along the Northwest African coast: circulation and oceanographic properties at the sea surface. Prog. Oceanog. 26: 307–355.

Moerbeek, D. J., 1982. De pelagische verspreiding van Kuhl's Pijlstormvogel, een methodisch onderzoek naar de mogelijkheden van automatische verwerking van vogelwaarnemingen op zee. Thesis, University of Amsterdam, 77 pp.

Murphy, R. C., 1936. Oceanic Birds of South America. Vols. 1 & 2.

Nelson, J. B., 1978. The Gannet. Poyser, Berkhamsted, 336 pp.

Peters, H., 1976. The spreading of the water masses of the Banc d'Arguin in the upwelling area off the northern Mauritanian coast. 'Meteor' Forschungsergebnisse (A) 18: 78–100.

Ryan, P. G. & C. L. Moloney, 1988. Effect of trawling on bird and seal distributions in the southern Benguela region. Mar. Ecol. Prog. Ser. 45: 1–11.

Tasker, M. L., P. H. Jones, T. J. Dixon & B. F. Blake, 1984. Counting seabirds at sea from ships: a review of methods employed and a suggestion for a standardized approach. Auk 101: 567–577.

Trotignon, E., J. Trotignon, M. Baillou, J. F. Dejonghe, L. Duhautois & M. Lecomte, 1980. Recensement hivernal des limicoles et autres oiseaux aquatiques sur le Banc d'Arguin (Mauritanie) (Hiver 1978/1979). L'Oiseau et R.F.O. 50: 323–343.

Urban, E. K., C. H. Fry & S. Keith, 1986. The Birds of Africa, Vol. 2. Academic Press, London, 552 pp.

Van Camp, L., L. Nykjaer, E. Mittelstaedt & P. Schlittenhardt, 1991, Upwelling and boundary circulation off Northwest Africa as depicted by infrared and visible satellite observations. Prog. Oceanog. 26: 357–402.

Van Dijk, A. J., E. C. L. Marteijn, F. E. de Roder, T. M. van Spanje & H. W. Spiekman, 1988. Rapport préliminaire des recensements des oiseaux aquatiques estivant sur le Parc National du Banc d'Arguin, Mauritanie. WIWO-rapport, 5 pp.

Van Dijk, A. J., F. E. de Roder, E. C. L. Marteijn & H. W. Spiekman, 1990. Summering waders on the Banc d'Arguin, Mauritania: a census in June 1988. Ardea 78: 145–156.

Von Westernhagen, W., 1970. Durchzugler und Gäste an der westafrikanischen Küste auf den Inseln der untiefe Banc d'Arguin. Die Vogelwarte 25: 185–193.

*Hydrobiologia* **258**: 211–222, 1993.
*W. J. Wolff, J. van der Land, P. H. Nienhuis & P. A. W. J. de Wilde (eds),*
*Ecological Studies in the Coastal Waters of Mauritania.*




# The functioning of the ecosystem of the Banc d'Arguin, Mauritania: a review

W. J. Wolff[1], J. van der Land[2], P. H. Nienhuis[3] & P. A. W. J. de Wilde[4]
[1]*DLO – Institute for Forestry and Nature Research, P.O. Box 46, 3956 ZR Leersum, The Netherlands & Dept. of Nature Conservation, Wageningen Agricultural University, The Netherlands;* [2]*National Museum of Natural History, P.O. Box 9517, 2300 RA Leiden, The Netherlands;* [3]*Netherlands Institute for Ecological Research, Vierstraat 28, 4401 EA Yerseke, The Netherlands;* [4]*Netherlands Institute for Sea Research, P.O. Box 59, 1790 AB Den Burg, The Netherlands*

*Key words:* Banc d'Arguin, Mauritania, ecosystem

## Abstract

The Banc d'Arguin, Mauritania, is an area of over 10000 km$^2$ of shallow water and tidal flats between the Sahara and the upwelling system off the Mauritanian coast. The results of the Dutch-Mauritanian project 'Banc d'Arguin 1988' as well as the relevant literature are reviewed in order to develop a preliminary conceptual model of the functioning of this ecosystem. Hydrographically the Banc d'Arguin can be characterized as a large-scale negative estuary with higher salinities near the shore. It is concluded that ultimately the Banc d'Arguin ecosystem is fueled by nutrients and organic matter derived from the upwelling area. The inner part of the Banc d'Arguin system is dominated by a detritus-based benthic foodweb in which seagrasses are the principal primary producers. Little of the seagrass production seems to be exported to other parts of the system or other areas. Zooplankton and zoobenthos biomasses are relatively low, but nevertheless support high densities of consumers. Aquatic birds are especially numerous. The study identifies many research topics for the future.

## Introduction

The Banc d'Arguin in Mauritania, West-Africa, is an area of over 10000 km$^2$ of tidal flats and shallow inshore waters bordering the sandy desert of the Sahara (Fig. 1). Since 1976 about 12000 km$^2$ of shallow water, tidal flats, and adjacent desert are included in the Parc National du Banc d'Arguin (PNBA).

Research of the marine part of the Banc d'Arguin has proceeded along two main lines for many years: one ornithological and another one related to fisheries.

De Naurois (1959) was the first to call the attention of ornithologists to the huge numbers of birds breeding on the Banc d'Arguin, especially pelicans, herons, flamingos, spoonbills and terns. Von Westernhagen (1968) discovered the importance of the Banc d'Arguin for migrating and wintering waders. His findings were confirmed by several other ornithologists (e.g. Pététin & Trotignon, 1972; Knight & Dick, 1975; Trotignon *et al.*, 1980; Altenburg *et al.*, 1983; Ens *et al.*, 1990).

Fisheries research on the Banc d'Arguin has been carried out for several decades from what is now the Centre National de Recherches Océanographiques et des Pêches (CNROP) at Nouad-

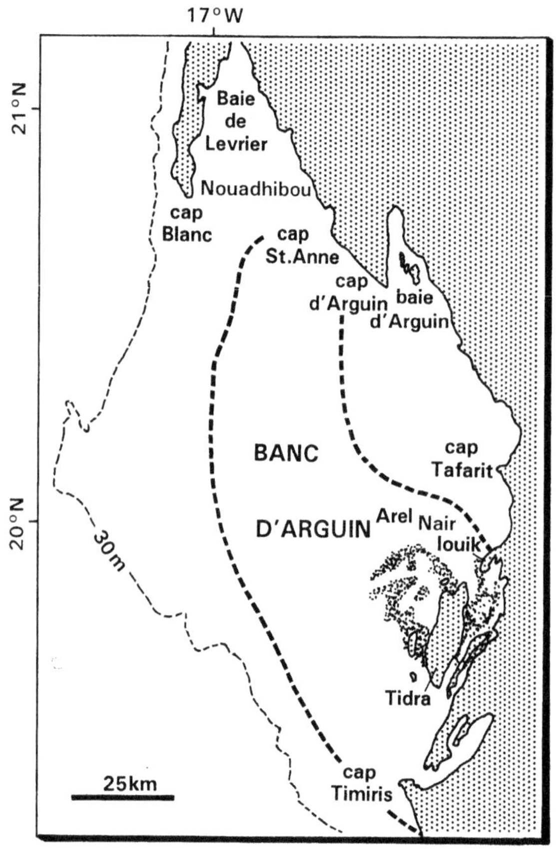


*Fig. 1.* The Banc d'Arguin, Mauritania. The thick broken line approximately encloses the area with depths less than 5 m.

hibou. On one hand it provided knowledge about the fishes of the shallow waters of the Banc d'Arguin (for example Gaudechoux, 1984; Sevrin-Reyssac & Forges, 1985), but on the other hand valuable information on the hydrography and the plankton of the area has also been collected by Maigret (1972), Brulhet (1974), Reyssac (1974, 1977), and Sevrin-Reyssac (1982, 1984). However, oceanographic and fisheries research at the Banc d'Arguin has been and still is hampered very much by the extremely shallow nature and hence the inaccessibility of the area.

The research on the Banc d'Arguin proper has to be viewed in relation to other research carried out offshore and in the coastal regions of Mauritania. The area offshore the Banc d'Arguin has been the site of many campaigns by the CNROP and also has drawn the attention of the international oceanographic community (e.g. Peters, 1976; Hempel, 1982). The terrestrial environment has been studied by geologists, botanists, and zoologists.

An ecosystem approach of the aquatic part of the Banc d'Arguin developed in the eighties. Sevrin-Reyssac (1983) tentatively sketched the foodchains of the Banc d'Arguin ecosystem. The French group 'GIS-Posidonie' in collaboration with the Parc National du Banc d'Arguin developed what they called a 'pre-model' of the Banc d'Arguin ecosystem (Boudouresque *et al.*, 1988) (Fig. 2). The Dutch-Mauritanian project 'Banc d'Arguin 1985–86' resulted in another simple model of the Banc d'Arguin ecosystem (Wolff & Smit, 1990) (Fig. 3). The next Dutch-Mauritanian project 'Banc d'Arguin 1988' aimed at the description and if possible a quantification of many functional relationships of the Banc d'Arguin ecosystem. During this project both the offshore and the inshore parts of the Banc d'Arguin were investigated. To these international studies the continuing research effort of CNROP and PNBA has to be added.

In this volume most of the results of the Dutch-Mauritanian project 'Banc d'Arguin 1988' have been brought together. The aim of this paper is to review the preceding papers as well as the previous literature and to develop a picture of the Banc d'Arguin ecosystem as we perceive it today. The picture still shows many white spots which have to be coloured in. But already now some conclusions are drawn which are very pertinent for the management of this extremely interesting coastal area.

*Hydrography*

The offshore upwelling system has been described in considerable detail (see for example Hempel, 1982). Few authors, however, have considered the relation with the shallow waters of the Banc d'Arguin.

The publications of Maigret (1972), Brulhet (1974), Peters (1976), Reyssac (1974, 1977) and

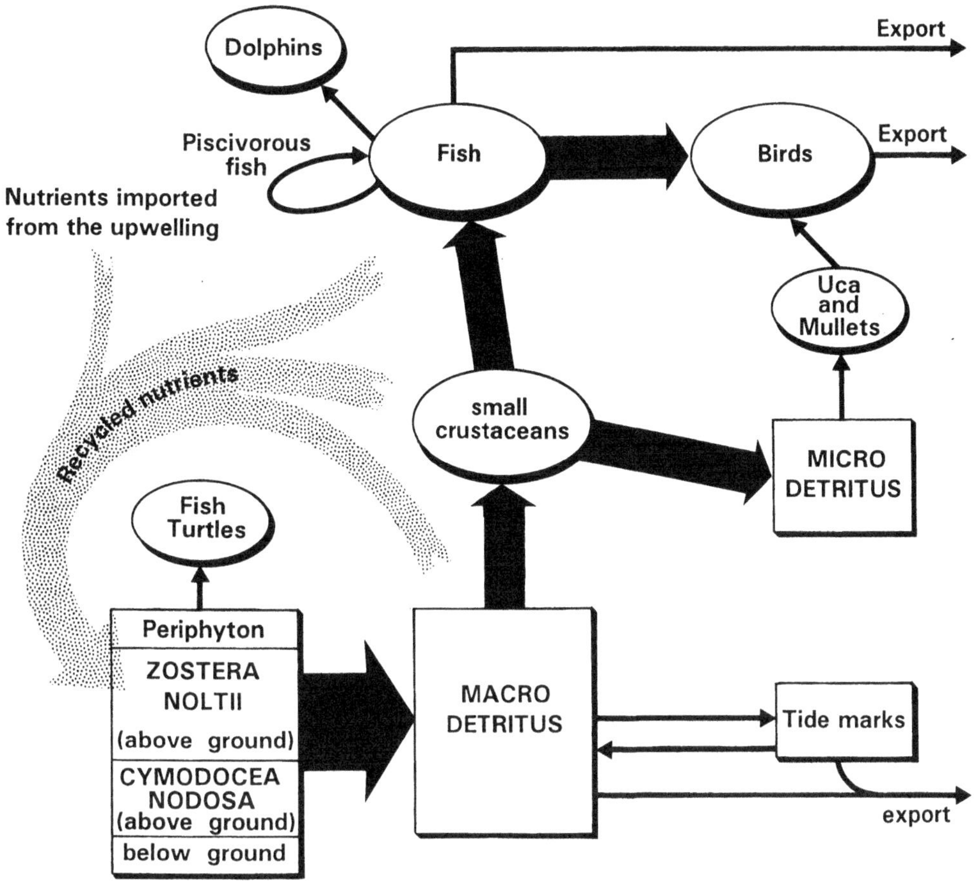


*Fig. 2.* Preliminary model of the functioning of the Banc d'Arguin ecosystem. After Boudouresque *et al.* (1988).

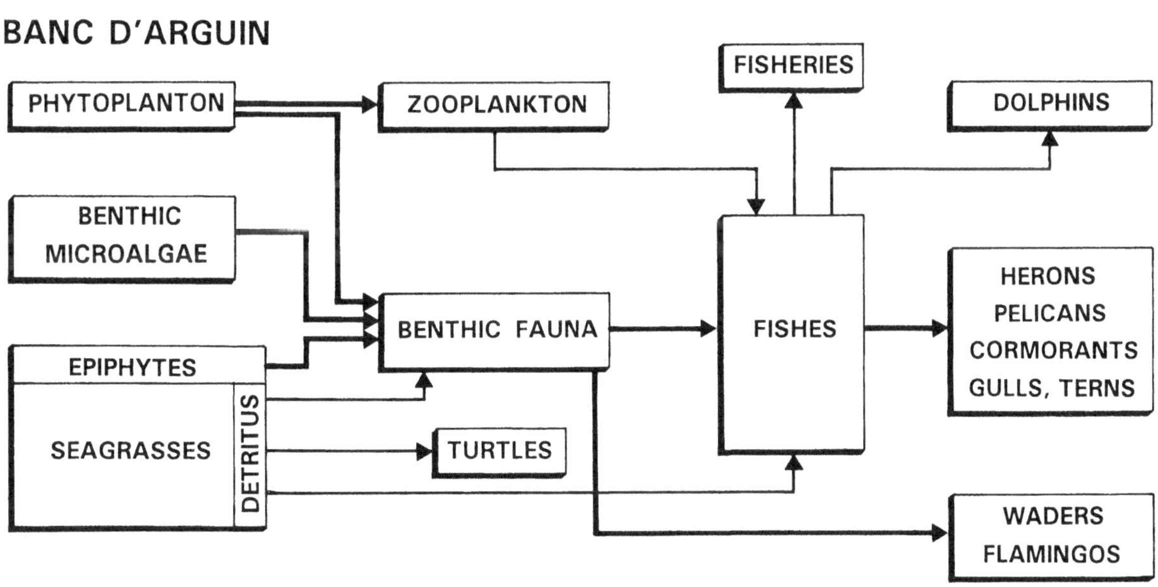


*Fig. 3.* Preliminary model of the ecosystem of the Banc d'Arguin. After Wolff & Smit (1990).

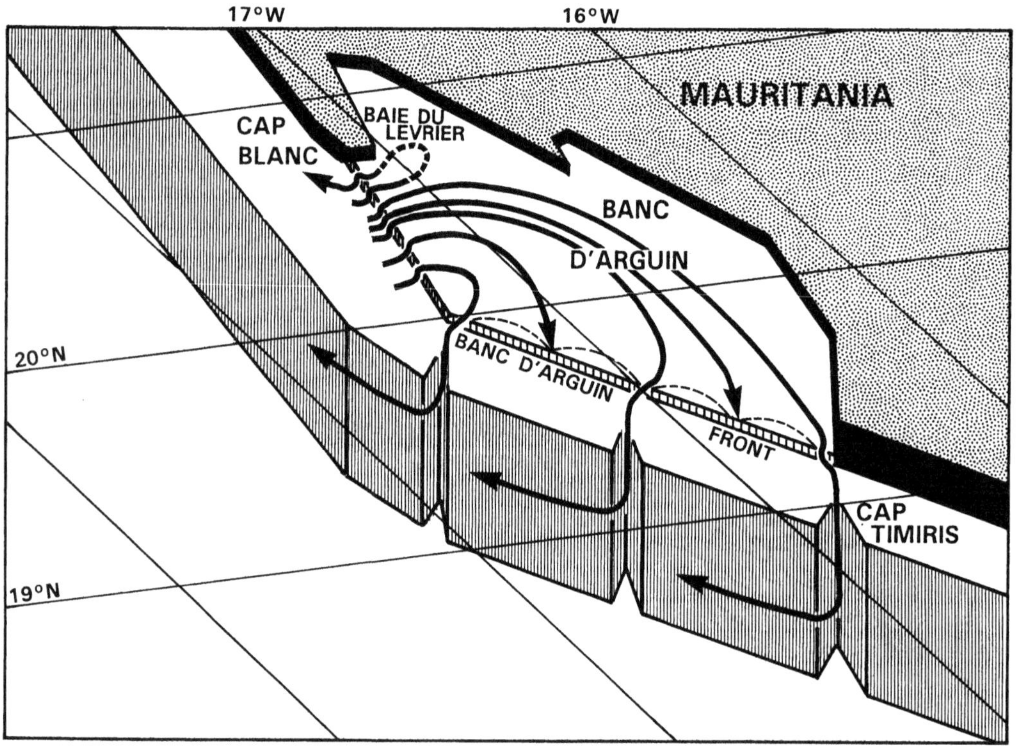


*Fig. 4.* Conceptual model of the water circulation at the Banc d'Arguin. After Peters (1976).

Sevrin-Reyssac (this volume) show that usually oceanic water enters the shallows of the Banc d'Arguin South of Cap Blanc (Fig. 4). Part of it moves northwards into the Baie du Lévrier, but the majority flows behind the Banc d'Arguin proper southwards into the direction of the Isle of Tidra. This water increases in temperature, salinity and density on its way South and finally flows across the southern shallow part of the Banc and cascades through numerous submarine canyons back to the ocean. Near the southwestern edge of the Banc masses of greenish to brownish water can be seen flowing to the West and subsequently disappearing under the less saline surface water (Van der Land, unpubl. observ.). The Banc water is very turbid and as a result sediments on the narrow shelf off the southwestern part of the Banc d'Arguin are very muddy and consist of sticky clay in the deeper parts. During our studies Loktionov (1993), however, was unable to trace this dense downward and northward flow of Banc water along the continental slope. He ascribed this to the relatively small scale of this process and to the cloudy conditions during the preceding period, which negatively influenced evaporation. Cuq (1993), on the other hand, confirms by means of satellite images how oceanic water enters the shallows South of Cap Blanc. Undoubtedly the bay acts as a tidal pump. The tide forces Canary Current water into the bay South of Cap Blanc, but a large part of this water does not return and is pushed southward by the prevailing northerly wind. Kuipers *et al.* (1993) suggest that the major mechanism bringing ocean water to the shallows is the variation in wind strength (see also Ould Dedah, 1993). Our data are largely in agreement with the model proposed by Peters (1976) (Fig. 4).

Sofar it has not been possible to quantify the water transport across the Banc d'Arguin. However, based on the transport time over the Banc d'Arguin of about one month (Peters, 1976; Reyssac, 1977) and the dimensions of the coastal topography we estimate that the southward flow is in the order of magnitude of 0.5 $km^3$ $d^{-1}$. The changes in the south-flowing water mass have been described only in terms in salinity and temperature. However, a description of the changes of the nutrient status and the concommitant changes in the phytoplankton is urgently required. Reyssac (1977) already showed that the phytoplankton East of the Banc d'Arguin proper is very different from the phytoplankton in the upwelling area offshore. Diatoms characteristic for the upwelling area do hardly or not occur East of the Banc d'Arguin. A factor contributing to this phenomenon may be the occurrence of high densities of filter feeders in between the two areas (see below).

The hydrographic conditions in the tidal flat region of the Banc d'Arguin have only summarily been described (Reyssac, 1977; Sevrin-Reyssac, 1993; Smit *et al.*, 1989; Wolff & Smit, 1990). The area can be characterized as a negative estuary with hypersaline and warm water near the coast. More detailed, year-round investigations are required.

*Primary production*

Reyssac (1977) reports for the Baie du Lévrier phytoplankton average chlorophyll-*a* values ranging from 1.6 to 22.0 mg $m^{-3}$ and productivity figures ranging from 192 to 929 mg C $m^{-3}$ $d^{-1}$. For the entire water column at a location off Nouadhibou this results in on average 1722 mg C $m^{-2}$ $d^{-1}$ for the period September–November 1973. The same author reports for the inshore region east of the shallow part of the Banc d'Arguin and North of the Isle of Tidra chlorophyll-*a* values ranging from 0.5 to 4.1 mg $m^{-3}$ and phytoplankton production values ranging from 41 to 164 mg C $m^{-3}$ $d^{-1}$. In the tidal flat region of the Banc d'Arguin phytoplankton biomasses are again much lower; Sevrin-Reyssac (1984) records chlorophyll-*a* values up to 0.11 mg $m^{-3}$. Hence, there is a clear gradient of decreasing phytoplankton densities from the open ocean towards the tidal flat area with very low phytoplankton biomasses in the latter area. Most probably the same is true for phytoplankton productivity.

Anybody visiting the Banc d'Arguin will be impressed by the vast expanses of seagrasses. Wolff & Smit (1990) estimate that of the 491 $km^2$ of tidal flats about 193 km are covered by dense seagrass beds and about 219 km are covered by less dense seagrass vegetations. Nienhuis *et al.* (1990) report that *Zostera noltii* is by far the most common species on these tidal flats. The species also occurs below the tidal zone down to depths of 5 m. *Halodule wrightii* is less abundant; it occurs from about mid-tide level down to 6 m below low-tide level. In general *Halodule* occurs in more sandy habitats than does *Zostera*. *Cymodocea nodosa* is the third species which is essentially sublittoral, although it occurs also in the lower part of the intertidal zone. Nienhuis *et al.* (1990) conclude that *Cymodocea* covers several hundreds of $km^2$ of sublittoral seabottom; they also present a distribution map. However, its average coverage is usually below 50%. Van Lent *et al.* (1991), who studied some of these seagrass beds in May 1988, record above-ground biomass values for *Zostera noltii* ranging from 60–119 g ash- free dry weight (AFDW) $m^{-2}$ and for *Cymodocea nodosa* from 41–76 g. Below-ground biomasses were roughly two times higher. Vermaat *et al.* (1993), who studied seagrass beds in April and September 1988, record above-ground biomasses of 38–76 g AFDW $m^{-2}$ for *Z. noltii*, 91–100 g for *C. nodosa* and a total biomass of 24–31 g AFDW $m^{-2}$ for *Halodule wrightii*. Below-ground biomasses of *Zostera* and *Cymodocea* were higher in April, but roughly equal or even lower in September. Van Lent *et al.* (1991) also give the total productivity rates for both species in May 1988: 1.6–3.4 g AFDW $m^{-2}$ $d^{-1}$ for *Z. noltii* and about 1.2 g for *C. nodosa*. Vermaat *et al.* (1993) determined a production rate of 1.1–1.2 g AFDW $m^{-2}$ $d^{-1}$ for *Zostera* in April and September and 2.8 g for *Cymodocea* in September. Since leaf loss in the seagrass beds investigated by Van Lent *et al.* (1991) was at least 1–3 g AFDW $m^{-2}$ $d^{-1}$ in May 1988 net productivity was only −1.4– +0.6 g AFDW $m^{-2}$ $d^{-1}$. Vermaat *et al.* (this volume) record leaf loss rates of 1.2–2.1 g for *Zostera* and

0.9 g for *Cymodocea*, and arrive at net production rates of $-0.05--1.0$ g AFDW $m^{-2}$ for the former species and $+1.8$ g for the latter. De Jong *et al.* (1991) determined seagrass productivity by microelectrode and bell jar measurements. They arrived at gross production rates of 2.0–4.2 g carbon $m^{-2}$ $d^{-1}$.

Periphyton amounts ranged from 0.1–0.6 g AFDW $cm^{-2}$ seagrass leaf area (Hootsmans *et al.*, 1993) Compared to other marine periphyton data this is a small amount.

Benthic macro-algae do not seem to be very important. Only small areas are covered by *Vaucheria* species (Altenburg *et al.*, 1982; Wolff *et al.*, 1993) and within the dense seagrass beds macro-algae are unimportant. The importance of micro-algae has been estimated by De Jong *et al.* (1991). Benthic diatoms occurred mainly on the upper tidal flats and on sandy beaches. Gross production ranged from 0.2–0.5 g C $m^{-2}$ $d^{-1}$. At 4–6 m depth subtidal microalgae, dominated by green algae, were responsible for a gross production of 0.9–2.6 g C $m^{-2}$ $d^{-1}$.

Sevrin-Reyssac (1983), Boudouresque *et al.* (1988) and Wolff (1991) conclude that in the shallow areas the Banc d'Arguin ecosystem is dominated by benthic primary production. Although this conclusion seems warranted by the low phytoplankton biomasses (Sevrin-Reyssac, 1984) and the large areas covered by seagrasses and other phanerogames (Wolff & Smit, 1990; Cuq, 1993; Wolff *et al.*, 1993), it is not supported by sufficient measurements of phytoplankton and seagrass productivity. This is another topic for future research.

### *Zooplankton*

The information on zooplankton of the Banc d'Arguin is very limited. Sevrin-Reyssac (1983) observed a very low zooplankton density in the tidal channels around the island of Tidra, and also Kuipers *et al.* (1993) concluded that zooplankton biomasses in Banc water are low.

## Direct consumption of benthic primary production

Wolff *et al.* (1987, 1993) draw attention to the fact that benthic suspension feeders are uncommon at the tidal flats of the Banc d'Arguin. Suspension feeder are even completely absent from the more seaward tidal flat regions, e.g. around the Isle of Arel. It is tempting to conclude that phytoplankton supply is insufficient to support benthic suspension feeders. Indeed, Sevrin-Reyssac (1983) concludes that phytoplankton is relatively sparse and unsuitable for (pelagic) suspension feeders in the tidal flat region of the Banc d'Arguin.

Invertebrates feeding directly on seagrasses, e.g. the isopod *Idotea chelipes*, have been observed at the Banc d'Arguin (Wolff *et al.*, 1993), but their numbers are very low. Their consumption will be of the order of magnitude of at most 1% of the seagrass production.

Some fish species, e.g. *Stephanolepis hispidus*, feed directly on plant material. Although their consumption has not been quantified, their importance seems to be limited when their numbers and biomass are compared to the seagrass biomasses (Jager, 1993).

This holds also true for the green turtle (*Chelonia mydas*). Compared to other coastal areas these animals are relatively plentiful at the Banc d'Arguin, but their number nevertheless is quite low (a few thousands?) considering the enormous potential food source provided by the seagrass biomass.

Ducks, geese, manatees or other vertebrate species feeding on seagrass do not occur at the Banc d'Arguin.

All taken together, it seems that direct consumption of seagrasses takes only a small part of the total production. However, this conclusion still has to be supported by sufficient quantitative data.

Periphyton consumption seems to be much more important than consumption of the seagrasses themselves. On the seagrass leaves several grazing gastropod species occur (Altenburg *et al.*, 1982; Wolff *et al.*, 1993). Grazing isopods and amphipods do occur as well but seem numerically less important. Quite important may be the fiddler crab (*Uca tangeri*) which strips the seagrass leaves very efficiently from periphyton growth (Hootsmans *et al.*, 1993). Also the mullets (Michaelis, 1993) feed on the periphyton.

The inevitable conclusion from the observations above is that the larger part of the seagrass primary production is not consumed directly, but instead is transported to other places or processed in a detritus food chain. This is in line with the observations of Hemminga & Nieuwenhuize (1991) who found large quantities of decaying seagrass leaves in the sediment of the seagrass beds.

### *Benthic fauna*

With at least 130 species the macrobenthic fauna of the tidal flats is rather rich in species (Wolff *et al.*, 1993), although the species diversity of the flats cannot be compared with the high diversity offshore. Its biomass is higher than was assumed originally (Altenburg *et al.*, 1982), but with about 17 g ash-free dry weight per $m^2$ in the area sampled not strikingly abundant. Moreover one species, the bivalve *Anadara senilis*, is responsible for about 8 g. Since this species has only very few predators (Swennen, 1990; Wolff *et al.*, 1987), it plays hardly any part in the food chains.

On the other hand large biomasses of benthic filter-feeders were found at shallow depths at the ocean facing side of the Banc d'Arguin (Duineveld, Lavaleye *et al.*, 1993). These concentrations apparently are dependent on the high phytoplankton production in the offshore area. No data have yet been published on the benthic fauna of the inshore subtidal area of the Banc d'Arguin.

Although benthic filter feeders are virtually absent from large parts of the tidal flat area, predatory species on the other hand are quite numerous with, for example, at least 11 predatory species of gastropods and at least 17 predatory species of polychaetes. Of the remaining species several gastropods and some crustaceans may be considered as grazers, but the majority, especially of the polychaetes and crustaceans, should be classified as deposit feeders (Wolff, unpubl. data), thus confirming the hypothesis of mainly detritus-based foodwebs at the Banc d'Arguin.

The productivity of the benthic fauna of the tidal flats is still unknown (Wolff & Smit, 1990; Wolff *et al.*, 1993). The population of *Anadara senilis* consisting of very old specimens has a very low productivity (Wolff *et al.*, 1987), implying that the remaining part of the benthic fauna should produce enough to supply the plentiful predators. Ansell *et al.* (1978) compared macrofauna production of tropical and temperate beaches, and concluded that an equivalent biomass in the tropics produces at a rate ten times faster than in temperate habitats. Indeed Alongi (1990), in a review of tropical soft-bottom benthos, lists P/B ratios ranging from 0.3 to 54.1. Wolff (1991) tentatively concludes that the disappearance of benthic biomass due to predators will be at least 20 g ash-free dry weight per $m^2$. Therefore and because of the year-round high temperatures a P/B ratio of about 3 seems very well possible for the benthic macrofauna. Interesting is the possibility that a large part of this production is realized by very small benthic fauna or even meiofauna (Wolff & Smit, 1990; Zwarts *et al.*, 1990). In this field many research questions still have to be answered.

### *Fish and shrimps*

Fish and shrimps are plentiful at the Banc d'Arguin (Brulhet *et al.*, 1974; Gaudechoux, 1984; Wolff & Smit, 1990; Jager, 1993). The variety of species is quite large and many different feeding types do occur. However, due to the shallowness of the Banc d'Arguin and the absence of charts our information on the nektonic fauna of the Banc d'Arguin proper is still very limited. We are able to classify most species with regard to their position in the foodweb, but in all cases we are unable to quantify their importance. For some species the Banc d'Arguin acts as a nursery (Jager, this volume), but the quantitative role of the Banc d'Arguin as a nursery is still very much in the dark.

*Birds*

Of all groups the numbers of birds probably have been quantified best. The breeding population counts some 40 000 pairs of mostly piscivorous species (Campredon, 1987). The piscivorous species feed mainly in the shallow water of the Banc d'Arguin area (Campredon, pers. comm.; Leopold, 1993) and based on physiological relationships we estimate that these take about 10 000 tons fresh weight (2000 tons AFDW) of fish and other nektonic organisms from the Banc d'Arguin ecosystem.

On the tidal flats about 2 250 000 waders, all feeding on benthic fauna, spend the northern winter (Trotignon *et al.*, 1980; Altenburg *et al.*, 1982, 1983). Summer numbers have been determined by Van Dijk *et al.* (1990). The same authors present information on the numbers of flamingos feeding on the tidal flats. Based on these numbers Wolff & Smit (1990) have attempted to quantify annual consumption of benthic fauna by birds and arrived at about 7000 tons AFDW of benthic animals. The latter value is equivalent to about 14 g AFDW $m^{-2}$.

Although the part played by birds quantitatively is now reasonably well known, still many questions await us with regard to the factors which determine these high densities of birds.

Much less well known are the pelagic birds which do not breed at the Banc d'Arguin and which in many cases spend here the northern winter or visit the area while on migration. Numbers of several species are impressive (van Spanje, pers. comm.; Leopold, 1993).

*Mammals*

The monk seals (*Monachus monachus*) living at Cap Blanc and further North only rarely visit the shallows of the Banc d'Arguin. The only marine mammals ocurring regularly are Cameroon river dolphins (*Souza teuszi*) and bottlenose dolphins (*Tursiops truncatus*). Other species occur mainly or exclusively offshore. The total population of both species is probably less than 100 individuals (Campredon, pers. comm.). Both species take fish and their annual food intake will be in the order of magnitude of 500–1000 tons wet weight (100–200 tons AFDW).

*Fisheries*

The local Imraguen are the only fishermen legally allowed to fish in the larger part of the shallow water area of the Banc d'Arguin. Their annual catch amounts to nearly 2000 tons wet weight of fish, mainly mullets (*Mugil cephalus*) and several hundreds of turtles (*Chelonia mydas*). They do not catch any shrimps or benthic fauna (Mengin & Van Spanje, 1989).

Some of the offshore trawlers as well as Senegalese artisanal fishermen sometimes penetrate illegally into the inshore waters. Their catch is very hard to quantify.

*Transports of matter*

With the upwelling ocean water penetrating into the shallows of the Banc d'Arguin large quantities of nutrients and plankton will be imported into the Banc d'Arguin ecosystem. Quantification is not yet possible but all evidence suggests that much of the nutrients and phytoplankton soon disappears (see also section 2). This may be explained by high densities of filter feeders, notably molluscs, in shallow water (10–15 m) in the area South of Cap Blanc. Here the Canary Current water, enriched by upwelling, penetrates into the shallow region. Filtering may be very effective because the bottom slopes very gently and the zone with filter feeders is several miles wide. Dense shell beds. e.g. *Venus* spp., were already observed in 1978 (Van der Land, unpubl. observ.), and more recent observations by the Centre National de Recherches Océanographiques et des Pêches at Nouadhibou confirm this. From the observations in 1988 we have the impression that the average size of the molluscs has become considerably smaller which may be interpreted as an effect of exploitation.

Even more obscure are the amounts of nutrients and plankton exported from the Banc d'Arguin by the dense and saline bottom water flowing back to the ocean.

An interesting aspect is the input of nutrients by windblown dust from the Sahara (Nienhuis, unpubl. obs.). Quantification is probably relatively easy, but still has to be done. Also the part played by the sebkhas, the extremely saline tidal flats just above high-tide level, which are only inundated during the equinoctial springtides, deserves further study. Do they act as a source or a sink of nutrients?

It is likely that the Banc d'Arguin forms a sink of nutrients. The decreasing phytoplankton biomasses behind the Banc proper from North to South and in the same direction the increasing area covered by seagrasses point to a transport to the bottom. Also the vast level bottoms covered with very soft sediments East of the Banc d'Arguin proper indicate retention of sediments and hence of nutrients. The dense concentrations of filter-feeders at the seaward edge of the Banc d'Arguin form another mechanism to remove particulate material from the water column. Thus the Banc d'Arguin may form an efficient nutrient trap (Fig. 5).

Transport was believed to be important for seagrass leaves. Large quantities are seen floating in the tidal channels and at sea. Very large quantities end up at the beaches, but possibly even larger quantities could drift to the open ocean where they ultimately sink to the bottom. Hemminga & Nieuwenhuize (1991) attempted to quantify transports of seagrass leaves in May 1988. They found very few seagrass leaves suspended in the water as well as only small amounts of floating leaves, although Van Lent *et al.* (1991) observed considerable leaf loss in the same period. On the beach between Iouik and Ten Alloul Hemminga & Nieuwenhuize (1991) observed that 27–87 g AFDW of seagrass leaves washed ashore per m of beach length per tide. This amount is sufficient to induce dune formation (Hemminga & Nieuwenhuize, 1990), but in view of the large area covered by seagrasses (according to Wolff & Smit (1990) about 400 km$^2$) and the production and leaf loss rates observed by Van Lent *et al.* (1991) and Vermaat *et al.* (this volume) Hemminga & Nieuwenhuize (1991) conclude that only a tiny amount of the seagrass ends up on the beaches. At the same time other observers (De Wilde, unpubl. observ.; Van der Land, unpubl. observ.) failed to observe seagrass leaves in the offshore area. Berghuis *et al.* (1993) also failed to find seagrass phytopigments in offshore bottom material.

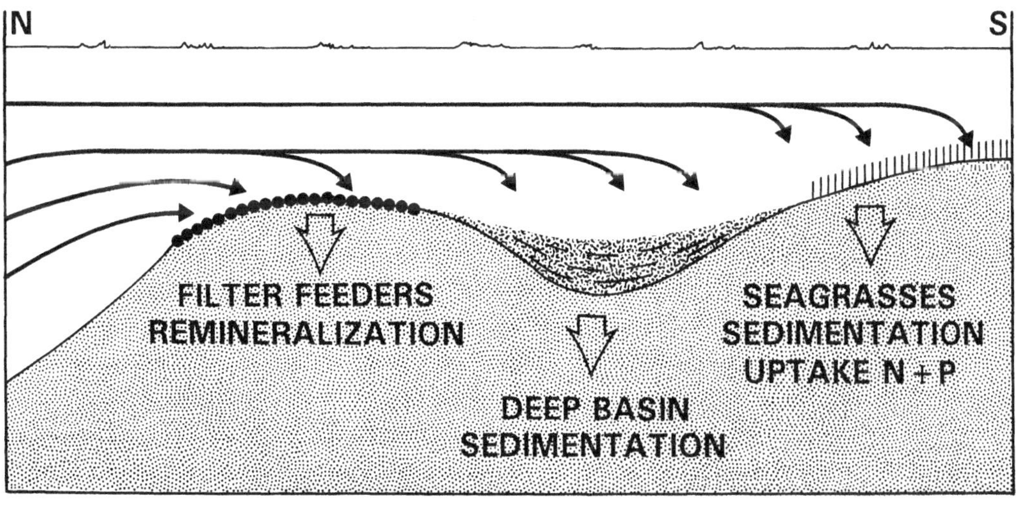


*Fig. 5.* Removal of dissolved and suspended particulate matter from the water over the Banc d'Arguin during its transport from the open ocean to the tidal flat region.

These observations are in line with the observations by Hemminga & Nieuwenhuize (1991) that the major part of the seagrass leaves remains trapped in the seagrass beds and decomposes *in situ*. This results in a detritus-based foodweb.

Finally transport by migrating animals has to be mentioned. However, net transports for birds are quite small and of the order of magnitude of 1% of the consumption by the various species. For mammals the situation probably is similar; for fishes we have no information.

The investigation of imports and exports of nutrients and particulate matter is still a large field of study, especially in relation to the possible retention of nutrients by the Banc d'Arguin.

*The ecosystem of the Banc d'Arguin*

Ecosystems can be described in many ways. Here we have tried to give a food chain picture. The complete picture is still far removed, but with the data at hand we can improve the schemes presented by Boudouresque *et al.* (1988) (Fig. 2) and Wolff & Smit (1990) (Fig. 3).

Figure 6 gives the picture which has emerged from the papers in this volume and other relevant literature. It is a picture of what we know, but at the same time it also shows very clearly what we do not know. Here we find the objectives for future research at the Banc d'Arguin.

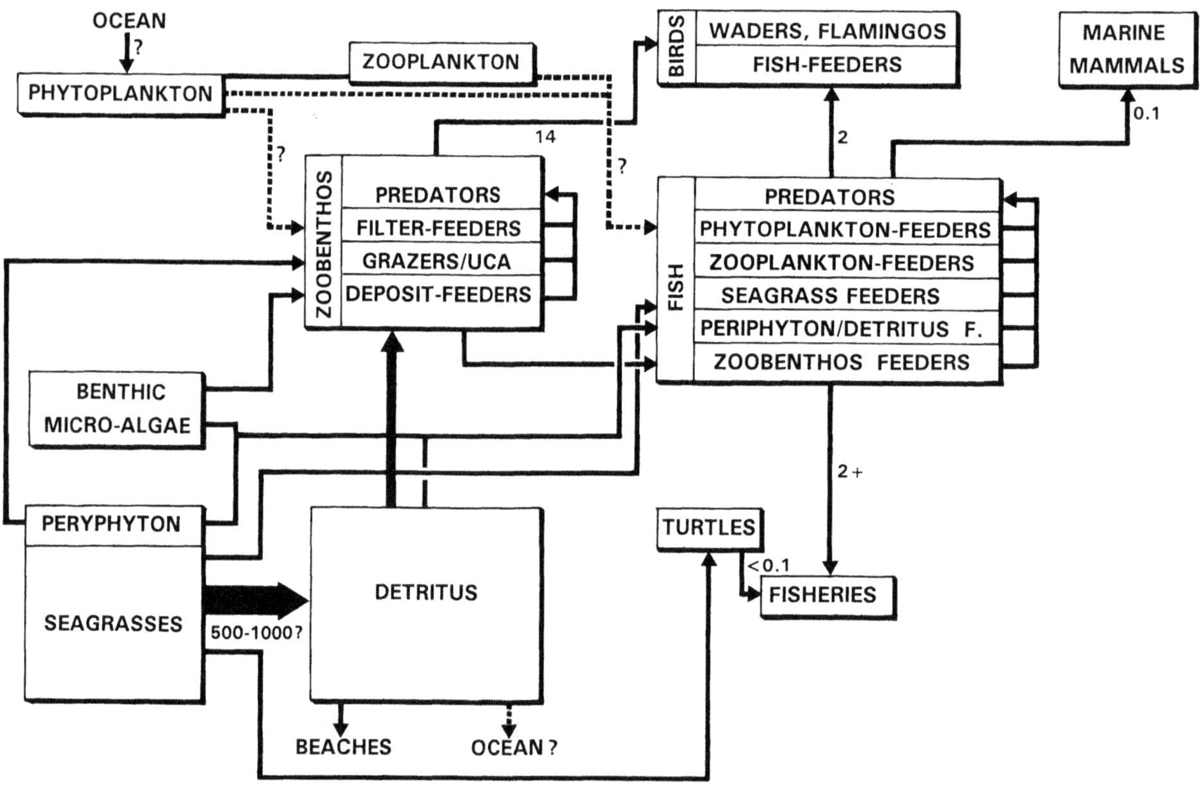


*Fig. 6.* Conceptual model of some aspects of the ecosystem of the Banc d'Arguin. The thickness of the arrows denotes the estimated size of the transport rates; in some cases a preliminary quantification (as g ash-free dry weight per m$^2$ per year) has been added for an area of about 1000 km of shallow water and tidal flats around the island of Tidra. The so-called small foodweb, probably responsible for the mineralization of the major part of the detritus, has not been shown because of lack of information.

## References


Alongi, D. M., 1990. The ecology of tropical soft-bottom benthic ecosystems. Oceanogr. Mar. Biol. Annu. Rev. 28: 381–496.

Altenburg, W., M. Engelmoer, R. Mes & T. Piersma, 1982. Wintering waders at the Banc d'Arguin, Mauritania. Report of the Netherlands Ornithological Expedition 1980. Stichting Veth tot Steun aan Waddenonderzoek, Leiden, 283 pp.

Altenburg, W., M. Engelmoer, R. Mes & T. Piersma, 1983. Récensement des limicoles et autres oiseaux aquatiques au Banc d'Arguin, Mauritanie. Gerfaut 73: 243–264.

Ansell, A. D., D. S. McLusky, A. Stirling & A. Trevallion, 1978. Production and energy flow in the macrobenthos of two sandy beaches in South West India. Proc. Roy. Soc. Edinburgh 76B: 269–296.

Berghuis, E. M., G. C. A. Duineveld & J. Hegeman, 1993. Primary production and distribution of phytopigments in the water column and sediments on the upwelling shelf off the Mauritanian coast (Northwest Africa). Hydrobiologia 258: 81–93.

Boudouresque, C. F., P. Campredon, P. Francour, D. Marchessaux & G. Pergent, 1988. Un pré-modèle du fonctionnement de l'écosystème du Banc d'Arguin. Parc National du Banc d'Arguin, Nouadhibou & GIS Posidonie, Marseille, 3 pp.

Brulhet, J., 1974. Onze années d'observations hydrologiques à proximité du Cap Blanc. Bull. Lab. Pêches Nouadhibou 3: 33–42.

Brulhet, J., J. Maigret & Sy Moussa Arouna, 1974. Résultats de la campagne de chalutage expérimental autour du Banc d'Arguin, 2e partie 1972/1973. Bull. Lab. Pêches Nouadhibou 3: 77–111.

Campredon, P., 1987. La reproduction des oiseaux d'eau sur le Parc National du Banc d'Arguin (Mauritanie) en 1984–85. Alauda 55: 187–210.

Cuq, F., 1993. Remote sensing of sea-surface and coastal features in the area of the Golfe d'Arguin, Mauritania. Hydrobiologia 258: 33–40.

De Jong, S. A., P. A. G. Hofman, A. J. J. Sandee & P. H. Nienhuis, 1991. Community oxygen budgets of seagrass beds and microbial mats on the Banc d'Arguin, Mauritania. In: P. M. J. Herman & C. H. R. Heip (eds.) – Report of the workshop Modelling the Benthos, Yerseke, The Netherlands, 20–22 March 1991. Delta Institute for Hydrobiological Research, Yerseke: 67–68.

De Naurois, R., 1959. Premières recherches sur l'avifaune des îles du Banc d'Arguin (Mauritanie). Alauda 27: 241–308.

Duineveld, G. C. A., M. S. S. Lavaleye & G. J. van Noort, 1993. The trawlfauna of the Mauritanian shelf (Northwest Africa): density, species composition, and biomass. Hydrobiologia 258: 165–173.

Ens, B. J., T. Piersma, W. J. Wolff & L. Zwarts (eds.), 1990. Homeward bound: problems waders face when migrating from the Banc d'Arguin, Mauritania, to their northern breeding grounds in spring. Ardea 78: 1–364.

Gaudechoux, J. P., 1984. Evaluation des ressources demersales dans la Baie du Lévrier et dans la partie nord du Banc d'Arguin en situation de saison froide (Décembre 1983–Janvier 1984). Bull. Centre nat. Rech. Océanogr. Pêches (Nouadhibou) 12: 114–126.

Hemminga, M. A. & J. Nieuwenhuize, 1990. Seagrass wrack-induced dune formation on a tropical coast (Banc d'Arguin, Mauritania). Estuar. Coastal Shelf Sci. 31: 499–502.

Hemminga, M. A. & J. Nieuwenhuize, 1991. Transport, deposition and *in situ* decay of seagrasses in a tropical mudflat area (Banc d'Arguin, Mauritania). Neth. J. Sea Res. 27: 183–190.

Hempel, G., 1982. The Canary Current: studies of an upwelling system. A symposium held in Las Palmas, 11–14 April 1978. Cons. int. Explor. Mer. Rapp. Proc.-verb. Réun. 180: 455 pp.

Hootsmans, M. J. M., J. E. Vermaat & J. A. J. Beijer, 1993. Periphyton density and shading in relation to tidal depth and fiddler crab activity in intertidal seagrass beds of the Banc d'Arguin (Mauritania). Hydrobiologia 258: 73–80.

Jager, Z., 1993. The distribution and abundance of young fish in the Banc d'Arguin, Mauritania. Hydrobiologia 258: 185–196.

Knight, P. J. & W. J. A. Dick, 1975. Récensement des limicoles au Banc d'Arguin (Mauritanie). Alauda 43: 363–385.

Kuipers, B. R., H. J. Witte & S. R. Gonzalez, 1993. Zooplankton distribution in the coastal upwelling system along the Banc d'Arguin, Mauritania. Hydrobiologia 258: 133–149.

Leopold, M. F., 1993. Seabirds in the shelf edge waters bordering the Banc d'Arguin, Mauritania, in May. Hydrobiologia 258: 197–210.

Loktionov, Y., 1993. Hydrographical observations west of the Banc d'Arguin, Mauritania, in May 1988. Hydrobiologia 258: 21–32.

Maigret, J., 1972. Campagne expérimentale de pêche des sardinelles et autres espèces pélagiques Juillet 1970–Octobre 1971. Tome 1: Observations concernant l'océanographie et la biologie des espèces. Lab. Pêches Nouadhibou & Soc. centr. Equip. Terr. int., Paris: 23–51.

Mengin, D. F. G. & T. van Spanje, 1989. Les pêcheurs Imraguen du Parc National du Banc d'Arguin en Republique Islamique de Mauritanie. Report Stichting Den Helder Wereldwijd & Fondation Internationale du Banc d'Arguin, Nouadhibou, 163 pp.

Michaelis, H., 1993. Food items of the grey mullet *Mugil cephalus* in the Banc d'Arguin area (Mauritania). Hydrobiologia 258: 175–183.

Ould Dedah, Sidina, 1993. Wind, surface water temperature, surface salinity and pollution in the area of the Banc d'Arguin, Mauritania. Hydrobiologia 258: 9–19.

Peters, H., 1976. The spreading of water masses of the Banc d'Arguin in the upwelling area off the northern Mauritanian coast. Meteor Forsch. Ergebnisse A 18: 78–100.

Pététin, M. & J. Trotignon, 1972. Prospection hivernale au Banc d'Arguin (Mauritanie). Alauda 40: 195–213.

Reyssac, J., 1974. Observations sur le phytoplancton et la production primaire de la région du Banc d'Arguin (Mauritanie) en avril et mai 1972. Bull. Inst. Français Afrique Noire (Dakar), ser. A, 36: 51–61.

Reyssac, J., 1977. Hydrologie, phytoplancton et production primaire de la baie du Lévrier et du Banc d'Arguin. Bull. Inst. franç. d'Afrique noire (Dakar), Sér. A, 39: 487–550.

Sevrin-Reyssac, J., 1980. Chlorophylle *a* et production primaire dans les eaux de la Baie du Lévrier et du Parc national du Banc d'Arguin (Septembre–Octobre 1980). Bull. Centre nat. Rech. Océanogr. Pêches (Nouadhibou) 9: 56–65.

Sevrin-Reyssac, J., 1982. Caractéristiques du phytoplancton de la région de l'île Tidra (Mauritanie) en Septembre et Octobre 1980. Bull. Centre nat. Rech. Océanogr. Pêches (Nouadhibou) 10: 41–47.

Sevrin-Reyssac, J., 1983. Quelques particularités de la chaine alimentaire marine dans la région des îles du Banc d'Arguin (Parc National du Banc d'Arguin, Mauritanie). Bull. Centre nat. Rech. Océanogr. Pêches (Nouadhibou) 11: 41–52.

Sevrin–Reyssac, J., 1984. Conditions hydrologiques et phytoplankton dans la partie méridionale du Parc National du Banc d'Arguin (Mauritanie) en mars et avril 1983. Bull. Centre nat. Océanogr. Pêches (Nouadhibou) 12: 107–113.

Sevrin-Reyssac, J. & B. Richer de Forges, 1985. Particularités de la faune icthyologique dans un milieu sursalé du Banc d'Arguin (Mauritanie). Océanogr. tropicale 20: 85–90.

Sevrin-Reyssac, J., 1993. Hydrology and underwater climate of the Banc d'Arguin, Mauritania: a review. Hydrobiologia 258: 1–8.

Smit, C. J., A-M. Blomert, A. Meijboom, W. J. Wolff & L. Zwarts, 1989. Hydrographical measurements. In: B. J. Ens, T. Piersma, W. J. Wolff & L. Zwarts (eds), Report of the Dutch-Mauritanian project Banc d'Arguin 1985–86: 43–56. WIWO-report 25/RIN-report 89/6, Texel.

Swennen, C., 1990. Oystercatchers (*Haematopus ostralegus*) feeding on giant bloody cockles (*Anadara senilis*) on the Banc d'Arguin in Mauritania. Ardea 78: 53–62.

Trotignon, E., J. Trotignon, M. Baillou, J.-F. Dejonghe, L. Duhautois & M. Lecomte, 1980. Récensement hivernal des limicoles et des autres oiseaux aquatiques sur le Banc d'Arguin (Mauritanie) (Hiver 1978/79). L'Oiseau et R.F.O. 50: 323–343.

Van Dijk, A. J., F. E. de Roder, E. C. L. Marteijn & H. Spiekman, 1990. Summering waders on the Banc d'Arguin, Mauritania: a census in June 1988. Ardea 78: 145–156.

Van Lent, F., P. H. Nienhuis & J. M. Verschuure, 1991. Production and biomass of the seagrasses *Zostera noltii* Hornem. and *Cymodocea nodosa* (Ucria) Aschers. at the Banc d'Arguin (Mauritania, NW Africa): a preliminary approach. Aquat. Bot. 41: 353–367.

Vermaat, J. E., J. A. J. Beijer, R. Gijlstra, M. J. M. Hootsmans, C. J. M. Philippart, N. W. van den Brink & W. van Vierssen, 1993. Leaf dynamics and standing stocks of intertidal *Zostera noltii* Hornem. and *Cvmodocea nodosa* (Ucria) Ascherson on the Banc d'Arguin (Mauritania). Hydrobiologia, 258: 59–72.

Von Westernhagen, W., 1968. Limikolen-Vorkommen an der westafrikanischen Küste auf der Banc d'Arguin (Mauritanien). J. Orn. 109: 185–193.

Wolff, W. J., 1991. The interaction of benthic macrofauna and birds in tidal flat estuaries: a comparison of the Banc d'Arguin, Mauritania, and some estuaries in The Netherlands. In: M. Elliott & J. P. Ducrotoy (eds), Proc. ECSA 19th Symp.: Estuaries and coasts: spatial and temporal intercomparisons. Caen, France, 4–8 Sept. 1989. Internat. Symp. Ser.: 299–306.

Wolff, W. J., A. G. Duiven, P. Duiven, P. Esselink, A. Gueye, A. Meijboom, G. Moerland & J. Zegers, 1993. Biomass of macrobenthic tidal flat fauna of the Banc d'Arguin, Mauritania. Hydrobiologia 258: 151–163.

Wolff, W. J., A. Gueye, A. Meijboom, T. Piersma & M. A. Sall, 1987. Distribution, biomass, recruitment and productivity of *Anadara senilis* (L.) (Mollusca: Bivalvia) on the Banc d'Arguin, Mauritania. Neth. J. Sea Res. 21: 243–253.

Wolff, W. J. & C. J. Smit, 1990. The Banc d'Arguin, Mauritania, as an environment for coastal birds. Ardea 78: 17–38.

Zwarts, L., A-M. Blomert & R. Hupkes, 1990. Increase in feeding time in waders preparing their spring migration from the Banc d'Arguin. Ardea 78: 237–256.

GPSR Compliance

*The European Union's (EU) General Product Safety Regulation (GPSR) is a set of rules that requires consumer products to be safe and our obligations to ensure this.*

*If you have any concerns about our products, you can contact us on ProductSafety@springernature.com*

In case Publisher is established outside the EU, the EU authorized representative is:

Springer Nature Customer Service Center GmbH
Europaplatz 3
69115 Heidelberg, Germany

**Batch number: 10607564**

Printed by Printforce, the Netherlands